PRAISE FOR GRANTLEE

'Engagingly written ... one of the most
nuanced portraits to date'
The Australian

'Vivid, detailed and well written'
Daily Telegraph

'A staggering accomplishment that can't be missed by
history buffs and story lovers alike'
Betterreading.com.au

'A free-flowing biography of a great Australian figure'
John Howard

'Clear and accessible ... well-crafted and
extensively documented'
Weekend Australian

'Kieza has added hugely to the depth of knowledge about
our greatest military general in a book that is timely'
Tim Fischer, *Courier-Mail*

'The author writes with the immediacy of a fine
documentary ... an easy, informative read, bringing
historic personalities to life'
Ballarat Courier

ALSO BY GRANTLEE KIEZA

Knockout: Great Australian Boxing Stories

The Remarkable Mrs Reibey

Hudson Fysh

The Kelly Hunters

Lawson

Banks

Macquarie

Banjo

The Hornet (with Jeff Horn)

Boxing in Australia

Mrs Kelly: The Astonishing Life of Ned Kelly's Mother

Monash: The Soldier Who Shaped Australia

Sons of the Southern Cross

Bert Hinkler: The Most Daring Man in the World

The Retriever (with Keith Schafferius)

A Year to Remember (with Mark Waugh)

Stopping the Clock: Health and Fitness the George Daldry Way
(with George Daldry)

Fast and Furious: A Celebration of Cricket's Pace Bowlers

Mark My Words: The Mark Graham Story
(with Alan Clarkson and Brian Mossop)

Australian Boxing: The Illustrated History

Fenech: The Official Biography (with Peter Muszkat)

FLINDERS

GRANTLEE KIEZA

ABC
BOOKS

Aboriginal and Torres Strait Islander people should be aware that this book contains descriptions, images and names of people now deceased. This book also contains descriptions of frontier violence that may be confronting or disturbing for some readers.

 The ABC 'Wave' device is a trademark of the Australian Broadcasting Corporation and is used under licence by HarperCollins*Publishers* Australia.

HarperCollins*Publishers*
Australia • Brazil • Canada • France • Germany • Holland • India
Italy • Japan • Mexico • New Zealand • Poland • Spain • Sweden
Switzerland • United Kingdom • United States of America

HarperCollins acknowledges the Traditional Custodians
of the land upon which we live and work, and pays respect
to Elders past and present.

First published on Gadigal country in Australia in 2023
by HarperCollins*Publishers* Australia Pty Limited
ABN 36 009 913 517
harpercollins.com.au

A catalogue record for this book is available from the National Library of Australia

ISBN 978 0 7333 4155 7 (hardback)
ISBN 978 1 4607 1349 5 (ebook)

Cover design by Louisa Maggio, HarperCollins Design Studio
Cover images: *Portrait of Captain Matthew Flinders, RN, 1774–1814*, 1806–07, Mauritius, oil on canvas, 64.5 x 50.0 cm, by Toussaint Antoine De Chazal De Chamerel, born 1770, died 1822. Gift of David Roche in memory of his father, J.D.K. Roche, and the South Australian Government 2000, Art Gallery of South Australia, Adelaide (20005P22); Cliff face and the sea 1896 by James Fawcett, courtesy State Library of Victoria; HMS *Investigator* by Art Collection 2/Alamy Stock Photo; Trim the Cat © Celia Pike; Silhouettes of Matthew and Ann Flinders courtesy Rachel Lewis Flinders, photography © Clive Aylard; all other images by shutterstock.com
Case: Sir Joseph Banks Papers, Series 65.03: Letter received by Banks from Matthew Flinders, 26 January 1801 (FL3149943/FL3149948) courtesy State Library of New South Wales
Endpapers: *A Voyage to Terra Australis* by Matthew Flinders (FL3750492) courtesy State Library of New South Wales
Author photograph by Milen Boubbov
Index by Garry Cousins
Typeset in Bembo Std by Kelli Lonergan
Printed and bound in Australia by McPherson's Printing Group

For Bruno Rizzo,
who has so often shown me the right track

Prologue

'[N]o difficulty could stop his career, no danger dismay him: hunger,
thirst, labour, rest, sickness, shipwreck, imprisonment, Death itself,
were equally to him matters of indifference if they interfered
with his darling Discovery.'

ANN FLINDERS, ON HER HUSBAND'S UNQUENCHABLE
THIRST FOR EXPLORATION.[1]

THE SMALL, KEEN, dark eyes of Matthew Flinders focused
through the rain beating down on his handsome face and
his careworn ship as he prepared to set sail on one of history's
great voyages.

Flinders was looking out towards the English Channel and
preparing to tackle the thousands of kilometres beyond – the vast
mountainous seas and icy gales – before beginning to unravel
the truth of a vast land stretching from the Indian Ocean to the
Pacific, which he would soon name Australia.

At just twenty-seven, Flinders had already proved himself to
have courage matching his unquenchable drive. As a teenager, he
had witnessed death and destruction on a terrifying scale when
he'd fought against Napoleon's great warships, and now he was
aboard a weary old leaking tub that he would steer to the far side
of the Earth to unlock the mystery of a whole new continent.

As he stood on the quarterdeck of his ship, the *Investigator*,
anchored at Spithead, on the Solent in the south of England,
Flinders barked out orders to his men while inside he ached with

emotional pain. On the wet and miserable morning of 18 July 1801, he was a young man torn apart by sadness, and yet driven by an insatiable ambition.

Flinders was on a mission to be the first man to circumnavigate and chart what he believed was a huge continent, and at the same time prove Australia was also an island. The voyage would also further Britain's imperial interests, advancing colonisation and the British expansion projects in areas new to Europeans.

As he looked through the driving rain, he did so with a heavy heart. He had been married only a few weeks when he was separated from his delicate wife, on orders from the British Admiralty, and he was leaving England with ill feelings between himself and his ailing father. He wondered if he would ever again embrace the country surgeon who filled his diaries with affectionate tributes, usually with fractured spelling and punctuation, to the astonishing achievements of his firstborn.

Flinders was ambitious, arrogant and audacious, but at the same time tender, romantic and passionate. He could be manipulative, annoying, envious, sometimes brash and undiplomatic, yet he also had a wicked sense of humour, could be self-effacing and could laugh along with Indigenous people who had never seen Europeans before. Always, he was brave, resolute and headstrong.

At Spithead, as the *Investigator* was loaded with 360 pounds of fresh beef, two barrels of peas and four tons of water – and three tons of beer in case the water went bad – Flinders jettisoned troublemakers from his crew at the last minute, judging them a waste of space on his little wooden ship. Cattle, sheep, pigs, goats, and poultry were brought on board and penned to provide dinner for the eighty-eight men during the first few weeks of their journey to Sydney Cove.

The young captain's beloved cat, Trim, had a much more comfortable existence than the other animals on the ship. Trim would be Flinders' constant companion through all manner of dangerous trials and thrilling victories.

Matthew Flinders would help complete the map of the world. But, agonisingly for a man who lived for adventure, much of his life would be spent in confinement, with long separations from his family and his ships. He spent years in a foreign prison, longing to see his wife once more before he died. From the depths of his tormented soul, he composed love letters to her, sometimes romantic songs as well, praying that, God willing, he would one day sing them to her when they were reunited.

Flinders' spirit was tied to the water long before his birth in a wide, flat, marshy area of England known as the Lincolnshire fens. As a boy he was intrigued by the voyages of James Cook, and while still a teenager he decided to pit himself against the raging seas after reading the novel *Robinson Crusoe*. But his life would prove even more thrilling than that of the fictional hero who inspired him.

Chapter 1

'My Boys [first] Birth Day, I bless God the child is better in his health, than for some time past.'
MATTHEW FLINDERS SR, ON THE SURVIVAL OF HIS FIRSTBORN.[1]

MATTHEW FLINDERS grew up in Lincolnshire's fens, marshlands where grass rose out of the water, and where the water stretched to the North Sea and across to the coast of Flanders, where his family had originated. Flinders could trace his ancestry back to the tenth century in Europe, while his English line went back at least to the birth of William Flinders in 1520 in the village of Carlton, Nottinghamshire.[2]

As early as 1108, a great part of the lowlands area of northern Belgium known as Flanders was 'being drowned by an exudation or breaking in of the sea', and there was a mass migration of Flemish-speakers across the North Sea to England, 'beseeching the king to have some void place assigned them, wherein they might inhabit'.[3] Some of these Flemish-speakers took the surname 'Fleming'; others adapted 'Flanders' into the surname 'Flinders'. A century and a half later, during the reign of King Edward Longshanks, merchants from Flanders based themselves in the Lincolnshire town of Boston, where they engaged in large-scale trading on behalf of clothmakers in the Flemish cities of Ypres and Ostend.[4]

Over the next few centuries, there were trickles of Flemings and Flinders into England, and occasionally floods of them, depending on the religious and political climate, the wars and persecutions.

In the early years of the eighteenth century,[5] Matthew Flinders' great-grandfather John Flinders[6] was a farmer and grazier who had lived in the villages of Gedling and Ruddington near Nottingham[7] before settling seventy kilometres to the east, in the fertile grazing country of the fenlands around Donington, Lincolnshire.

Donington was a small market town 200 kilometres north of London in one of the three low-lying divisions of Lincolnshire known as the Parts of Holland. It lay about twenty kilometres south-west of the bigger port town of Boston, and a little further from the wide, brown, muddy shore of The Wash, a rectangular blending of earth and the North Sea, with estuaries of the rivers Welland, Witham, Nene and Great Ouse.

Matthew's paternal grandfather, also named John Flinders,[8] had been a churchwarden in Donington, and through his own studies to become an apothecary and surgeon had formed a friendship with Thomas Secker, who was a medical graduate before being appointed the Archbishop of Canterbury.

Landowners around Donington, with the help of Dutch experts, were constantly draining the surrounding marshes to create more farmland for their tenants. Large amounts of the rich, dark soil surrounding the town were the result of this work, and it was given over to patchworks of green fields growing hemp that the Royal Navy used to make rope and rigging. A major fenland drainage scheme had started in 1765, and Donington had a small canal port to transport goods to Boston and The Wash.

Matthew Flinders' father, Matthew Sr,[9] described himself as a 'surgeon, apothecary, and man-midwife', who was 'gratefull to Providence' for the blessings of his life.[10] He performed minor surgeries, though he had not graduated from a university, and like many medical men with a similar background was known as Mr Flinders, not Dr.

Flinders Sr served his apprenticeship as an apothecary in 1770 with Richard Grindall, a surgeon to the London Hospital and the Prince of Wales.[11]

He returned to Donington, then a village of about 1000 people, to set up his business in 1771. A brother, John Flinders, had a similar business in the nearby town of Spalding. Professional men such as him wore ruffled shirts, waistcoats, breeches, frockcoats with plated buttons, and buckled shoes. Sometimes they even wore powdered wigs. But most of the homes Flinders Sr visited on his rounds, in what was still regarded as a rural outpost of Britain, were small, airless cottages made of mud and straw with thatched roofs, bare-earth floors and a few candles providing feeble light.

Flinders Sr found it hard to make a profit in his first year. Many of his patients struggled to feed their families; more than a few survived only by his benevolence. He took £72 and a few pence in his first year of business but paid out £58 15s for the ingredients for his medicines, leaving him with just £13 and change. Things improved as his reputation grew.

Practical and methodical, Flinders Sr was a pious man with close ties to Donington's Anglican community, centred on the centuries-old Church of St Mary and the Holy Rood, where his mother was buried. He had a reputation for being generous and affectionate towards his relatives, though at times he could also be stern and intractable.[12]

He kept up to date with the latest developments in medicine by subscribing to various journals, though he was particular about finding those within his budget. By 1773 he could gleefully record that he had made £100 during the preceding twelve months.[13]

Flinders Sr was then twenty-three years old, and a short, clean-shaven and active young man with brown hair and grey eyes.[14] With a comfortable income and good prospects, he married Susanna Ward, two years younger, in the Lincolnshire village of Bolingbroke (now called Old Bolingbroke), home of the Ward family, on 6 May 1773. Susanna had a pleasant, amiable disposition, as well as dark hair and soft eyes.[15]

The young surgeon borrowed £80 at 5 per cent interest from a moneylender in nearby Wigtoft, and bought from his father a comfortable two-storey brick house with an attic fronting the western end of Donington's cobblestone marketplace. It had three large windows on the second floor looking out over the town square, and dormer windows for the living space in the attic.

A ground-floor annexe was tacked onto the side of the house, and this Flinders Sr used as a surgery and apothecary shop. Inside the door of the surgery he had a slate hanging on a nail, on which he would chalk reminders and appointment times each day.

A traveller through Donington at the time said the town had fallen from the grace of previous years, when it was a major commercial centre in medieval England, and was now ill kept and dirty.[16] It was certainly remote,[17] and there was localised rioting as common land was acquired by the wealthy and fenced off, while the reclaiming of the marshes affected the livelihood of fishermen.

But Flinders Sr and civic-minded townspeople like him were making it the best community they could. While Boston was a bigger and more affluent commercial centre, Donington held four annual fairs, where horses, cattle, hemp and flax were traded.

Physicians of the time often prescribed cures that were worse than the illness – such as bleeding out sickness through cutting or the application of leeches. But skilled apothecaries combined minor surgeries with the use of potions and medicines, which they mixed and learnt to use effectively through work in the field.

And while infant mortality rates remained high, by the late eighteenth century British women were looking more and more to trained man-midwives, rather than entrusting the delivery of their babies to unskilled relatives or friends. Soon Flinders Sr would be plying his trade in his own household. He and Susanna were hardly married before she fell pregnant.

THE FLINDERS' FIRST CHILD, who would become the great mariner, was born at their Donington home on Wednesday, 16 March 1774, and was given over to a nursemaid, as was the custom of the time. At six weeks old, Matthew Flinders Jr was baptised on 28 April 1774 at the Church of St Mary and the Holy Rood. His father expected him to follow his forebears into medicine, and began plotting out his education almost immediately.

Young Matthew was not a hardy child, though, and for a long time his health was poor – and this was an age when one in three children under the age of five died.[18] His parents despaired about his prospects for longevity, but the baby boy had spirit. What's more, he had arrived in the world at a momentous time in human history.

Across the Atlantic Ocean, thirteen British colonies were throwing off the yoke of King George III, and with a series of rebellions were preparing for a full-scale war that would lead to the creation of the United States of America.

Britain was also still agog with the three-year circumnavigation of the world by the towering Yorkshireman James Cook and the wealthy young patron of science Joseph Banks, aboard the converted coal carrier HMS *Endeavour*. Their perilous voyage had introduced Europe to new lands, peoples, plants and animals from the furthest parts of the globe.

The discoveries were akin to finding civilisations on distant planets today. Banks' fortune was centred at his Revesby estate in Lincolnshire, about thirty kilometres north of the Flinders' home, and he was one of the Lincolnshire landholders at the forefront of

Flinders' birthplace at Donington. The ground floor annexe to the right was used by his father as a surgery and apothecary shop. *State Library of South Australia B-1785*

draining the marshlands. He had helped to finance the voyage of the *Endeavour*, and the small but sturdy ship had returned to England in 1771, having completed the greatest scientific expedition to that time, bringing home thousands of specimens of unique flora and fauna along with captivating tales of cannibals in New Zealand and of strange bounding creatures – the 'kongouros'[19] – from a vast land Cook had named New South Wales. Cook had also named a shallow body of water there 'Botany Bay' after Banks' studies, which were so important to the voyage.

On his return to England, Banks, a burly, pony-tailed playboy in his youth, had turned his London mansion in New Burlington Street, Mayfair, into a multi-storey 'Academy of Natural History', welcoming fellow scientists and collectors from around the world.

Matthew Flinders' young father was also a keen student of natural history. Life fascinated him, and his devotion to saving it and preserving it kept his business flourishing throughout

The Kongouro from New Holland, painting by George Stubbs, 1772. *National Maritime Museum, London ZBA5754*

the 1770s. He began to charge rates that reflected the growing status of the man-midwife's profession, though he invariably gave discounts to the poor.

Although Flinders Sr was part of a small group of middle-class professionals, his hard work was always performed with one eye on his accounts, he was anxious about the ever-present threat of financial hardship during turbulent times.

He and Susanna employed two servants, a maid and a boy, and when business was good their household workforce would stretch to a gardener. But they had to count every farthing. They let go their maid, Rebecca Knights, because Susanna thought her wages of £3 13s a year plus board and lodgings too high. Instead, Susanna employed her niece Polly Stanney, although Flinders Sr rightly feared that the seventeen-year-old girl would not be strong enough for the work.[20] Within six months, a new maid, Mary Goodyear, was living in the home. Servant Thomas Barton was let go after two years because Flinders Sr thought he had become 'careless, saucy and very ill behaved'.

Flinders Sr leased two hectares of land from his father at £7 10s a year, though he tried repeatedly to have this reduced to £5. Ultimately they fell out over the arrangement, which Flinders called 'exorbitant and cruel', and he blamed his father's 'tyrannical' second wife, Mary Baxter.[21] John Flinders had wasted no time before remarrying following the death of his first wife, Elizabeth,[22] and he fathered five more children with Mary.

Flinders Sr had enough land to keep a horse and a cow and to grow vegetables. He built lattice structures to tame the branches of his Nonsuch apple trees, which produced sweet, crisp red fruit, and he hired an itinerant gardener to sow purple broccoli and speckled dwarf kidney beans.[23] He shaped a lush and thick hedge around this bountiful garden.

His property had various small outbuildings too, including a coalhouse, a brewhouse and a barn, and before long he was replacing a rotting paling fence around his animals with a new post-and-rail one, although he judged the cost of the work, forty shillings, 'very expensive'.

WHEN MATTHEW WAS JUST four months old, the return to London of HMS *Adventure*, which had left England with Captain Cook's HMS *Resolution* two years before, was being talked about across England.

Cook had led a second voyage to the Pacific, and this time his two ships had sailed south into the Antarctic Circle, past huge icebergs. Cook had come within about 120 kilometres of the Antarctic coast in January 1773, before ice forced him back. The *Adventure*'s skipper, Tobias Furneaux, surveyed the southern and eastern coasts of Van Diemen's Land. Tasmania's Furneaux Group of islands and Adventure Bay are reminders of his visit. At Queen Charlotte Sound in New Zealand, Furneaux sent a large cutter with a master's mate, a midshipman and eight hands to gather fresh greens for the voyage. A fight broke out with the Maori at 'Grass Cove', and the sailors were killed and parts of their bodies eaten.[24]

Diminishing stores eventually forced Furneaux to return home. The *Adventure* reached Spithead, on the Solent, in July 1774. On board was a young and handsome Polynesian man named Mai, a native of Raiatea, who had escaped to Tahiti – or Otaheite, as the British called it then – after Bora Bora warriors invaded his home. Mai had asked Cook if he could accompany the British back to their land across the sea.

Joseph Banks made Mai right at home at his mansion in New Burlington Street, dressing him in the finest from London's best tailors, and he organised his friend the Earl of Sandwich to present him to King George.[25]

The muscular Polynesian became a celebrity throughout England for the two years he lived among the rich and aristocratic, stirring in the heart of many British youngsters a desire to one day voyage to the distant corners of the world.

On 30 July 1775, Cook triumphantly sailed the *Resolution* into Spithead.[26] In next to no time he was planning a third voyage around the world.

THE YEAR AFTER MATTHEW was born, his father attended forty-three births. No mother died and he lost just one baby, a commendable record for the time.

January 1775 was bitterly cold but Flinders Sr and Susanna kept up their social life as best they could, often dining with friends or having them over for supper by the fire as baby Matthew, overcoming one illness after another, kept them fascinated with his attempts to walk and talk.

Flinders Sr had three midwifery cases that month, prescribing large doses of opiates in one of them for Mrs Fairbanks of Gosberton West to ease her pain and violent colic.[27] He was also called to the home of his lawyer friend Jonathan Gleed[28] to treat his newborn, Thomas.[29] The baby's scalp was 'tumified' or swollen at birth, and since then a tumour had grown on the back of his skull. Flinders Sr consulted with Edward Blythe, an ex-army surgeon in Spalding. Blythe advised to open a small puncture, and as little Thomas screamed the house down, Flinders Sr extracted two large spoonfuls of thick, black blood. He plugged the wound with some lint and smeared alcohol over the whole area. In quick time the child became well.[30]

On the nights he was not working or entertaining, Flinders Sr sometimes drank with friends at the Red Cow Inn, a popular establishment then almost a hundred years old. Sometimes he'd be at the Black Bull, which had been constructed in the 1500s. He would play cards and win or lose a shilling here and there,[31] and he gambled on the state lotteries with his brother John.[32] Just as often he was visiting the homes of his many clients or friends, sharing a meal and drinking tea, an expensive delicacy which the East India Company's ships were bringing from China.

Little Matthew was not quite a year old when his parents took him outside on 21 February 1775 to observe the phenomenon of the aurora borealis in the Lincolnshire night sky, a dazzling show known as the 'northern lights'. Matthew's father remembered it as 'remarkably splendid ... extending a great way, and possessing uncommon and extensively waving motions towards the centre ...

seeming as if the atmosphere was in violent lucid agitations or convulsions'.[33] The infant was far too young to commit the event to memory, but from the earliest moments of his life his young, malleable mind was being encouraged to have a sense of awe for the world around him.

On 16 March, Flinders Sr noted in his diary that it was Matthew's first birthday and that, thank God, the little fellow was better in health than for some time past.[34] Susanna was pregnant again but there was little time for celebrations for either event.

The following day, Flinders Sr assisted with a natural birth at Bicker Gauntlet, three kilometres north-east of Donington, and arrived home in the cold and dark at 7 p.m., hoping for a pleasant dinner by the fire and a good night's sleep. Instead he received an urgent message that Sarah Cottam at Swineshead Fenhouses, eight kilometres north-east of Donington, was in the midst of an agonising life-or-death struggle.[35] Another midwife had been with her two days.

No matter how hard Mrs Cottam pushed, the baby would not come out, and after screams that echoed through the spring night like claps of thunder, Flinders Sr was obliged to use his forceps for the first time. The Chamberlen family had created these instruments to help midwives safely extract a child during difficult births.[36] The forceps consisted of two hollow metal arms that clamped around the baby's head.

In the past, devices such as hooks shaped like crochet needles had been used to deliver stillborn babies, but these invariably ended the life of a living child stuck in the birth canal. Midwives thought that preferable to watching the mother die as well.

The forceps helped the exhausted Sarah Cottam deliver a healthy child, although there were 'several slippings',[37] and little John Cottam did not utter his first cries until days after the labour began. Flinders Sr rode his mare home across the fenlands to Susanna and his own infant, exhausted but delighted that he had brought a child into the world that might otherwise have died – and hopeful that news of his success with the forceps would bring him more work. He reached Donington late in the afternoon and collapsed into bed at 5 p.m. He had not had his boots off for forty hours.[38] He charged the new parents £1 1s for his services.

Flinders Sr had his forceps with him late in the evening on 18 April, when John Lee asked him to help his wife with the delivery of their daughter Ann. He delivered the baby but it was a difficult night. 'I wish I could always remember to turn the head with the forceps 'ere extracting, as I think it would be accomplished easier,' he wrote.[39] The child's head proved to have a tumour similar to that which had affected young Thomas Gleed. Flinders Sr opened it and drew blood, with similar benefit to the newborn.

He was kept on the move constantly with other medical cases too. On 1 May 1775 he was back in the village of Bicker Gauntlet gently examining the fractured arm of William Storey's seven-year-old daughter, Mary, who had been injured by a 'leap and fall'. Some 'volatile Spirituous Embocation' relieved the pain and swelling, and 'strict bandage with splints compleated the cure'.[40] Flinders Sr frequently mended elbows dislocated by falls, and once even had to fix the dislocated jaw which Mrs Cook of Quadring had suffered through yawning too hard.[41]

When he became ill every autumn with what was most likely malaria, young Mr Flinders treated himself for the 'debility and pain and *chilness*'.[42] Sometimes he would take three of Dr James's Analeptic Pills, which promoted purging through vomiting and multiple bowel motions. Once, when he experienced a brief bout of deafness, he had Susanna syringe his ear while he bled himself from a wound in his arm.

On 24 September 1775, Flinders Sr helped with the delivery of his first daughter, Elizabeth – known as Betsey[43] – after a difficult labour. He breathed a sigh of relief because he knew how quickly things could go wrong.

Six weeks later, Flinders Sr rode through thick snow as a hard winter neared, and arrived at the home of Mrs Garner in Donington Fen at 4 a.m. He was immediately gripped by a sense of dread, later writing that it was his second visit to the home, and he had previously been 'obliged to use the Crochet' – that is, needed to destroy the baby to preserve the mother's life. Gladly, he later scrawled into his diary that while he spent close to forty hours with Mrs Garner in that little house, 'the birth proved natural'. He got home by 8 p.m. to collapse into bed.

The weather was fierce at that time of year. Flinders Sr had covered the potatoes and asparagus in his garden with straw and manure as a protection against the frost, a bitter chill that lasted for more than a month. He couldn't remember a time in his twenty-six years when the cold was 'so severely sharp' and the countryside was frozen everywhere, the ice being near thirty centimetres thick in many places. His accounts of the time show him paying out 1s 5d for seventeen leeches needed to bleed patients, £1 14s 9d for 'Druggs' required to make his medicines, 2s 3d for a new blanket for Matthew's bed and 1s 6d for the boy's little chair.

Susanna took great care to make sure Matthew and Betsey were wrapped in thick woollen blankets. Her family regarded the young mother as 'clever in housekeeping and managing children'.[44]

On some of the freezing nights, Flinders Sr would sit beside the fire and transcribe into a bound book the love letters between Susanna and himself written in the days when they were courting. He also swapped books at the local fairs; on one occasion when Matthew was still very small, he bought a recent release, *Institutes of Natural and Revealed Religion, Vol. 3*, written by Joseph Priestley, who had had just discovered a gas called 'oxygen'.

Flinders Sr was also taken with the writings of the popular author Jonathan Swift, and was a keen reader of news and theatrical magazines. He had joined with his father to pay for a subscription for a Cambridge journal, though he preferred the Stamford newspaper because it was 'impartial', whereas the Cambridge rag was 'very barren of entertainment' and, with the war against America exploding, 'partial on the furious and patriotic side'.[45] He kept his books and periodicals atop his desk in a new bookcase, which he called 'a neat peice [sic] of furniture and very usefull'.[46]

Flinders Sr would encourage his son to read widely from a young age, and there were always intellectual tomes to stir the boy's mind as he grew.

ON THE FOURTH OF JULY 1776, the Second Continental Congress, meeting in Philadelphia, approved a Declaration of Independence from Great Britain. The declaration would change the world forever. In Lincolnshire, Flinders Sr was more interested in the heavens above than in what was happening on the other side of the world.

On 30 July, when Matthew was just two, his father sat up with him till well past midnight to gaze upon a long anticipated lunar eclipse. Susanna was heavily pregnant again and needed her rest. Flinders Sr had already enrolled Matthew at Mrs Moor's local nursery school,[47] but out here there was a whole universe to study. One of Matthew Sr's friends, John Birks, had an eighteen-foot telescope, which made the spectacle of the eclipse even more 'amazing, grand and striking'.[48] The 'globular appearance' of the moon was breathtaking, and as Flinders Sr pointed out the extraordinary sight to his little son, there was a 'beautiful diversity of light and dark parts, with great reason said to be water and land'.[49]

While that miracle was astonishing, Susanna produced another just seven weeks later when she delivered the family's second son after a difficult birth. For three days before the child's delivery on 28 September 1776, Susanna was in the grips of a violent fever, with a screaming headache and a fast but weak pulse. Sometimes she shivered with cold, and at others dripped with sweat. Flinders Sr gave his ailing wife tartar emetic in the hope that it would purge her body of the poison swirling through her, and he administered two anodyne pills to relieve her pain.

After Susanna spent a day in hard labour, Flinders Sr broke his wife's amniotic sac at 6 p.m. and Matthew's baby brother, Jackey Flinders, arrived into the world at 7.30. He was an 'exceedingly fine child', but within a couple of weeks looked poorly and was unable to keep milk down. He gradually lost what little flesh he had. His father wrote: 'On Wed. Nov 13th at 4 in the Morn, Death made his first approach in our little Family.'

He saw no point in sentimentality as he prepared to bury the little chap:

> We ought to account of this a mercifull Dispensation in that Providence made choice of the Youngest; to have parted with either of the other two would have afflicted us much more; and as we have nought in a natural sense, but my industry in Business to depend on, we ought to think the non increase of our family a blessing.[50]

Death made another visit to Flinders Sr on the day after Christmas, when it took his father, John, just short of his sixty-fourth birthday.[51] 'He departed this existence for one I hope more perfect,' Flinders Sr wrote, adding that Matthew's grandfather had been suffering from the retention of fluid associated with dropsy for some time. All the efforts by medical men, using expectorants, bleeding and heat packs to create blisters, had provided little relief.

> He was sensible of his death, and to appearance resigned to the will of God, and expressed not the least regret at leaving the world. He continued quite rational as long as he could speak, which was within half an hour of his death, after which he gradually sunk. I left him before expiration as the scene was too affecting.[52]

John Flinders was buried at St Mary's on 28 December, leaving a young second family in what Matthew Sr described as 'scanty' circumstances.

Then, on Saturday, 19 July 1777, as three-year-old Matthew fretted over his mother's condition, Susanna delivered tiny twin girls. It was two months before she was due, and – as Flinders Sr feared – the little girls were both stillborn. Flinders Sr again consoled himself that it was perhaps the will of kind 'Providence' to spare his little family the 'expence' of more hungry children,[53] but before long Susanna was pregnant once more.

FLINDERS SR INOCULATED Matthew and Betsey against smallpox on 7 December 1777, the year after the disease first reached his parish.

In the eighteenth century in Europe, some 400,000 people died annually from the 'speckled monster', and a third of the survivors went blind.[54] The case-fatality rate in children was even higher, approaching 80 per cent in Britain.[55]

A pioneering medical family, the Suttons of Suffolk, claimed they had made the inoculation procedure 'almost painless' and much safer than previous treatments against the disease.[56] But it was still an ordeal. Robert Sutton Sr had modified the inoculation technique so that it involved only a tiny jab through the skin with a sharp lancet and the placement of a small amount of smallpox pus inside the wound to induce immunity in the patient.

After their father inoculated Matthew and Betsey, he brushed away their tears to monitor their reactions. The injection site resembled a tiny red fleabite. It began to itch on the fourth day. The children were fed mercury to make them salivate, and then given laxatives calculated to produce at least four stools a day. A pustule with a domed top was fully formed seven days after inoculation.

On the last day of 1777, Flinders Sr noted that Matthew and Betsey had 'passed through that calamitous disorder in the most favourable and easy manner. I have inoculated several others with the greatest success and expect more Business of this kind before we stop.'

Donington may have been far from the universities and academic societies of London, but it was certainly no intellectual backwater. Little Matthew Flinders, now almost four, was growing up surrounded by a thirst for knowledge.

Chapter 2

'I burned to have adventures of my own. I felt as I read that there was born within my heart the ambition to distinguish myself by some important discovery.'

<small>MATTHEW FLINDERS, ON READING *ROBINSON CRUSOE* IN HIS YOUTH.[1]</small>

DONINGTON WITNESSED another gobsmacking phenomenon a few weeks after Matthew Flinders' fourth birthday. Robert Powell, the world's most famous fire-eater, performed at the Red Cow Inn on 18 May 1778.

As a hush descended over the crowd, the old showman appeared in long, flowing robes, with an elaborate turban on his head. He displayed some sleight-of-hand magic tricks and there were a few murmurs, but Matthew's father had seen those capers from other magicians. Then the fire-eater gave his audience what they came for. With theatrical flourish, a drum roll and a nod to the supernatural, Powell drew gasps from the pub patrons with his burning breath, keeping young Matthew and the others on the edge of their wooden pews as he swallowed fire, chewed on hot coals and munched burning wax.

Powell came to Donington with an enormous reputation built upon more than half a century of astonishing crowds, but he had fallen on hard times and was now travelling around the countryside without the fine carriage and servants which he once commanded. Matthew's father felt that the performer had rather oversold himself, and that at more than seventy years old he was getting beyond such hijinks. Donington's surgeon and man-midwife noted that the famous fire-eater did not 'absolutley lick the Iron Red Hot', while the charcoal he mouthed 'very tenderly'. Still, the crowd had loved it, and Flinders Sr conceded that the show was 'amazing',[2] though he was sure that the old man often drank away his troubles.

Young Matthew was spellbound by all the astonishing sights that Donington could offer. At the local fair a week after Powell's visit, a cassowary from Java strutted about the Lincolnshire grass. Matthew's father, unaware of the emu, reckoned the cassowary to be the second-biggest bird in the world, after the African ostrich.

Matthew's mother did not attend the fair. Two days later, with her husband's assistance and six weeks ahead of time, Susanna Flinders delivered two more sons, John and Samuel. They died within hours of each other and were interred in a single small coffin on 30 May.[3] Susanna had delivered four children in eleven months, and all had perished. She would soon try again.

Matthew was already painfully well educated in the uncertainties of life, and he was further exposed to the great complexity of the universe when all of Donington stood outside to watch a solar eclipse just three weeks later.

In the autumn of 1778, Matthew's father travelled to Boston to see what he called 'that enormous and in these parts of the world uncommon Fish', a whale. It had been chased onshore from an area of The Wash called Boston Deeps by the crew of one of the king's small fishing boats. Its body had then been transported up the Witham and dragged onto land. Matthew's father was astonished by the sight of the great creature, but wished he had arrived in town two or three days sooner, before the locals began to cut it up:

[The] sight of him entire must have been an amazing curious Spectacle. Even the mess we now saw was curious from its amazing Bulk – when entire he measured 52 feet in length [17 metres], his tongue as large as a Calf and each finn joining at the shoulder about 2 yards long.[4]

When Flinders Sr returned home, he regaled his family with the tale of the great creature, explaining that the sea beyond Boston contained many mysteries and dangers. Matthew wanted to learn all he could about the oceans of the world and life beyond the low, flat marshes in the land of his birth.

He soon became enamoured with the foreshore of Boston, where he watched the dock workers, and the ships departing for and arriving from exotic ports around the world. Boston was dominated by the ornate lantern tower of St Botolph's Church, but Matthew was more interested in the River Witham and the way sailors navigated their way into The Wash, and how they picked up speed as their sails were unfurled and they roared out across the slate-grey sheet of the North Sea to those places of promise and wonder beyond the horizon.

SUSANNA WAS PREGNANT AGAIN and the family prayed at St Mary's for her safety. She had already borne seven children and only Matthew and Betsey had survived.

On 22 May 1779, Matthew's mother delivered a daughter, whom she named after herself, and the Flinders family was now five. Flinders Sr sent the new baby out to a nurse, paying three shillings a week for the care, and seven pence for milk. The baby did well and stayed with the nurse for six months.

'It saved us a great deal of troubles,' Matthew Sr wrote, 'but is expensive.' Not long after that he forked over 17/6 to Mrs Moor to settle the account for Matthew's schooling. Betsey was enrolled at Mrs Codling's nursery school, and then her father paid two shillings for a new pair of buckle-up shoes for Matthew because his little boy would soon be going to 'big school'.[5]

As the newest addition to the Flinders family began to crawl about the Donington home, a pall was cast over Britain's scientific community. Disturbing reports about the fate of Captain Cook during another voyage into the Pacific began to reach Joseph Banks in London in January 1780. Dutch East Indiamen and Swedish ships coming from China brought news that the man who had mapped the world had met with a violent death.

St Botolph's Church in Boston, Lincolnshire. Engraved by T. Jeavons, after Joseph Mallord William Turner, published 1835. *Tate Gallery, London T06115*

Cook had been leading his third great voyage of exploration and was again in command of the *Resolution*, while Captain Charles Clerke was the skipper of an accompanying ship, the *Discovery*. The voyage had returned the celebrated Polynesian Mai to Tahiti before Cook and Clerke headed along the west coast of Canada, hoping to find a 'north-west passage' that would connect the Atlantic and Pacific oceans. Cook named the archipelago now known as Hawaii 'the Sandwich Islands' after the Earl of Sandwich, at the time the acting First Lord of the Admiralty.

While Cook had befriended many of the Hawaiians, when some of them stole a small boat from the *Resolution*, he decided to take a local chief hostage until it was returned. On the morning of 14 February 1779, Cook came ashore with a company of armed marines, woke the local chief, Kalani'o-pu'u, and marched him towards the beach. A melee ensued, and Cook shot and killed a man.[6]

On the deck of the *Resolution*, the ship's young master, William Bligh,[7] the ill-tempered, foul-mouthed son of a Plymouth customs officer, watched the carnage unfold through a spyglass. Bligh saw Cook knocked to the sand and stabbed in the neck. Thousands of miles from his wife and children – a family who had barely seen him during his last decade of exploration – Cook fell face first into the water and bled to death. Four marines were also killed, and two wounded. Cook's men fired into the crowd before fleeing to their boat, leaving blood on the beach and in the water.

A FEW WEEKS BEFORE HIS sixth birthday, Matthew was enrolled at the Donington Free School under teacher Jeremiah Whitehead. The school was sixty years old, founded by local landowner Thomas Cowley, who had bequeathed part of his estate to fund a charity school to teach twenty poor children to read and write. Although the Flinders family was not poor, there was a friendship between Whitehead and Flinders Sr, and the school extended something of a scholarship to his boy. Flinders sent the school 10/6 to seal the deal. He bought his son some new stockings, a new hat and some gloves,[8] and before long was handing over 2/6 for Matthew's school feast.

But while Flinders Sr imagined that Matthew would one day follow him into the world of medicine, a new career path showed itself for the boy soon after, when Matthew's thirteen-year-old cousin, Jackey Flinders,[9] 'being wholly bent to try his fortune at Sea', procured a place as a captain's servant on the 32-gun frigate HMS *Apollo*. The youngster's family saw it as the first step to becoming a naval officer, but the teenager's early career was more about hard work and cheating death.

Jackey had hardly joined the *Apollo*'s crew when, during hostilities associated with the American Revolution, the ship's captain, Philemon Pownoll,[10] was killed by a cannonball from a French ship off Ostend, in Flanders, in June 1780. Five other crewmen were killed and twenty wounded before the French ship was driven ashore and captured. Yet Jackey had a heart for adventure. Rather than wait for the *Apollo* to be repaired, he joined the crew of the HMS *Amphion* to fight what Flinders Sr called 'the perfidious Dutch'.[11]

By this time, Joseph Banks was, at just thirty-seven, the president of the Royal Society. Despite a financial crisis caused by Britain's war with the thirteen United States of America, Banks' collection of artefacts from around the world, and the staff required to preserve them, had grown so much that he had to move from his mansion in Mayfair to an even bigger abode in Soho Square.

Shortly after his wedding in 1779, Banks and his new wife, Dorothea, had travelled north from their London home to supervise renovations at his Revesby Estate, just north of Boston. It would be an almost annual migration for the next thirty-eight years, as Banks and his household holidayed in the fens between August and October. Banks became a central figure in Lincolnshire life, and eventually in the life of young Matthew Flinders.

Banks was a major wool grower in Lincolnshire and, now recognised as the chief spokesman on all scientific matters in Britain, determined to expand the agricultural interests of Britain to new horizons. England had been sending convicts and prisoners of war to the British colonies in North America for almost 200 years, but the American Revolution had curtailed that, causing massive overcrowding in British jails. Rotting ships crowded the Thames as floating prisons, and still London was running out of room.

The American war also meant that Britain's supplies of hemp, tar and timber from the Baltic – vital for its ships – were being dried up by the League of Armed Neutrality, an alliance between Russia, Holland and other northern European powers.

England now looked to establish a new penal colony and a supply chain for raw materials. The government had started having discussions with Banks about New South Wales and the possibilities it offered to house prisons and naval, military and trading outposts. Sir Charles Bunbury[12] was asked to chair a parliamentary committee to look into the question of convict transportation. Bunbury interviewed hard, cruel men who had been shipping enslaved people from Africa to the Caribbean, and those who had transported prisoners to the Americas.

Bunbury asked Banks about the best place to establish a colony of convicted felons in a distant part of the globe, from where their escape would be difficult – and where, 'from the fertility of the soil', they might be enabled to maintain themselves with little or no aid from 'the mother country'.[13]

Banks advised that Botany Bay, 'on the coast of New Holland, in the Indian Ocean, which was about seven months' voyage from England', would be the place most 'eligible for such settlement'.[14] Banks said there would be 'little probability of any opposition from the natives' as during his stay there in the year 1770 he saw very few, and did not think there were 'above fifty in all the neighbourhood'. He said he believed the country was 'very thinly peopled'.

Banks explained that he 'was in this bay in the end of April and beginning of May, 1770, when the weather was mild and moderate. The climate, he said, was similar to that 'about Toulouse, in the south of France'.

As a landowning scientist motivated by profit for Britain, Banks viewed Botany Bay as much more than just an outdoor prison: he saw his appearance before the Bunbury Committee as the first step towards establishing a self-sufficient colony in New Holland that could not only generate great agricultural and mineral wealth for the British economy, but also, through exploration and scientific research, enrich European knowledge about unique fauna and flora. Crucially, it would also allow Britain to dominate the South Pacific – before the French, Dutch or Spanish moved in.

BEFORE MATTHEW WAS SEVEN years old, his father could gladly report that the boy had started to learn writing at Mr Whitehead's and could now read 'tolerably well'. Matthew's sixteen-year-old cousin Henrietta – or 'Henny',[15] as she was known – came to stay for two weeks and took Matthew back to her parents' home in nearby Spalding for a short holiday.

Flinders Sr bought Matthew a book of Latin grammar, and as the winter of 1781 gave way to spring, the little boy spent his nights by an oil lamp receiving lessons in that language,[16] though he always hankered to learn more about the world outside his county. In 1781 Susanna presented Matthew with a baby brother[17] after 'a severe but natural Labour', though the child would not have a happy life; his father would eventually come to call him 'my unfortunate son John'.

Early the following year, Susanna was pregnant again with her tenth child. At thirty, she was weary beyond her years. She developed a severe fever and lost a great deal of weight, but she delivered another brother for her children, Samuel Ward Flinders.[18] Their father noted that Samuel was a 'very fine boy' and it had been 'a good Labour', but Susanna had been 'very poorley for a long time'.

The baby was sent to suckle on a wetnurse, Mary Redhoof, while Susanna stayed in bed for six long weeks. She celebrated Christmas with her family, but during a bleak winter, she was white as the snow all around and before long was forced back to her bed. By the end of 1782, Flinders Sr was in the 'the utmost distress for her safety'.[19] Susanna became delirious, and Flinders despaired that the state of 'my poor wife' made it the most distressful time of his life. His thirty-third birthday came and went without him even noticing it, such was his anxiety, but he read everything he could on what Susanna was experiencing and noted half a dozen cases when the unfortunate women recovered after a change of air and scenery.

On 19 February 1783, with Matthew and the other children watching ashen-faced, Flinders Sr carried his emaciated wife to the carriage of his friend John Harvey, a grocer. Flinders Sr then drove Susanna to her brother John Ward's home at Bolingbroke, forty kilometres to the north-east, in the hope the visit might lift her spirits. Susanna remained there for a month, then Flinders Sr received an urgent letter from the Wards that she had taken a turn for the worse.

He set off immediately and found her 'bad indeed': he was not certain if Susanna even recognised him. She was 'greatly emaciated, & almost incapable of taking sustenance'. She was fed herbal potions for her nerves and another to stimulate the flow of menstrual blood, and while she recovered a little, it was all too brief.

Susanna Flinders died on the morning of 23 March 1783, aged just thirty-one. Her husband was beside himself with grief, writing that he had lost 'the dearest & most valued friend I had on Earth'. Matthew and the two older children watched on as their mother was buried in the churchyard of St Peter and St Paul's at Bolingbroke, in what her devastated husband described as 'a decent but not pompous manner'.

Fighting back tears, he wrote:

> My situation is truly horrible and unhappy on my own account, my comfort being gone, but doubly so on account of my 5 children, two very small and out at nurse … My tears are plentifully shed each day, and when I shall regain any peace I cannot tell … This world has now no charms left for me, there appears nothing for me but care & trouble. However God is infinitely Wise and Good and to him I resign myself all my concerns.[20]

MATTHEW'S COUSIN HENNY came to stay with the family to help them in their distress. She provided support and consolation for Flinders Sr, but still he wore his grief like a shroud and admitted that he was 'miserably unhappy'.[21]

He was a practical man, though, and a few weeks later, just as he had done following the death of five of his babies, he told himself that a 'continual grieving could be of no avail, but rather injurious'. He must now '[n]ote a circumstance [that] will perhaps appear somewhat odd in my records, after the real and extraordinary Grief which I have manifested for my late valuable partner & whom I shall regret to my latest hour'. Flinders Sr was thinking about marrying again, and admitted that he had 'pitched on the amiable' and attractive young widow Mrs Elizabeth Ellis.[22]

Since the recent death of her husband, Elizabeth had been living with her married sister Hannah Franklin in Spilsby, about forty-five kilometres to the north-east of Donington. John Ward told Flinders Sr that it was no good to keep moping about and that Elizabeth was a good woman who was lonely too. Flinders Sr wrote to the young widow suggesting that since they had both been bereaved, they might find some solace with each other. Elizabeth agreed. So he wrote to her again, and after a few letters, 'by which I was rather assured of an agreeable reception', he decided to meet her in person.

On 20 July 1783, just four months after Susanna's death, Flinders Sr, accompanied by John Ward, rode to Spilsby to meet Elizabeth for Sunday lunch at the Franklin house. They were greeted by 'most friendly treatment' and 'indeed had a most delightful visit'. Flinders Sr bought 'a new suit of Cloaths' from Hannah's husband, a shopkeeper named Willingham Franklin, because he would need to look his best during this new courtship.

His mind was in a whirl on the long ride home. Life was often short in the 1700s, and Flinders Sr did not want to waste a minute. Within two weeks he had decided that Mrs Ellis would be his wife, and stepmother to nine-year-old Matthew and the four other children.[23]

Flinders Sr was always careful with his money – he was a single man with five children and servants to feed – but he splashed out £10 to make renovations on his house. Young Matthew watched as workmen created a more comfortable bedroom for him and the older children, with an improved ceiling, a better window and a new closet. Two new hearthstones and fire screens were installed, and wallpaper was hung in the best room, which would accommodate Flinders Sr and his bride.

Despite his relatively speedy remarriage, Flinders Sr wanted Matthew and the rest of his family to understand that he would never forget his 'ever dear departed friend and wife':

God knows I have yet many bitter hours on her account, and I'm not without some fears, that it will be impossible for me ever to be so happy as I have been, but what can I do? I find my life uncomfortable and hope in the Society of the agreeable Woman I have chosen in some degree at least to regain my former happy state.[24]

On the first day of winter in 1783, Flinders Sr rode alone on horseback through a thick mist across his now well-worn track to Spilsby, carrying his marriage licence in his pocket. He arrived at mid-afternoon, and he and Willingham Franklin then met with Richard Vesey, the curate of Spilsby, who was to perform the wedding ceremony the next morning.

Flinders and Elizabeth Ellis were married after breakfast the next day, and at 10 a.m. Flinders and Franklin jumped on horseback, while the new Mrs Flinders, her sister and a maid climbed into a chaise carriage. The wedding party spent four hours on the rambling journey to Boston for dinner at the White Hart Inn, which boasted the best food in town. By a shining moon, the group then journeyed home to Donington to tell Matthew and the other children about their grand day and their prospects for a happy family life. Flinders Sr noted that his wedding 'expences' were £8 13s 2d.

Four weeks later, on the first day of 1784, frost was all around the Donington home and there were frequent severe winds, heavy snow and pools of ice. The poor of the town were 'most distressed', begging for charity. Flinders Sr, though, was still floating on air, 'having again the comfort of a kind and Bosam [sic] friend'.

'I can with truth assert,' he wrote, 'that every reasonable prospect of a Happiness in a 2d marriage I do experience as far as a month can make me a judge.'[25]

After Elizabeth recovered from a severe illness that included excessive menstruation that lasted for weeks, young Matthew journeyed with her back to the Franklin home in Spilsby for eight days of convalescence. The trip worked wonders and Matthew made a firm friend of his cousin Tommy Franklin,[26] who would visit him for holidays in Donington.

Whenever Matthew was in Boston, looking out over the wide expanse of The Wash, the prospect of travelling across the water stirred his enquiring mind. From early childhood he manifested a taste for discovery. Once he was missed from home, and after a search lasting some hours, he was finally found in a marshy field tracing a brook in the fens – 'to see,' as he said, 'where it came from'. His little pockets were stuffed with pebbles and plants, and all sorts of curiosities for later study.[27]

A few weeks after Tommy Franklin's first visit to the Flinders home in 1784, Matthew's father borrowed a pony for his boy and together they rode more than thirty kilometres south-east to the town of Tydd St Mary, to visit some Flinders relatives living there. Matthew was no horseman yet, but his father conceded that he 'performed tolerably'. That year Matthew's seagoing cousin Jackey returned from fighting the Dutch and the Americans, with thrilling stories of battles on the sea, and peoples and places on the far side of the world.

As his oldest child neared his eleventh birthday, Flinders Sr was thrilled at the way Matthew's study of Latin was reaping dividends. Matthew was only going to school at Mr Whitehead's for half-days, and was working on his Latin at home in the afternoons, and Flinders decided to keep it that way until the boy was twelve. Then he would send him for two years' tuition at the best nearby grammar school he could afford.

In the first week of 1785, Flinders Sr prayed that 'the Good God' would allow him to remain in fine health so that he could continue to provide a comfortable home and life for his 'Young ones'.[28]

That was not always easy. In the summer of 1785, measles made its mark on all the Flinders children, as well as their ten-year-old cousin Mary Franklin,[29] who was staying with them in Donington. Mary was known as 'Polly', and sometimes Matthew would refer to her as his 'dear little Molly'. All the children recovered well except for Matthew, who suffered terribly with fever and cough.[30]

When he was at last restored to health, he continued well in his studies, especially Latin, which he was now learning from Philip Lound, one of Whitehead's assistant teachers. Flinders also put Matthew and Betsey into a dancing school: he wanted them to have the social skills that would eventually help them make suitable marriages.

AS THE PROBLEM OF overcrowding on England's prison hulks worsened, the Irish statesman Edmund Burke and Lord Beauchamp[31] set up a select committee to investigate and report on the implementation of the *Transportation Act* for sending prisoners to overseas colonies. The Beauchamp Committee heard from Sir Evan Nepean,[32] Lord Sydney's[33] permanent undersecretary at the Home Department, on 27 April 1785. Nepean revealed that the government had already chosen Lemain, on the Gambia River, as the next convict settlement.[34] (It is now MacCarthy Island, Gambia.)

On 10 May, Banks appeared before Beauchamp's committee. Banks said he had no doubt that the soil in many parts of the eastern coast of New South Wales was sufficiently fertile to support a considerable number of Europeans who could cultivate it in the 'ordinary modes' used in England.[35] Botany Bay, he said, was 'in every respect adapted to the purpose'. He did not know enough about Aboriginal language or government to advise about negotiating the takeover of an area, he said, but fish were plentiful on the coast, and there were no dangerous wild beasts. He thought European cattle would thrive there, as would grains and legumes. Women might be brought from the Pacific islands to keep the men happy.[36]

IN JULY 1786, WHEN MATTHEW was twelve, his father sent him to a new school just opened by the Reverend John Shinglar in the village of Horbling, twelve kilometres west of Donington. Flinders Sr thought it a fine set-up with 'genteel people', and was confident his boy would do well boarding there, though it wasn't cheap: £18 18s for a year's tuition and living expenses. Then there was the worrying expense of clothes for a growing boy who had to keep up appearances for the sake of his well-known father.

Matthew would be one of only seven boys at the establishment, and Flinders Sr was happy that he wasn't too far away. Shinglar had just been made the curate at Horbling, but he had spent seven years as the assistant teacher at the Huntingdon School, and Flinders Sr was confident in his abilities. Matthew had now started in Greek, and there were high hopes he would do well in literature too. Flinders Sr estimated that, academically, Matthew was the 'first Boy except one'.[37]

As important as Matthew's schooling was, Flinders Sr had other pressing matters on his mind. Elizabeth delivered their first child – a daughter – at 4 a.m. on 30 August 1786. But it was a 'most tedious and severe Labour – pain excessive'. Flinders tried the forceps, but they were ineffective, and reluctantly he brought out the crochet hook instead. The baby did not survive; the mother barely did.[38]

BY THIS TIME, the British government had settled on Botany Bay as its next penal settlement rather than Gambia. Lord Sydney gave orders to the Admiralty for a fleet to be made ready. In consultation with Banks, Prime Minister William Pitt and his government put the plan for Botany Bay into action on 21 August. A warship was to be made ready as an escort for vessels carrying 'seven or eight hundred convicts' and 'two companies of marines to form a military establishment on shore'.[39]

Lord Sydney put Captain Arthur Phillip[40] in charge of the expedition. A small, olive-skinned man with a pear-shaped head, a large, fleshy nose and a smile missing a front tooth, Phillip had gone to sea at fifteen aboard the *Fortune*, stripping blubber from whale carcasses in the Barents Sea. He had fought in the Seven Years' War and the Battle of Havana. He became a captain in the Portuguese navy, serving in the fight against Spain and ferrying Portuguese convicts to Colonia del Sacramento, Uruguay. He was recalled to service for Britain in the war against America, commanding the 64-gun HMS *Europa*. From October 1784, Nepean – in charge of the British secret service – used him to spy on the French navy.

By August 1786, when the 'Heads of a Plan' for Botany Bay was devised, Phillip was seen as a cool, calm and competent leader.

MATTHEW HAD GROWN into a bright, active and dependable boy, who at thirteen was trusted to escort his stepmother on a two-week vacation with the Franklins in Spilsby. He then rode with Tommy Franklin for another twenty-five kilometres north to visit his Aunt Mary Carr[41] and her husband, Richard, who were both druggists.

Matthew's travels on land often brought him near the ocean, and he longed to explore it. His cousin Jackey was now serving with the navy in Jamaica, and Matthew was envious of his adventures.

At the end of 1787, Arthur Phillip's First Fleet – eleven ships carrying the first convoy of prisoners and their jailers – had reached Cape Town via Rio de Janeiro, and was heading towards Botany Bay. They reached their destination on 18 January, but six days later were startled to find two French ships[42] approaching. These 500-ton vessels were under the command of the Comte de La Pérouse,[43] who had tormented the British Navy during America's War of Independence, and who was now on a mission to circumnavigate the Pacific for King Louis XVI.[44]

Phillip found Botany Bay a different place from that described by Joseph Banks. The waterway was open and unprotected against storms, and the water was too shallow for his ships to anchor close to shore. Much of the surrounding countryside was of barren, sandy soil. Looking for a better place to establish his penal settlement, Phillip sailed a few kilometres north. Entering Port Jackson, he 'had the satisfaction of finding the finest harbour in the world, in which a thousand sail of the line may ride in the most perfect security'.[45] He wrote:

> The different coves were examined with all possible expedition. I fixed on the one that had the best spring of water, and in which the ships can anchor so close to the shore that at a very small expence quays may be made at which the largest ships may unload.[46]

He named the cove after Lord Sydney.

On the evening of 26 January 1788, Phillip oversaw the erection of a flagpole at Sydney Cove and the raising of the Union Jack. His marines fired volleys, and he saluted the health of the king, George III, and the future of this new colony.[47]

Phillip then sent stocky, round-faced Lieutenant Philip Gidley King to lead a party of fifteen convicts and seven free men to Norfolk Island, 1700 kilometres to the east.[48] King arrived in March and quickly established a settlement, and a family. He formed a relationship with convict Ann Innett and they had two sons, Norfolk and Sydney, who would both become naval officers.

Closer to home, Governor Phillip dispatched Scottish sea captain John Hunter[49] on an expedition to explore the Parramatta River, which fed into Sydney's harbour. Hunter was a fairly tall man for his times at 178 centimetres, fifty years old with a strong face, kind blue eyes, thick eyebrows and white hair. He had grown up at Leith, the port area of Edinburgh, and his Scottish brogue was delivered with authority. He was also a keen natural historian and disciple of Joseph Banks, and a talented artist whose sketchbook grew to contain some of the earliest artistic impressions of people, birds, flowers and fish found in and around Sydney, and on Norfolk and Lord Howe islands.

Hunter had been in the navy for more than thirty years and was a veteran of wars with the French and the Americans, and he had served in the East Indies. He knew sailing and navigation inside out.

On the voyage of the First Fleet, Hunter had commanded the flagship HMS *Sirius* from England, but now, in a small open launch with Lieutenant William Bradley and a guard of red-coated marines, he made soundings along the Parramatta River. On 5 February, while Hunter was having breakfast at a place that from then on was known as Breakfast Point, he made what is believed to be the first contact between Europeans and the traditional owners of the land, the Wangal people. Bradley recorded:

We made signs to them to come over & waved green boughs, soon after which 7 of them came over in two

Canoes & landed near our Boats, they left their Spears in the Canoes & came to us; we tied beads &c. about them & left them our fire to dress their Muscles [mussels] which they went about as soon as our Boats put off.[50]

Hunter and two others went up to meet them, holding up their hands to show they were unarmed.

We mett them & shook hands but they seemd a good deal Alarm'd at four or five Marines under Arms by the Boats, upon which they order'd to ground their Arms & stay by them, they then came up with great Chearfulness & good humour and seated themselves by our fire amongst us, where we eat what we had with us, & invited them to partake, but they did not relish our food or drink.[51]

During the time Hunter was employed surveying the Parramatta River, he 'had frequent meetings with different parties of the Natives, whom we found at this time very Numerous'. It was difficult for British naval officers to see these new interactions through Aboriginal eyes. Hunter only had a skewed sense of what he and his men were experiencing as they had no understanding of cultural traditions of the Wangal.

I confess I was a little surprised … after what had been said of them in the Voyage of the *Endeavour*, for I think it is observ'd in the Account of that Voyage, that at Botany Bay they had seen very few of the natives and that they appeard a very Stupid & incurious people. We saw them in very Considerable numbers & they appear'd to us to be a lively & inquisitive Race; They are a Straight, thin, but well made people rather small in their limbs but very active, they Examin'd with the utmost attention & great … astonishment at the different Coverings we had on, for they certainly consider'd our Cloaths as so many different Skins, and the Hatt a part of the head …[52]

Although many of the Europeans at the time wrote of First Nations people as curiosities under study, Aboriginal people were certainly studying the white people too. Although there was a power imbalance, these encounters were never completely one-sided.

While Hunter tried to have peaceful dealings with the First Peoples of the land, he was also extending an olive branch to the French, despite Britain's years of naval conflict with them. La Pérouse, still in Botany Bay, gave his journals and letters to Hunter to take back to the Cape of Good Hope on the *Sirius*, from where they could be conveyed to France. La Pérouse expected to be back in Paris and reporting to Louis XVI by June 1789.

The French ships remained in Botany Bay until 10 March 1788, before heading for New Caledonia. The two great vessels and the 220 men on board were lost at sea.[53]

MATTHEW WAS HEADING towards his fourteenth birthday when his father decided he'd had enough formal education. He took Matthew out of the Reverend Shinglar's school after the boy had made what Flinders Sr called 'a Proficiency in Learning' that exceeded his greatest hopes. It was now time for Matthew to be schooled in the family business. Heaven knows Flinders Sr was busy enough for two men with the exhausting work, which stretched again to his own bedchamber.

On 22 February 1789, Elizabeth delivered a healthy daughter named Hannah Flinders after a four-hour labour. Elizabeth took ill with a violent fever after the baby arrived safely, and Matthew was ordered to race on horseback to Spilsby and urgently return with Hannah Franklin in case the end was near for her sister. Mercifully, it wasn't, but it was five weeks before Elizabeth could leave her bed.

Matthew knew that his father did important work, and that many of the residents of Donington and surrounding hamlets owed their lives to his skill and willingness to rush to an emergency. But Matthew did not share his father's passion for medicine or midwifery.

Flinders Sr kept a slate in his apothecary shop in the annexe beside his house, on which he wrote notes for his cases. Matthew knew how much his father wanted him to follow his same career path, and was dreading breaking the heart of a man he loved and respected. So he wrote a brief note to his father on the slate, explaining that he had his heart set on joining the Royal Navy.[54]

Flinders Sr wasn't convinced and continued seeking out medical opportunities for his son. Matthew wrote to Jackey, who had now spent nine years in the Navy but had not yet sat for the examination to become a lieutenant. Jackey told him that making a good career in the Navy was more about who you knew than what you knew, and that the boy would do well to find a patron with powerful connections. He also told Matthew to read up on important texts: the works of the ancient Greek mathematician Euclid, the 1754 book *Elements of Navigation* by the English mathematician John Robertson, and the 1772 work *The New Practical Navigator and Daily Assistant* by the Scot John Hamilton Moore.

Jackey then wrote to a friend, Robert Laurie, who was an acting lieutenant on HMS *Alert*, a two-masted fourteen-gun brig-sloop, then in the West Indies. In the late 1700s, entrance to the Navy was often secured by the nomination of a senior officer, who could take young recruits on board as a favour to relatives, to curry favour with wealthy friends or even in return for cancelling a tradesman's bill.[55] Jackey asked Laurie to accept Matthew as a 'lieutenant's servant' and put him on the ship's muster list, even though Matthew remained at home in Donington.

Though illegal, the practice of listing unseen servants on naval muster lists was not uncommon. It gave ambitious young men official 'sea time', as they had to complete six years of naval service before sitting for their lieutenancy exams. The practice had been in vogue for decades. Thirty years earlier, for instance, William Bligh had been signed onto a vessel at age seven.

Laurie was only too happy to register Matthew on the ship, because if his servant was absent, the officer pocketed his pay. Matthew's name remained on the *Alert*'s muster list until 4 October 1790.[56]

In Donington, the teenager devoted himself to studying navigation, geometry and trigonometry. He was also studying the politics of naval advancement. His cousin Henny was now the governess in the family of the 'Tough Old Commodore' Thomas Pasley,[57] commander of the 64-gun HMS *Scipio*. She gushed to him about Matthew's brilliance.

Years later, Matthew told stories, perhaps apocryphal, of how he was invited to dine with the Pasleys. They insisted he stay overnight. He had brought no nightshirt with him, though, and when he was shown to his bedroom, he found sleeping attire laid out on the bed, folded and apparently for his use. When he put the garment on he thought it rather odd, as it had frills and ribbons at the front and delicate lace at the wrist and collar. He was mortified at breakfast the next morning to discover that one of Pasley's daughters had given up her bedroom for him – and forgotten that she'd left her nightdress in the room.[58]

Despite Matthew's bedroom miscalculation, Pasley thought him a clever lad and promised him a place on the *Scipio*.

Matthew's father tried one last time to convince him to pursue medicine. On 26 April 1790, a month after Matthew's sixteenth birthday, he and his father journeyed fifty kilometres south from Donington to the town of Lincoln, to meet with Alderman Joseph Dell, a local apothecary and surgeon. As they travelled, Flinders Sr sensed he was fighting a losing battle.[59] He felt Matthew was already showing the vanity and arrogance not uncommon among British naval officers of the time.[60]

Alderman Dell agreed to take Matthew on as his apprentice and promised to pay him 10 guineas a year, well beyond Flinders Sr's budget. Flinders Sr needed Matthew as an assistant and said he would miss his boy a great deal, but he felt the offer was the best thing for the lad: it would put money in his pocket and new clothes on his back. Matthew's father would have to hire an assistant, and when his eldest daughter Betsey was not at school, he would make her as 'usefull in the Shop' as he could.[61]

But on the way back to Donington, Matthew told his father that he was planning to travel much further than Alderman Dell's shop in Lincoln. Not only was he on the muster for the *Alert*, he had read a book by one of his father's favourite authors, Daniel Defoe. The book was *Robinson Crusoe*,[62] the thrilling survival tale of a shipwrecked mariner who had defied his father's wishes and followed his heart to go to sea. The tale ignited Matthew's maritime ambitions, 'against the wishes of his friends',[63] like a spark falling on straw.

On Friday, 14 May 1790, just three weeks after their journey to see Alderman Dell, a crestfallen Flinders Sr went with Matthew to Spalding. There the small, active man with kindly dark eyes embraced his sixteen-year-old firstborn and bade him safe travels, wherever he might go. He put £20 into his son's pocket – about half a year's wages for a labourer – to help him launch his career, and later picked up the account of £7 7s for the young man's smart new uniform of white breeches and stockings and a long blue linen-wool frockcoat with brass buttons.

From Spalding, Matthew took a coach and saw London for the first time. St Paul's Cathedral and Westminster were as astonishing as the new lands and peoples he would soon encounter. From London, he journeyed east to Rochester, on the Medway in Kent, where, at the vast naval base at Chatham, great warships with their yellow and black hulls bobbed about under a veritable forest of towering masts, while dozens of smaller craft swarmed about them.

Even though Matthew would remain on the *Alert*'s muster list for five more months, he joined the crew of Pasley's *Scipio* on 17 May 1790.

Flinders Sr lamented that his dear son's desire to join the Navy 'has long been his choice, not mine'. He had become a stubborn lad, and his father despaired that with another war looming in France, Matthew would find life dangerous on the high seas.

'Henny got him this situation, pray God it may be to his advantage,' he mused. 'I suppose he will be in the capacity of a Midshipman, the stipend of which is but bare subsistence.

'I shall heavily miss him.'[64]

Chapter 3

*'He is going with Capt. Bligh in the Providence to circumnavigate
the Globe … and will be near 3 years performing this great under
taking … God only knows what may be the event
of such a long voyage.'*

MATTHEW FLINDERS SR, ANXIOUS ABOUT HIS SON'S FIRST
GREAT ADVENTURE.[1]

THE WORLD WAS CHANGING rapidly around the young
Matthew Flinders in 1790 and Britain was desperate to
preserve its empire.

Britain had lost thirteen American colonies after they declared
their independence. Spain was now in control of Florida, and
the French had taken the Caribbean island of Tobago, as well as
Senegal in Africa. Another war loomed with France, where a civil
uprising would soon depose the monarch and see thousands of
heads roll from the guillotine.

In October 1788, with the penal colony at Sydney Cove
struggling to feed itself, Phillip sent John Hunter on a voyage in
the *Sirius* to the Cape of Good Hope. Hunter returned in May
1789, having circumnavigated the globe, but the ship was leaky
and constant pumping was needed to keep it afloat. By 1790 a
Second Fleet had arrived, bringing more hungry prisoners to an
infant colony that was again short on food. Before long, Governor
Phillip was writing to Joseph Banks in London to say that his
health was 'gone': he was 'worn out'[2] and desperate to go home.

Colonisation remained lucrative for Britain, though, with the East India Company flying the British flag in much of the Asian subcontinent. George III was determined that Britannia would continue to rule the waves, and millions of pounds were being spent to ensure he had the most powerful navy and army in the world.

Off the east coast of Kent, sixteen-year-old midshipman Matthew Flinders applied himself tirelessly to learning all he could about the *Scipio*, pacing out its 160-foot, five-inch (forty-nine-metre) length, and studying its sixty-four cannons, twenty-six of which fired massive twenty-four-pound (10.9-kilogram) cannonballs. The ship was of a design considered optimal for war: fast and manoeuvrable under sail, and well armed, with heavy guns on two decks.

As the summer breezes whipped up the waves of the English Channel, Matthew developed his sea legs, growing accustomed to the rise and fall of the vessel as it sailed up and down the sheltered waters of the Downs. He began memorising every part of the complex piece of machinery that was a fully rigged British warship, and he literally 'learnt the ropes', studying how to perform every task, from working on rigging high up on swaying masts while hanging on for dear life, to catching rats and cleaning the crowded, reeking living quarters, all the time recoiling from the salty language that was unheard in his father's home. He learnt from the more experienced sailors how to study both the sky and the sea for approaching peril.

Within a few weeks, Pasley transferred to an even bigger ship, the four-year-old HMS *Bellerophon*, which most of the sailors called the 'Billy Ruffian'. More than fifty metres long, it carried seventy-four guns, with twenty-eight of them firing thirty-two-pound (14.5-kilogram) cannonballs. It would become one of the most famous ships in the world, and Matthew was entered on the *Bellerophon*'s books as a midshipman.

He had only been in the Royal Navy for three months when he was allowed leave to return to Donington. In the short time he had been away, he had altered physically and psychologically, and already saw himself as a man of the sea.

Thomas Luby painted the HMS *Bellerophon* lying at anchor off Berry Head, Torbay, in 1815 with the defeated French emperor Napoleon Bonaparte onboard. Matthew Flinders sailed on the ship as a midshipman in 1790.

The dark-haired young officer who arrived unannounced for a surprise visit at his father's home on 6 August 1790 was dressed in his crisp blue coat, white britches and black bicorne hat. His thin, pale face, aquiline nose and jutting chin gave him a determined bearing, and he had the keen-eyed visage of an intrepid adventurer with his dark, bright eyes and small, firm mouth. Matthew was now 168 centimetres tall,[3] a little below average height for the times, with a slim, wiry body and brisk mannerisms.

'He is grown, and much altered by his Uniform and dress,' his father wrote, 'and he appears satisfied with his Situation, I pray God he may be prosperous.'[4]

Matthew spent four days with his family before returning to duty, but he had already grown tired of sailing up and down the English Channel. He was transferred to HMS *Dictator* for a brief period, but the work and scenery remained the same. Matthew wanted to see the world, and he wanted the sort of adventure that men such as Cook and Banks had encountered on the *Endeavour*.

Then into Matthew's life came the pugnacious William Bligh, who at thirty-six had just completed one of the most audacious adventures in maritime history.

FOLLOWING THE DEATH of expedition leaders James Cook (in 1779) and Charles Clerke (a few months later, from tuberculosis in the Bering Strait), William Bligh had led the *Resolution* and the *Discovery* back to England in August 1780.

Bligh married the following year, and soon after fought against the Dutch in the North Sea. He then joined the fight against the French and Spanish at Gibraltar in 1782, under Lord Richard Howe,[5] known among his sailors as 'Black Dick' – a reference, we are told, to his dark complexion.

Five years later, Bligh's patron, Joseph Banks, endorsed him for a special mission to take breadfruit – a starchy food staple in Tahiti – to the British-controlled islands in the Caribbean, as a cheap way to feed slaves. Banks was an expert on breadfruit, having studied it in Tahiti on the *Endeavour* voyage, and he chose a converted collier for the mission, which had been renamed HMS *Bounty*.

Despite Bligh's reputation for surliness and a harsh tongue, Banks had been impressed by his help in preparing an official account of Cook's last voyage,[6] and assured him of the command for the breadfruit expedition. Bligh wrote to Banks thanking him for his 'great goodness'.[7]

The *Bounty* set sail from Spithead for Tahiti on 23 December 1787. By the time it arrived in Cape Town on 28 May 1788, after a difficult voyage, the skipper had appointed the dark and muscular Fletcher Christian[8] as acting lieutenant. After a roundabout journey of 40,000 kilometres, the ship arrived in Tahiti on 26 October 1788.

The crew stayed there for five and a half months, collecting 1005 breadfruit trees, while also falling in love with the tropical paradise – and with the local women. Eighteen of the *Bounty* men, including Christian, received treatment for venereal infections, a disease they spread in this tropical paradise, apparently without care or remorse. While Bligh remained faithful to his adored

William Bligh, the master mariner, captured by artist John Webber in about 1776 while serving with Captain Cook on the *Resolution*. Bligh taught Matthew much of what he knew about map-making though Matthew clashed with his skipper. *National Portrait Gallery, Canberra*

wife, Betsy, he understood the excitement of the men in a land of beautiful women, because 'the allurements of dissipation are beyond any thing that can be conceived'.[9]

Despite Banks' reputation as a bed-breaking ladies' man in his younger days,[10] Tahiti had been a sexual eye-opener even for him. While he would later regale the American scientist Benjamin Franklin with boasts of how he taught the Tahitian woman to kiss on the lips in 1769, they had some surprises for him too. Bligh was appalled by what he called the 'sensual and beastly act of gratification', where 'even the mouths of women are not exempt from the pollution'.[11]

When the *Bounty* left Tahiti on 5 April 1789, it was a powerful wrench for the crew, who resented more and more Bligh's harsh tongue and aggressive manner. Early on the morning of 26 April, off the Tongan island of Tofua, Fletcher Christian and his supporters mutinied, tying Bligh's hands behind his back and casting him and eighteen others into a boat just seven metres long. Although the mutineers taunted him from the deck of the *Bounty*, no doubt believing he had no chance of survival, Bligh then commenced one of the most heroic seafaring feats of all time, leading his loyal band in their tiny craft on a 6500-kilometre, forty-seven-day journey across the ocean to Kupang, Timor.

While some of the mutineers died in the Pacific, and others were eventually returned to London and hanged, Bligh was exonerated at a court martial in October 1790. Two months later, Banks wrote to his old schoolfriend Lord Auckland,[12] Britain's ambassador at The Hague, to tell him King George had approved a second breadfruit expedition, with Bligh again at the helm.

As well as transporting fodder for enslaved people, Bligh saw the expedition as vital for further exploration of sea lanes to the Pacific.[13]

Bligh and Banks both saw that mapping a safe way through the 150-kilometre stretch of water and hazardous reefs separating Cape York and New Guinea was crucial to the survival of the penal colony in Sydney. A navigable route would cut the travel time from New South Wales to Britain, allowing ships to bypass the great icy waves of the Southern Ocean.

On 23 February 1791, Bligh wrote to Banks to say he had gone to the Blackwall Yard on the Thames and purchased a new three-masted sloop-of-war, HMS *Providence*, for the voyage. It had three decks and was slightly bigger than the *Bounty* at 420 tons (381 tonnes) and ninety-eight feet, eleven inches (thirty metres) long. It would be crewed by a hundred men. It carried twelve four-pounder guns on carriages, and fourteen half-pounder swivel guns for fast action.[14] Bligh had also chosen a small vessel which could act as a tender, a stout little brig aptly named the *Assistant*, fifty-one feet (15.5 metres) long and 110 tons (100 tonnes).

Bligh appointed his friend Nathaniel Portlock,[15] a twenty-year veteran of the Navy and a survivor from Cook's final voyage, as the skipper of the *Assistant*. Francis Bond,[16] the son of Bligh's half-sister, would be first lieutenant on the *Providence*, and James Guthrie the junior lieutenant.[17]

In their sumptuous and imposing boardroom facing Whitehall, surrounded by dark oak panelling and floor-to-ceiling windows, the Lords of the Admiralty approved Bligh's plan to 'make a complete examination of Torres Strait'.

As Bligh was preparing to sail, the imagination of Matthew Flinders was working overtime, fired by the story of Robinson Crusoe and the voyages to exotic, thrilling and often dangerous destinations by Cook, Bligh and the French count Louis de Bougainville.[18] He pleaded with Commodore Pasley for the chance to serve on the *Providence*, and Pasley and Bligh agreed.

On 8 May 1791, Matthew saw the ship for the first time at the sprawling Deptford yard, twelve hectares of brick warehouses and workshops on the Thames, seven kilometres downstream from London. On one side of the yard were rolling green hills and lush pastures, and on the other a forest of towering masts. A steep wooden plank led Matthew onto a new vessel that still smelled of sawn timber and the pitch used to seal the deck. He doffed his hat in a salute to Bligh. Matthew knew the tough skipper was regarded as a tyrant by some, but he believed the great mariner could teach him all there was to know about ships and the sea.

Matthew stowed his gear below deck. Having just turned seventeen, he was about to set off on a mission that was as thrilling and daunting as a voyage to the moon would be for later generations.

As he prepared for the voyage – and faced the prospect that he might never return, even with Bligh's reputation as a master mariner – Matthew visited his family in Donington on 11 May 1791, this time staying for ten days. Flinders Sr had just become a father again, to a daughter, Henrietta. He hoped she might be the last child Elizabeth bore. The labours had taken a terrible toll on his delicate wife, and with so many dependants, money for the family was always tight. Matthew told his anxious father that he would probably be at sea for 'three years performing this great undertaking', as the plan was to 'Circumnavigate the Globe'.[19]

When the visit ended, Matthew farewelled Elizabeth, his siblings and his little step-sisters. He and his father embraced tightly, unsure whether they would ever see each other again.

'God only knows what may be the event of such a long Voyage,' his father wrote ruefully.

> May [God] prosper and befriend [Matthew] in every Country, Climate and People. I have desired [Matthew] to keep an exact journal as if it please God we live, and he returns safe, I have some idea a publication possibly may be advantageous. He has made much improvement in his knowledge of Navigation and is thanks be to God well and in good Spirits. If he is Successful this Voyage may be a great means to promotion.[20]

Not long after his return to Chatham, and as the *Providence* began taking on the large wooden boxes and pots in which the crew would transport the breadfruit plants from Tahiti, Matthew wrote to Commodore Pasley of his visit home. Pasley replied, noting the grand impression the young midshipman had made since joining the Navy a year earlier:

> I am favored with your letter on your return from visiting your friends in the Country and am pleased to hear that you are so well satisfied with your situation on board the *Providence* – I have little doubt of your gaining the good opinion of Capt. Bligh if you are equally attentive to your Duty there as you were in the *Bellerophon*.[21]

WITH MATTHEW'S HEART pounding, the *Providence* and the *Assistant* left Deptford on 22 June 1791, bound for Jamaica via Tahiti. His family would recall that from this time 'may be dated that passion for discovery, which never left him during life'.[22]

Bligh's men eased the two ships down to Gallions Reach, fifteen kilometres downstream, where they took on guns and ammunition. The two ships sailed on together, little and large, past Gravesend, heading for the open waters of the English Channel. But the *Providence* was causing problems and Bligh told Banks that while he was sure the ship would eventually 'answer

Lieutenant George Tobin's drawing of the *Providence* and the little *Assistant* on which Matthew sailed with William Bligh to Van Diemen's Land and Jamaica. *State Library of NSW FL1606671*

to my most sanguine expectation when I have taken in a little more iron ballast ... at present she is rather, (what we call) crank, that is, lays rather too much down when under Sail'.[23]

At Sheerness, where the mouth of the Medway met the mouth of the Thames, twenty more tons of iron ballast and some shingle was placed in the bilge to give *Providence* better balance.

Providence boarded fifteen marines, their lieutenant, a sergeant, two corporals and a drummer. Bligh was taking no chances with another mutiny. The ship also carried two skilled botanists, James Wiles[24] and Christopher 'Paddy' Smith, who would take care of the breadfruit plants bound for Jamaica. Wiles, six years older than Matthew, came from Holywell, near Stamford in Lincolnshire, and would become firm friends with his young neighbour.

The botanists were also tasked with helping protect the crew from the ravages of the sailor's constant curse, scurvy, a condition caused by a lack of vitamin C and which resulted in loose teeth, paralysis and sometimes death. Banks provided Wiles and Smith with nectarine trees and pineapple plants, and they would also busy themselves raising the 107 orange trees and twenty lemon trees on board.[25]

Each ship received a year's supply of food. Salt pork and salt beef were to be served on alternate days. There was hard biscuit, oatmeal, peas, cheese and butter. In the absence of fresh water, there was to be a gallon of beer per man per day, and wine and brandy for when the beer turned rancid. In the never-ending fight against scurvy, Bligh insisted also on supplies of sauerkraut, spruce beer and wort, a sugary liquid extracted from the mashing process during the brewing of beer or whisky.

There were twenty-seven men aboard the little *Assistant*, including four marines, and 100 men on *Providence*.[26] Bligh's crew included Lawrence Lebogue and John Smith, who had both been cast into the open boat with him after the *Bounty* mutiny. Matthew was fascinated by the way Bligh controlled *Providence* as it sailed at great speed along the English coast, past Brighton and Bournemouth to Spithead off Portsmouth.[27]

Early in the afternoon of 2 August 1791, and with a fine breeze blowing, Bligh ordered anchors aweigh and the ships finally left English waters. As the crew took their last sight of the hills of their native land,[28] Matthew wondered if he would ever see home again.

MATTHEW HAD HEARD horror stories about the wild seas in the Bay of Biscay, off the southern coast of France, but his introduction to ocean sailing was not unpleasant as the ships made slow but steady progress towards Africa. Bligh had a sentry guard a large fire in the brick-lined galley that burnt constantly and was used to dry the men and their clothes. Rain started pouring two weeks into the voyage, and nothing generated sickness on board a ship so much as wet or damp clothes.[29]

Matthew quickly became accustomed to the constant creak of straining timbers and the whistle and whoosh of wind in the sails, and the noise and smell of the livestock on deck.

The days were always busy as men kept watch and cleaned. On Sundays they were mustered and inspected. Bligh performed a divine service until his health began to deteriorate and he was confined to bed. Matthew, his pale skin scorched by the blazing heat, nevertheless impressed all on board with his sailing prowess and his navigation skills, balancing himself on the rolling deck as he used the heavy brass sextant to 'shoot the sun' and determine the ship's latitude. He and the other seven midshipmen also maintained the three chronometers used for astronomical observations, and to pronounce noon and the formal beginning of a new nautical day.

He copied for himself the ship's Standing Orders, the instructions designed to keep the crew and ship as healthy, clean and disciplined as possible, and he noted how the ship's log and other records were kept. Bligh had learnt these tasks from Cook, and Matthew was now learning them in his turn.

After twenty-five days at sea, and on a hazy, misty morning, Matthew heard the lookout perched high above deck shout 'Land ahoy' after seeing the sun glistening off the distant snow-capped volcanic peak Mount Tiede. The ships had reached Tenerife, the most populous of the Canary Islands, with about 68,000 inhabitants guarded by a small militia of 300 or so, who mustered every Sunday to go through their drills.

The *Providence* and the *Assistant* moored in the town of Santa Cruz after passing a barren, dreary landscape with 'not a vestige of vegetation ... but on the very summit of the mountains, and on a few spots in the valleys'.[30] Bligh sent Lieutenant Guthrie on shore to ask if the governor would return a cannon salute, but it was politely declined: the Spanish government frowned on the expense of such customs.

The request to procure water and refreshments, on the other hand, was 'most readily granted',[31] and Matthew was given leave to go ashore. He stepped on foreign soil for the first time as he and the botanist Paddy Smith went exploring with some others. Bligh decided to spend a week in Santa Cruz and sent his men off to find 125 gallons (570 litres) of the best local wine to bring back for Joseph Banks,[32] but the crew asked to leave after a couple of days, preferring the ocean breezes under sail to the sultry and oppressive sun on that small, rugged island. The temperature hit more than 100 degrees Fahrenheit (37.8 degrees Celsius) on the sand.[33]

Matthew had never experienced such heat, but he still kept busy investigating this new place and the swarthy, bony people he met. For a young man who had hardly been out of Lincolnshire a couple of years earlier, there were fascinating sights everywhere. He kept his promise to his father and made detailed notes in a journal about everything he saw. He also kept Commodore Pasley informed, penning regular updates that he gave to passing ships heading back to England.

Santa Cruz was not a large town but its streets were wide, 'ill pavd [sic] and irregular'. The houses of the wealthier inhabitants were large, and while they had little furniture, the homes were airy and pleasant, 'suitable to the climate'. Most had balconies, where the owners sat and enjoyed the ocean breezes. On the east side of the town was a small public garden containing two rows of poplar trees, and full of the marvel of Peru, a small, sweetly scented perennial that flowered in the afternoon with clusters of red, pink, white, yellow or striated tubular flowers.[34]

In the town of Laguna there was a large prison with inmates of both sexes, 'who looked thro' the Bars and begd [sic] very heartily'. Matthew and his companions met an old gentleman who, in exchange for the gift of a 'large ugly knife', asked the visitors to his house and 'entertaind [sic] us with Fruit and some excellent Wine'.[35]

We visited a nunnery of the order of St Dominic. In the chapel was a fine statue of the Virgin Mary, with four wax candles burning before her. Peeping through the bars, we perceived several fine young women at prayers. A middle-aged woman opened the door halfway, but would by no means suffer us to enter this sanctified spot. None of the nuns would be prevailed upon to come near us. However, they did not seem at all displeased at our visit, but presented us with a sweet candy they call Dulce, and some artificial flowers, in return for which Mr Smith [the botanist] gave them a dollar.[36]

On the north side of the town, next to the beach, was a small fort and barracks for the soldiers, but to Matthew they were dilapidated buildings 'as shabby and as despicable as my old Hatt [sic]'.[37]

Matthew drew a map of his journey so far, and made sketches of the islands and the African coastline. He sent all of them home to his father in Donington, along with notes about the ships he had seen on the journey. He reported that Bligh was well pleased with his work.[38]

CAPTAIN BLIGH'S HEALTH had suffered after the mutiny on the *Bounty*, when food and fresh water had been severely rationed during the gruelling month and a half of his open-boat voyage.

By the time *Providence* had reached Tenerife, he was seriously ill with 'a violent attack of a fever from being exposed to the sun ... and dreadfull headach'.[39] He complained that he had never experienced such extreme heat in all his time at sea, and that the winds at night blew like a furnace.[40]

With Bligh so ill, Portlock led the ships out of Santa Cruz on 2 September, bound for the Cape Verde islands and the Tropic of Cancer. Matthew and the crew suffered in the heat too – as did the deck timbers. Portlock insisted that the decks were washed down every evening with seawater to stop them cracking and to help cool the sleeping quarters below. Yet it was still so musty for the men trying to rest in their hammocks that wind sails had to be fitted on the hatchways to ventilate them with fresh air.

As the wind died, the vessels stalled, covering just seven kilometres an hour. It took nine days in oppressive heat for the ships to cover the 1700 kilometres from Tenerife to Cape Verde, off the coast of West Africa. Storm petrels – or 'Mother Carey's Chickens', as the sailors called them – were everywhere over the sea, often hovering near the surface in the wake of the two vessels, searching for food.

The ships reached the bay at Porto Praya, on Santiago in the Cape Verde islands, on 11 September. There was a whaling boat from Nova Scotia in port and an American schooner, and

Portlock gave the Americans Bligh's dispatches on the progress of his voyage to take back to London. Bligh was so sick, though, that the dispatches had to be written by the ship's surgeon, Edward Harwood.[41]

Matthew noted the 'Great Numbers of fine beautiful fish of various colours about the ship'; the Spanish mackerel were 'very conspicuous'. Flying fish abounded too: one that was fourteen and a half inches long (thirty-seven centimetres) flew onto the deck as a special dinner guest. The fish were similar to mullet, and the third lieutenant on the *Providence*, George Tobin,[42] who was compiling an illustrated record of the voyage, imagined it would be excellent sport trying to shoot them when they were on the wing.[43]

Most of the Portuguese inhabitants lived in 'straw coverd Hutts'. Matthew observed lush, green groves of coconuts, oranges, plantains and sugarcane, in sharp contrast to the rocky land and mountains surrounding them.

Lieutenant Tobin was sent ashore to ask the governor if the ailing Bligh could stay at his residence until his fever passed. The governor, who resided in 'a very sorry mansion indeed' – poor even by comparison with a small English cottage – told Tobin that Porto Praya was in the grip of its own fever pandemic, which was carrying off five or six people daily, and so it would be safer if Bligh stayed on his ship. He sent over a few green oranges and some decayed coconuts, and Portlock decided to sail on to cooler climes.

Lieutenant Bond was put in charge of the *Assistant*, which despite Bond's coaxing could not match the speed of the bigger vessel, and *Providence* often had to proceed under reduced sail so that it could keep the smaller ship in visual contact.

MATTHEW MADE DETAILED notes of the extraordinary creatures he encountered on this voyage. Butterflies and moths were constantly about the ship, some with light red and purple variegated wings. The *Providence* sailed through a sea of what the other sailors said was fish spawn, extending in a line, a few metres broad, as far as the eye could see and having the resemblance of sawdust on the surface of the water.[44]

On 16 September[45] heavy rains drove the heat away for a while, and the men brought some nectarine plants on deck to receive a good soaking. They caught what Flinders called the 'Hedge Hog fish',[46] a light blue creature, finely striated, spotted with black and covered with black spines. Its defence was to suck in large amounts of water and inflate its size to ward off predators. Porpoises played around the ship, and Matthew and the rest of the crew marvelled at the way they could leap two metres above the water, and double the distance across it. The dark-brown man-of-war birds (better known today as frigatebirds) astounded the sailors by the way they could dive at astonishing speed from a height of double the *Providence*'s highest mast straight into the water to scoop out a fish. One bird dropped a fish in mid-flight and then swooped down to catch it again before it hit the water.[47]

Bligh was slowly recovering from his fever when he was cheered on deck on 3 October 1791 to preside over 'crossing the line' festivities, held every time a British ship moved into the Southern Hemisphere. The crew had rigged up a 'ducking chair', a contraption to drop all those crossing the equator for the first time into the ocean, but Bligh prohibited it for reasons of safety. Still, in the spirit of maritime mayhem, he allowed these 'new chums' – including Matthew – to undergo a baptism of a different kind.

The initiates were ordered below deck. Tarpaulins were placed over the hatches and a sentry was posted to prevent any of the youngsters from seeing what Matthew called the 'dark Mysteries' going on above. Presently, they heard a voice call one of them out. Two sailors with blackened faces came down as 'constables', blindfolded a man and marched him on deck for a special ceremony.

Matthew could hear muffled sounds and splashing above. The ritual was repeated again and again until it was Matthew's turn, and he was blindfolded and escorted out into the light. When the blindfold was removed, he saw two sailors dressed as a hideous

King Neptune and his ugly wife. They had swabs for tails and oatmeal on their faces, and they were carrying tridents. They were seated on makeshift 'thrones' on the quarterdeck, mounted on a gun carriage that was rolled along the deck.

One by one, Matthew and the other newcomers had to pay a fine in liquor before they had their faces smeared with an oatmeal concoction and were then thrown backwards into a tub of water as the older sailors tossed buckets of seawater over them. Matthew picked up a bucket and started throwing water back at his superiors, and a huge water fight erupted. There was much laughter and bawdy language. Bligh called a halt before it all ended in tears, and the men went downstairs to put on dry clothes. The ceremony concluded with Bligh giving all men on board a dram of liquor.[48]

Three days later, the sun was nearly vertical over the ships and there was no shadow. A cool trade wind made conditions more pleasant. On 15 October, the crew caught and cooked three dolphins, which Matthew rated as 'fine fish, both with Respect to flavour and form', noting that 'their Colours change according to the angle you view them in, to blue, yellow, green, and brown'.

The same afternoon they caught a shark a little over two metres long, which, when cut open, contained five wriggling babies, each about forty-five centimetres in length. While Lieutenant Tobin thought sailors abhorred cruelty, he noted that the catching of a shark excited the crew, and that 'every kind of barbarity is too often exercised on these rapacious fish' – the sailors insisting that the shark would be just as vicious to them if roles were reversed.[49] Matthew was shocked by the savagery of the men. The shark had not been on board three minutes 'before she was in a Dozen Peices and perhaps before half an hour they were feasting upon him'.[50]

As the ships sailed through the Tropic of Capricorn and towards the Cape of Good Hope, Matthew watched as giant albatrosses stretched out their astonishing wingspans above, their flight rapid and majestic as they circled on high, watching for prey. Bligh told his men that during the passage of the *Bounty* near Cape Horn, a great number of albatrosses were caught using baited hooks. They were kept in hen coops and 'crammed' with oatmeal. After a few days they made surprisingly good eating.[51]

For several nights Matthew and the others on the *Providence* were astonished to behold a luminous phenomenon that was like constellations of stars in the water. Tobin called it a 'Sea of liquid fire'; sometimes it would stretch for a kilometre and a half in the ship's wake, and be so bright that it would bathe the *Providence* in an eerie white light as it cut through the glowing waves. As dolphin, porpoises and the tuna-like bonitos darted about the ship's bow, it was as though they were swimming in perfect daylight, and many fell to bullet and harpoon.

At first Tobin attributed the myriad of lights in the water to the vast numbers of 'blubbers', or jellyfish, that were about. The sea was home to all manner of these strange creatures, some as solid as house beams, others that resembled brilliantly coloured snakes six or seven metres long, some small and dangerous, such as the 'Portuguese man o' war' or 'bluebottle' with its poisonous sting. The most common blubbers were the size and shape of large mushrooms. The men on the *Providence* were puzzled that these mysterious creatures seemed impervious to pain when they were cut in half, as no spasm or convulsion took place.

On 3 November, the ship passed through a quantity of what the men still referred to as 'fishes spawn'. Matthew watched with fascination as some of it was scooped up in buckets to be studied through a magnifying glass. Tobin described this odd matter as 'transparent globules like jelly, of the colour of the water, about the size of a pea'; he thought it resulted in the sea of stars that followed his vessel.

Science would eventually reveal that the remarkable 'star show' was bioluminescence or 'water phosphorescence' – forms of plankton that emit a bright light when water is agitated.

CAPTAIN BLIGH REMAINED 'very infirm', but as the two ships approached the Cape he was able to tolerate 'some Noises and Bustle' and was giving orders again, confident the mountain breezes on the southern tip of Africa would restore him to his usual strength.[52]

On 6 November 1791, three months after the ships had left England, Table Bay came into view. The coast appeared barren and mountainous, and about fifteen kilometres from Cape Town they passed the prison fortress of Robben Island, where flag signals were hoisted to alert officials on the mainland that two ships were approaching. The *Assistant* continued to lag behind, and Bligh had to constantly slow the *Providence* to keep the ships together.

Matthew gasped at the majestic flat top of Table Mountain as it came into view, three kilometres long and more than a kilometre high. Adjoining it was the majestic peak of the Lion's Head, which the English then called 'the Sugar Loaf'.[53]

The winds at the cape were treacherous. Matthew was only seventeen, yet watching the skill of the experienced skippers and crew as they negotiated the dangers convinced him that he too could sail to the ends of the Earth. He wrote to his father to say that the two ships would likely reach Jamaica in fourteen months, and he could address a letter to him there.

Matthew had already travelled further than most men or women would in their lifetimes – and he was just getting started.[54]

Chapter 4

*'The anxiety of the Dutch government at Batavia to know how far
the South Lands might extend towards the Antarctic circle was the
cause of Tasman being sent with two vessels to ascertain this point;
and the discovery of Van Diemen's Land was one of the results.
It was not, however, the policy of the Dutch government to make
discoveries for the benefit of general knowledge ...'*
MATTHEW FLINDERS, ON THE NATURE OF EARLY SEA EXPLORATION.[1]

MATTHEW AND THE CREW of the *Providence* spent almost two months at the Cape of Good Hope as Bligh ordered the refitting of his ship and the *Assistant* in preparation for the next stage of their journey through the mountainous waves of the 'Roaring Forties' as they headed east towards Van Diemen's Land, 10,000 kilometres away.

Bligh made arrangements for a tent to be pitched on shore while the vessels were being thoroughly overhauled. He gave orders for his men to be served fresh mutton, greens and soft bread every day, and directed the ship's surgeon to send all sick persons to recover in the tent, so they could benefit from the 'land air'. Some of the crew had left England with 'virulent venereal complaints', and Bligh saw it as crucial to eradicate the disease from them.[2]

Bligh was still plagued by his own ailments, a fever and 'a convulsed motion' in his eyes that gave him great pain.[3] Portlock had the rig and sails on the *Assistant* modified so that it could better keep up with Bligh's bigger ship. In Table Bay there were several vessels from India, the Americas and Europe, and their men told such fantastic tales that there was a collective excitement among Matthew and his companions about the wondrous sights that lay ahead.[4]

The two botanists on the voyage, Paddy Smith and James Wiles, went about collecting exotic plants for Joseph Banks, and they gathered together fifty pots containing about 200 plants – figs, pomegranates, quinces and grapevines – to carry to Van Diemen's Land.

George Tobin and Matthew explored the Cape, documenting everything they could about this extraordinary place and the fascinating people who lived there. Cape Town had 'a pretty Appearance from the Bay', Matthew noted. '[A] great deal of Wood being dispersd [sic] all over it, indeed it is a Handsome regular Built Place, the Houses are in general low [set].'[5] Along the shoreline was a row of shabby little tents, where Tobin observed 'people of colour' selling liquor, vegetables, ostrich eggs and other articles to sailors.

Matthew was amused that the 'Dutch from having great Quantities of Animal food are rather corpulent … neither are they very polite … every Man is a Soldier, and wears his square riggd Hat, Sword, Epauletts and military Uniform, they never pass each other without a formal Bow'.[6] Self-denial in eating was 'not much practised' and the Cape Town men had hardly finished their meal when they reposed for two or three hours, only rising for coffee and to smoke their beloved pipes.

The Dutch had many enslaved people, Matthew noted, some from the 'Coast of Mosambique', who were the 'Tillers of Land' and they carted and cracked limestone for construction work. There were also slaves from the Khoekhoe people, the nomadic indigenous pastoralists of South Africa. Matthew described the Khoekhoe as copper-coloured, with high cheekbones. The Dutch referred to them by the pejorative term 'Hottentots'. Most of the slaves, Matthew saw, were domestic workers who had been shipped from the Dutch East Indies (modern-day Indonesia) and who had brought Islam to southern Africa. These 'Malays' had been forced to give up the habit of chewing betel nuts by their Dutch masters.[7]

Matthew saw several French ships arriving under national colours that represented liberty, equality and fraternity, but each vessel bore African slaves bound for the French colonies in the Caribbean. The arrival of these slave ships seemed absurdly at odds with the prevailing spirit of the people's revolution against tyranny then raging in France. The master of an English whaler at the Cape told Tobin that he had bought a slave for a bottle of brandy, and that he could have purchased many more at nearly the same value. Tobin prayed that recent agitation in Britain over the trade in human lives would improve the lot of slaves in Britain's West Indian colonies, and that this was the 'cordial wish of every feeling heart'.[8]

Some of the officers, on a visit to Table Mountain, observed a variety of chameleons, while some of the men, including Bligh, his health and vigour returned thanks largely 'to the Air & Exercise',[9] travelled through the vineyards of Stellenbosch, about fifty kilometres east of Cape Town.

Hiding out in Cape Town at the time, away from the prying eyes of Matthew and the rest of the British crews, was the Scottish artist Thomas Watling, a convicted forger who had escaped from the convict ship *Pitt* when it docked off Cape Town in late November. The *Pitt* had left England with 410 convicts (352 men and boys, and fifty-eight women and girls), as well as a company of the New South Wales Corps, who would replace the marines in the prison settlement at Port Jackson. Leading that corps was the new lieutenant governor of New South Wales: a portly 33-year-old Army major named Francis Grose,[10] who had twice been wounded in the war against the American rebels. Twenty-nine of the *Pitt*'s convicts had already died on the journey.[11] The fugitive Watling would eventually be handed over to the British authorities, and would arrive in Sydney in October 1792, on the *Royal Admiral*, which also carried the enterprising fifteen-year-old horse thief Mary Reibey, destined to become the richest businesswoman in early Australia.

On 16 December 1791, a Dutch ship, *Waaksamheyd* ('Wakefulness'), sailed into Table Bay, hoping to anchor near the *Providence* and the *Assistant*. The Dutch ship was a 'snow', a square-rigged vessel with two mainmasts, complemented by a 'snow' – or trysail-mast. It was flying signals of distress.[12]

On board was Captain John Hunter and his crew from HMS *Sirius*, which had been wrecked off Norfolk Island in March 1790 when caught in a violent storm and driven onto a coral reef. Hunter waited for nearly a year on the island before being rescued. The loss of the *Sirius* was a severe blow to the colony at Sydney, which was already short of rations and ships that might deliver more.

Hunter had hired the Dutch ship to take him and his crew back to Britain at the rate of £300 per month, but the trip to the Cape had already taken a wearying eight months, having sailed a difficult route east of New Guinea through Saint George's Channel.[13]

Five days after the ship had arrived in the Cape, strong winds drove it to the east side of Table Bay and forced it back out to sea. Bligh concluded that *Waaksamheyd* was a 'bad sailer', and he had the *Providence* fire twenty-two guns at intervals, signalling the need for assistance. Boats from various ships, including his own, helped tow the Dutch vessel to a secure situation.[14]

Hunter was on his way back to England to face a court martial over the wreck of the *Sirius*, but he would be exonerated and would soon be sailing back to New South Wales as its new governor.

MATTHEW SAILED FROM the Cape on 22 December 1791, as the *Providence* and the *Assistant* passed the *Waaksamheyd* and its cheering, waving well-wishers. Bligh set a course for the Indian Ocean and Van Diemen's Land six weeks away to the east, knowing it would likely be more than a year before they met Europeans again. The skipper had left letters with Hunter for delivery to Joseph Banks when the *Waaksamheyd* finally reached London, relating the experiences of the breadfruit voyage so far.[15] Two days later, on Christmas Eve, Matthew lost sight of the

African coast, and the next day enjoyed a celebration of mutton and grog with 'as much jollity as if on dry land'.[16]

The weather soon became boisterous, with damp, wet gales raging. Bligh insisted that a fire in the galley be maintained constantly, and that 'Brodie stoves' burn night and day in both the ship's cockpits to dry the air. On a Black Friday, 13 January 1792, the *Providence* was 'rolling very deep and taking water over the gunwales'.[17]

On 17 January, Matthew sighted the conical peak of the subantarctic volcanic island of St Paul, just six square kilometres jutting out of the southern Indian Ocean. After two and half more weeks of rolling through the big waves under dark skies into the Southern Ocean, he saw the sun rise fiery and full of threats on 3 February, but Bligh and Portlock kept their ships on an even keel in the rough seas. Then, five days later, as a cold wind whistled through the sails of the *Providence*, Matthew saw the faint coastline of the place the Dutch explorer Abel Tasman had named Anthoonij van Diemenslandt in 1642. By now it was known as Van Diemen's Land.

In cool breezes, the ships passed the Eddystone, a tower-shaped rock twenty-seven kilometres from the mainland, and the next morning they anchored off what is now known as Bruny Island, in fifteen metres of clear blue-green water in Adventure Bay. Bligh knew the area well, having been to Adventure Bay during Cook's third voyage on the *Resolution* in 1777, and then again on the *Bounty* in 1788.

Bligh took one of the ship's boats to the white sandy beach rimmed by a forest of tall gum trees and firs, and immediately began searching for the sites he had used for water and wood on the *Bounty*'s visit four years before. His sawpit had been partially filled in, and of the apple trees which he had planted in 1788 – the first apples in what is now Tasmania – he found only one. It was in a healthy state but still small; he guessed the others had been consumed by fire. He began marking two charts of the area.

Bligh sent out one party under Lieutenant Tobin to gather wood, and another under Lieutenant Guthrie to gather fresh

George Tobin drew Matthew's shipmates on the *Providence* resting in an abandoned Aboriginal hut, or 'wigwam', as he called it, while reprovisioning in Adventure Bay on Bruny Island in Van Diemen's Land. *State Library of NSW FL1606696*

water and catch fish. The crew dined on some small sharks that night.[18] Tobin's men killed different kinds of snakes, but without ascertaining whether their bite was poisonous or not, as it was judged imprudent to make 'any personal experiments on such a question'.[19] On the following days, the men collected mussels and caught small trout with hook and line. Sometimes they shot a parrot or seagull and tossed it into an iron pot to vary their diet. The two work parties ate their dinners together in a hut abandoned by the Nuenonne band of the Palawa people next to Resolution Creek and beside large piles of mussel shells, the sight of which convinced the visitors the mussels were safe to eat.

The hut – Bligh and his men called it a 'wigwam'[20] – was a little over two metres wide and half as high. It was constructed of branches of trees stuck in the ground and fastened with coarse grass, with bark arranged like tiles or shingles for a roof. It could house about six people.[21] Tobin wrote:

Wretched as such a habitation may appear, it sheltered us from many a 'pelting of the pitiless storm,' and hot ray of the sun. It had not been long deserted, the remains of native cookery being still fresh. Muscles [sic] have been before mentioned, the bones of animals were also strewed about, which were conjectured to be those of the Kangoroo. In another hut the vertebral ones of some large fish or animal were found; these from their magnitude were probably of the Whale, or Grampus.[22]

The only mammals they encountered were what Tobin called 'kangoroos', and 'a kind of Sloth, about the size of a roasting pig, with a proboscis two or three inches in length. On the back were short quills like those of the Porcupine.'[23] The 'kangoroos' were too fast for the British muskets – Bligh felt that a good dog was needed to run them down – but Lieutenant Guthrie killed the 'sloth-like creature' – an echidna – and it was roasted, providing a 'delicate' flavour.[24] There was all manner of fish in the bay, and birdlife too abounded – black swans, parrots, wild ducks, gulls, gannets and a variety of sand plover.

Although Matthew frequently saw smoke from Aboriginal fires, the crew only had one interaction of significance. The Indigenous people took some trinkets that were left in one of their huts, but kept their distance from these white invaders.

Matthew noted a 'remarkable circumstance … when presents wrapped up in paper were thrown' to some of the Palawa. They took the articles out and placed them on their heads in what Flinders took to be a 'a display of gratitude'. Almost a century before, Dutch explorers had recorded similar experiences with Indigenous peoples east of the Gulf of Carpentaria, 3000 kilometres to the north.[25]

Bligh saw canoes and considered them 'ill constructed'. On 19 February, however, some of the officers came suddenly on a party of the local people – sixteen men and six women. The men were bearded and the group clothed in kangaroo skins thrown over their shoulders. One was a mother with a baby at her breast.

All except one young man, curious about these strange invaders, ran away. The young man warily took some bread when it was offered to him, yet his terror soon made him bolt for the bush too. In the abandoned hut of these frightened people, Tobin found more kangaroo skins and short spears, sharp at one end and hardened by fire.[26]

Bligh's departure from Adventure Bay was delayed when it was discovered that one of the crew from the *Assistant*, a man named Bennett, was missing. Bligh dispatched search parties to find him, and Lieutenant Pearce of the marines discovered him cowering in a hollow tree about 800 metres from the beach. Bennett claimed that he had been unjustly accused of theft, and without the smallest portion of food with him, or anything with which to produce a fire, apparently preferred the risk of starvation in a distant land than remaining on board to face his accusers.[27] In defiance of his reputation for harshness, the skipper saw to it that the distressed man, his mind addled by anxiety, was treated kindly and returned to his vessel.[28]

Flinders and the others were impressed by the giant Tasmanian blue gums fringing the bay. Tobin measured one that was twenty-nine feet around (8.8 metres) and noted that while one of the ship's carpenters deemed the wood too thick and heavy for masts, it would someday probably provide good timber for building. The botanists, Wiles and Smith, collected eighty banksias to take home to the man after whom the plant was named.

Near a stream at the southern part of the bay, they planted, on Bligh's behalf, nine oak saplings and watercress seed, along with quince, strawberry, fig and pomegranate plants, a rosemary shrub and a Spanish chestnut tree. On nearby Penguin Island they planted fir seed, almonds, and apricot and plum stones.[29] An inscription was carved into a trunk: 'Near this tree Captain William Bligh planted seven fruit trees 1792:– Messrs. S. and W., botanists.'[30] Bligh turned a rooster and two hens loose into the forest of gums, hoping they might multiply for the benefit of future crews, though Bligh's men believed they would soon be eaten by the Indigenous people. Bligh also sent a small boat with the ship's goats to graze on nearby Penguin Island, and when two went missing they presumed they had most likely found their way over the rocks to the bigger island.[31]

The *Providence* and the *Assistant* weighed anchor and cruised out of Adventure Bay on 22 February, planning to skirt the southern tip of New Zealand, bound for the breadfruit trees of Tahiti.

Bligh wanted to find the entrance to Frederick Henry Bay but the wind became so difficult that the solid timber spar on the *Assistant* shattered. Bligh signalled a return to Adventure Bay so that urgent repairs could be made. In the distance there was snow on what the Palawa called kunanyi and which Bligh called Table Mountain. It would also later be known as Mount Wellington. Nearby, Bligh named Mount Nelson after David Nelson, Joseph Banks' botanist on the *Bounty* voyage, who died after the open-water journey to Kupang.

THE *PROVIDENCE* AND HER sturdy *Assistant* finally left Adventure Bay in heavy squalls on the cloudy evening of 24 February 1792, after the men had eaten a meal of porridge-like gruel mixed with dried kale – then known as 'borecole' and renowned for its antiscorbutic properties. They were escorted out of the bay by some seals and porpoises. Bligh set a north-east course for Cape Frederick Henry, on the north of Bruny Island. Matthew and the others saw several fires in the thickly wooded country around them, but the Palawa stayed out of sight, Bligh noting that they 'absented themselves and wished to have no communication or intercourse with us'.[32]

Bligh maintained the washing of decks and the regular scrubbing of hammocks as the ships turned east towards a point well below New Zealand's South Island. Matthew paid £1 2s 2d for tobacco from the ship's stores, and would take his long-stemmed pipe for a relaxing smoke beside the constant flame in the galley. He started his own log, copying Bligh's example by making detailed notes of the voyage, recording in minute detail the weather, sail adjustments, incidents on the ships and astronomical observations.

Matthew was impressed by the way Bligh preserved the health of his men through exercise and diet. Bligh ensured they were regularly given kale, wort, sauerkraut, peas and spruce beer to ward off scurvy. The sauerkraut could be cooked in many ways, but was mostly parboiled as a salad. The spruce beer was brewed on board from pine needles, molasses and hops.[33] Bligh's dietary enforcements, Matthew noted, 'were nearly a copy from Captain Cook, whose Attention in preserving the Health of his men was unremitted'.[34]

Gruel, salted pork and beef were staples, but fish were caught constantly. On 1 March the officers dined on the fried liver of a porpoise, remarking that it tasted like hog.[35] The following day they were at a latitude of almost 50 degrees south, skirting the southern tip of New Zealand. As the ships reached a longitude of 140 degrees west, Bligh steered them north-east towards Tahiti, 4000 kilometres away.[36]

The captain was still suffering 'Head ache & inclinable to low spirits',[37] but there was too much work for him to do to concentrate on his ills. Along with the marines, the sailors took turns at target practice, first with muskets and pistols and then with cannons, in case there was trouble in the paradise of Tahiti.[38] In his own show of strength, Bligh ordered the marine drummer James Davies lashed twelve times across his bare back 'for disobeydience & insolence to his Sergeant'.[39]

MATTHEW TURNED EIGHTEEN on 16 March 1792, as the ships charged east on the face of a strong gale. It was an unpleasant day, with the ships pitching violently, but the sights all around were fascinating.

Late on the evening of 25 March, Matthew and the crew saw a fiery meteor in the south-west, which Bligh said 'lasted a few seconds like a plate of burning brandy held in the wind'.[40] Matthew wrote that the 'blazing meteor ... enlighten'd the *whole* Atmosphere'.[41] By the end of March, the South Pacific air was so warm that awnings were fitted to shade the crew from the sun. Tropical birds began to fly about the ship, and the warmer weather brought cockroaches out of their lurking places in swarms. Bligh had boiling water splashed along the ships' beams and into any crevices or hiding holes to destroy the roaches and their eggs.[42] Thunderstorms and hailstones came with the hot weather, and in the tropics the colour of the ocean turned from cold grey to royal blue.

Flinders watched as Bligh repeatedly ordered the *Providence* to reduce sail so that the little *Assistant* could stay in visual contact, particularly when the winds were raging. Once, the *Assistant*, which usually struggled to keep pace with Bligh's ship, powered ahead and was 'three cable lengths', or about 200 metres, directly in front of the *Providence*, when a rapid change in wind direction saw the vessels come perilously close to each other.[43]

Bligh ordered that the two ships proceed with caution, especially at night, as there were uncharted rocks and atolls everywhere. On 5 April, Matthew saw 'a small low island just above the horizon … [with] a good surf breaking upon the Beach and over the Rocks near it'.[44] As they sailed closer, they saw many coconut palms and birds but apparently no human inhabitants. The land became known as Bligh's Lagoon Island; it is now the atoll of Tematagi – its original name – in French Polynesia.

The men on the two ships practised with their guns almost daily, and became so proficient in handling their weapons that Bligh had little doubt 'of their being steady' in case of an attack from hostile Polynesians.[45] Bligh again read the Royal Navy's Articles of War to his assembled crew, reminding them that the death penalty could be applied for mutiny or dereliction of duty.

Yet this skipper, whom history would paint as a harsh autocrat, did not want any trouble for the Tahitians. Bligh seemed to be regarded with great respect there and he wanted to maintain that reputation. As the temperature rose and the ships rocked about on the balmy nights, the crews of the *Providence* and the *Assistant* were champing at the bit to enjoy the sensual delights of Tahiti. The island and its fabled sexual freedoms had created bawdy legends around the Pacific paradise. One popular brothel in London's Pall Mall advertised live shows with naked dancers of both sexes recreating 'the famous feast of Venus, as it is celebrated in Tahiti'.[46]

In preparation for his crew's great release of sexual tension, Bligh ordered the ship's doctor to examine each man to see if 'any disease was among them, particularly the Venereal'.[47] Tobin wrote that the whole body corporate, including eighteen-year-old Matthew Flinders, 'passed through the hands of our worthy associate Ned Harwood, and never did the Doctor take a pinch of snuff, with more solemnity, or handle a subject with a less risible countenance … It was his duty now, to examine the affairs of men with a scrutinizing eye.'[48]

While the inspection of the men's private parts might not have been 'very agreeable to the Parties concerned', Matthew wrote, 'it was certainly done with a good intention' – to prevent any transmission of infection to the Tahitians.[49] Five of the crew, including Tobin, were found to be carrying syphilis, and Bligh ordered that their intercourse with the Tahitian women should be prevented 'if possible'. Many of the men ignored this directive.

After divine service on Sunday, 8 April, the two ships neared the lush volcanic island of Mehetia, a place Tobin thought 'one of the most beautiful spots that can be conceived'. Matthew made a sketch of it, showing the lush foliage on all sides of its peak, except the north side, which appeared broken off and washed by the sea. Bligh performed a divine service and then read out his regulations to the officers and ships' company for encouraging friendly relations with the Tahitians, as they would soon be at their destination. Guns were to be a last resort. Bligh had every piece of the crew's clothing itemised to prevent them swapping apparel for sex when they arrived. He gave strict instructions to the men about treating the Tahitians with respect, and threatened them with punishment if they disobeyed. Matthew copied out Bligh's rules in detail.

As Bligh believed that James Cook had been revered in Tahiti, he ordered that no one was to mention his violent death or the mutiny on the *Bounty*. No one was to tell the Tahitians that the purpose of this voyage was again to take breadfruit trees in case there were initial objections. Every crewman and marine was to win the esteem and goodwill of the Tahitians, to treat them with all kindness, and not to resort to violence even if it was to recover items stolen from them. If a man lost items, the value would be taken from his wages and he would face severe punishment for neglect. Trading was to be done only through an assigned person from the ship. No firearms were to be taken ashore without the captain's permission.[50]

That afternoon, while the ships passed along the north side of Mehetia, men in three canoes came paddling furiously after them. Bligh welcomed these men on board to barter breadfruit and coconuts for nails and other articles. One of the men, who was enjoying a local drink called kava, was dressed in a European shirt, procured from the whaling ship *Matilda*, which had recently visited the island after bringing convicts to Sydney with the Third Fleet.

The *Providence* and the *Assistant* kept under easy sail throughout a night of heavy rain, and at daylight on 9 April, Matthew beheld the astonishing beauty of Tahiti for the first time. Initially it was just a blur of distant mountains, but as Bligh steered the *Providence* closer, he saw that the heavy showers of the preceding night had given an additional glow to the vivid green of the island, with many white cataracts snaking their way amid the thick foliage. The beach, glistening with black sand, was crowded with Tahitians under the covering of the luxuriant breadfruit trees and towering coconut palms.

As Bligh guided the ships into Matavai Bay, they were soon surrounded by Tahitian canoes – and, surprisingly, a British whaling boat. But if Matthew and the others were worried about the powerfully built local men attacking them, as they had done twenty-five years earlier when the first Europeans arrived there on the HMS *Dolphin*, their concerns were immediately assuaged by friendly shouts and waves. 'Immediately we put into the bay,' Matthew wrote, 'the natives began to come off in their Canoes – they presently found out that Captain Bligh was on board, which seemed to give them much Pleasure – for they were not possitively certain whether he was alive.'[51]

The Tahitians called the 'King of the *Providence*' Captain 'Brihe', and news of his arrival generated great excitement across Matavai Bay. The Tahitian king, Pomare II, was on nearby Mo'orea when

Bligh arrived, but one of his queens and many local chiefs came on board Bligh's ship to express 'unfeigned joy and satisfaction at meeting their old friend' again.[52] The Tahitians had ulterior motives to express joy at Bligh's arrival. Foreigners were often brought into the power struggles among Tahitian peoples too.

Bligh's arrival also meant a supply of precious iron for the Tahitian people which was often used in barter for sex with local women.

Matthew watched on as the queen presented Bligh with a gift of hogs, fruit and Tahitian cloth. The Tahitians told Bligh that the *Bounty* mutineers had sailed that ship to Tahiti, where some of the crew had stayed to live with local women before they were captured by the men of the ill-fated HMS *Pandora*.[53] George Vancouver had recently visited Tahiti on another voyage arranged by Joseph Banks, and the whaling boat in Matavai Bay belonged to the twenty-one survivors, including a convict stowaway, from the wreck of the *Matilda*, which had broken apart on a shoal almost 1000 kilometres to the south of Tahiti a month earlier. Some of the other survivors had already been lost while trying to steer one of the boats back to Sydney.[54]

More canoes kept coming across Matavai Bay, laden with pigs, coconuts and breadfruit that the Tahitians hoped to trade for iron goods, especially hatchets. And then the moment the sailors and marines had been waiting for: canoes began arriving with near naked young women, as beautiful as their reputation in England had suggested. The women went through friendship ceremonies with these British visitors, rubbing noses and exchanging gifts as they chose their *tayo*, or companion. Tobin noted that, after sunset, as the canoes returned to shore, they left 'by far the most desirable part of their freight among our crew, which after the trying self-denial of a long voyage, shut out from the dearest solace life affords, could not but be truly acceptable'.[55]

Bligh kept a stern eye on these relationships, though, for the sake of the Tahitian women. Marine James Coombes took twelve lashes for 'having disobeyed my orders & having connection with a woman while he was infected with the venereal disease.

Nothing but severe punishment nor even that will prevent these wretches from committing this infamous act among these poor people,' Bligh noted.[56] Bligh wasn't just protecting the Tahitian women. He knew that the spread of venereal disease among his men would endanger his mission if too many of them were struck down and became unfit to sail.

Bligh himself still suffered from 'scarecely bearable' headaches and seizures, which paralysed the left side of his face, but after the failure of the *Bounty* mission he was set on making this transport of breadfruit trees a success. He established a base on a rise over the beach with huts for the men, a long, shady area for the plants and an observatory to help with his navigational calculations. The rest of his men busied themselves salting hogs and repairing the ships for the voyage to Jamaica, caulking the starboard bends, unbending sails and erecting a greenhouse on *Providence* to preserve their precious cargo.

They were busy in other areas too, and by mid-June Bligh reported that 'Our Sick List consists of 22 venereals'. Some of his men were not reporting their symptoms, and Bligh had his crew

George Tobin's sketch of the observatory site at Point Venus in Tahiti in 1792. *State Library of NSW FL1606708*

mustered and examined again by Dr Harwood. Bligh's suspicions were confirmed and he uncovered 'two wretches – the Boatswain & a Midshipman infected with the disease & one of them kept a woman constantly with him'.[57] The paybook of the *Providence* shows that Matthew Flinders was twice docked pay – thirty shillings – for mercury, a common treatment for venereal disease.[58]

In many ways, the British were envious of the seemingly free and easy lifestyle of the local people, and as they frolicked in the surf with their *tayos* they watched children aged just five or six 'amusing themselves in the heaviest surf with a small board on which they place themselves outside the breaking, whence they are driven with great velocity to the shore'.[59] But Tobin and the others had their suspicions about other aspects of this seemingly idyllic life. Free love resulted in many unwanted pregnancies, and, like James Cook and Joseph Banks before them, they heard rumours about 'the abominable crime of infanticide, and the paucity of females to males' – a paucity so extreme that the crew were told there were ten times as many males on Tahiti as women, and that many female babies were smothered at birth.[60]

When the crew moved to tents on shore, petty theft became a problem, as it had been during Cook's first visit there on the *Endeavour*. Tobin lost a valuable handkerchief on his first night in Tahiti, and Flinders noted that although four sentries were stationed around their tent, 'a large Bag of dirty Linnen of considerable value' was taken at night from beside Lieutenant James Guthrie's bed.[61]

Tobin and the surgeon on the *Assistant* lost their jackets and pistols when a Tahitian guide, who was taking them on an exploration through the wilds of the island, offered to carry them for the gullible visitors and then scampered off into the undergrowth, never to be seen again.[62]

More troubling were the rumours Bligh heard of human sacrifice during a war between different Polynesian groups. He recalled that he was shown the rotting body of a victim wrapped in plaited coconut leaves and tied to a pole, and he later saw the putrid corpse at the centre of a religious ceremony as it was offered to a Tahitian deity.[63]

Chapter 5

*'[T]o fire after they saw the poor Fellows so frightened, when they
had given up the Contest, was not much better than Murder –
at least it was Cruelty.'*
MATTHEW FLINDERS, AFTER A TORRES STRAIT ISLANDER WAS SHOT DEAD
DURING A SKIRMISH WITH BRITISH SAILORS.[1]

MATTHEW AND THE CREWS spent three months in
Tahiti, soaking up the sun and forging connections with
the local people. Bligh continued to suffer from headaches and
mood swings. Sometimes his temper appeared uncontrollable,
and his language was shocking even for men used to the cruelties
and hardships of life at sea. Matthew felt he had lost his skipper's
favour, but told his father he was unsure why.

In early July 1792, the botanists Wiles and Smith told Bligh that
the young breadfruit plants were healthy enough to be transported
from their nursery to the ships. The lower decks were stripped
and Matthew watched as men wearing thick protective coats
washed every surface and hole they could find with boiling water,
to eradicate cockroaches and other vermin that might damage
the trees.

Wiles and Smith wrote to Joseph Banks to say that 1156
pots, tubs and boxes had been loaded onto the ship to carry the
breadfruit trees to Jamaica, as well as other plants for Banks'
research back in England.[2] Bligh estimated he was carrying 2126
breadfruit trees and 2634 plants all together.

As the departure of the ships drew closer, the British encampment was deserted and the flagstaff bare. A pall came over Matavai Bay as romances and friendships were about to be broken, perhaps forever. The marines would ensure that, despite the deep emotional bonds some of the crew had formed over the preceding fourteen weeks, there would be no repeat of men demanding to stay on the island. Bligh would take back to England fifteen survivors from the *Matilda*. Five had elected to stay on the island with the Tahitians, and the *Matilda*'s stowaway convict, Samuel Pollend, remained hidden in the forest.

As Bligh prepared to set sail, both his ships were crowded with Tahitians, all carrying various farewell presents for their English *tayos*, who were equally generous in return. As the sun set on 18 July, many of the locals took their leave with tears in their eyes, while others, including some of the women, stayed on board for the whole night with their British lovers.[3]

The *Providence* and the *Assistant* sailed out of Matavai Bay the next morning, again surrounded by canoes, but this time the occupants were producing what Matthew took to be a chorus of lamentations. Some of the Tahitians made such a show of their grief that they tore at their heads with ornamental sharks' teeth until their scalps were covered with blood.[4]

Many of the Tahitians had asked Bligh if they could travel with him to England, after hearing the stories told by Mai on his British adventure, and some asked their friends to shut them up in chests and casks to be smuggled aboard.[5] Bligh had refused almost all the requests, even from his elderly Tahitian friend Tynah, to whom he instead presented a 'Musquet & 500 rounds of powder & shot' as a farewell gift.

Bligh agreed to take Tynah's *towtow*, or servant, Mydiddee, to Britain: as Mydiddee was 'a fine Active Person about 22 Years of Age at most', Bligh said, he was more likely to learn British ways and customs than 'a Chief who would be only led into Idleness and Dissipation as soon as he arrived in Europe'.[6] Mydiddee had hopes of meeting King George III, just as Mai had done.

To Bligh's astonishment, another Tahitian named Pappo was uncovered secreted between decks when the *Providence* was well away from the island. Pappo had met Bligh during the *Bounty*'s visit four years previously, and according to James Wiles had 'distinguished himself by his activity in supplying them with provisions and curiosities'. The *Bounty*'s crew had presented him with a seaman's jacket, which he constantly wore, and which earned him the nickname 'Jackets'.[7] When Fletcher Christian returned to Tahiti after the Mutiny, Pappo accompanied him to the island of Tubuai, before Christian sailed the commandeered ship to Pitcairn, where the *Bounty* was eventually set alight.

Pappo had spent weeks helping Wiles and Smith collect and care for the breadfruit trees, but Bligh was still furious at him for hiding himself on board. Bligh's face turned beetroot-red and blue language spewed from his mouth. But a gale was blowing too fiercely for the skipper to 'beat back and land him, without much loss of time'. As the salty spray lashed Bligh's angry countenance, he told the startled crew that he just didn't have the heart to make the uninvited guest jump overboard.

> While I was debating in my mind what was best to be done, the Botanists told me he had been a valuable Man to them, & would be of great use if I kept him ... I conceived he might be useful to our Friends in Jamaica in attending the Plants, about which he knew a great deal.[8]

The ships sailed west from Tahiti in hazy weather and with a cross-running sea, as they prepared for a voyage of 37,000 kilometres that would take them through the Torres Strait, into the Indian Ocean, around the Cape of Good Hope, on to Jamaica and eventually home to London. They passed Mo'orea and afterwards saw the southern shores of Huahine and Raiatea: 'high mountainous islands', Matthew wrote. To the north was Bora Bora.[9]

They reached Aitutaki, one of the Cook Islands, then passed the Tonga archipelago and sailed by Fiji – or Bligh's Islands, as some called them, as he had made the first European charts there during his voyage in the open launch after the *Bounty* mutiny. Each glowing dawn revealed more sublime sandy beaches, deep forests and towering volcanic peaks, just the sort of place for the adventures of Matthew's inspiration, Robinson Crusoe. Matthew was stunned by each glorious vista and the thrill of sighting places where few Europeans had ever ventured.

Matthew considered Gau island (then called Ngau) to be the most beautiful place he had yet seen on the voyage – which was saying something. 'This island,' Matthew wrote, 'was thought worthy to be called Paradise Island.'[10]

By 10 August, the ships were weaving their way through the rocky islets and dangerous breakers around Fiji's Kadavu Island, where they saw several fires and the flat top of Mount Buke Levu. Matthew helped the senior officers with the preparation of eleven official charts for the voyage, and he drew seven small plans himself, marking them with a small monogram, all the time improving his cartographic skills.

From Fiji the ships sailed to the northernmost of what Cook had called the New Hebrides – now Vanuatu – as Bligh wished to revisit the islands of the Banks Group, which he'd seen from the *Bounty*'s launch. He did not try to communicate with the inhabitants, though, as they were massed in strong, worrying numbers upon the beaches.

There was ten weeks of cautious sailing through the Pacific reefs, with the *Assistant* leading, using a lantern at night, and Portlock often at its masthead to ensure that his ship was on a safe course. At dusk on 31 August, there was the terrifying noise of breakers 'thundering on the reef',[11] as a great surge of water threatened to topple anything before it. The next day, Matthew, from the masthead of *Providence*, saw the first shoals of the Endeavour Strait, in the far south of the Torres Strait. 'In keeping to the Northward we had expected to have gone clear of the many rocks and Shoals, which extend out to some Distance all along the North East coast of New Holland,' Matthew wrote, 'but it appears those from the South coast of New Guinea are equally troublesome and dangerous'.[12]

Two days later, the ships arrived at a clear space of water twenty-five kilometres wide – still known today as Bligh Entrance – between New Guinea and the land Cook had named Cape York. Bligh had been here in the open boat, and knew it was a maze of treacherous reefs and rocks rarely navigated by Europeans since Luís Vaéz de Torres led a Spanish voyage through there in 1606.

MATTHEW WORKED ON HIS own chart of the Torres Strait, to accompany his log, signing it 'M. Flinders'. At the same time, Bligh cautiously pushed the ships west towards Kupang, sending out the small boats – the cutter and the whaleboat – to search for safe passage through the hazards. It was too dangerous to sail at night, so the *Providence* and the *Assistant* would anchor after sunset. Surf broke alarmingly on the coral reefs and rocks nearby.

Although his relationship with Captain Bligh had become strained, Matthew could think of no one better to navigate the ships through dangerous waters, to 'extricate us from all Difficulties and bring us safe thro' the Strait'.[13]

On 5 September, the third day of this slow going, the whaleboat and cutter were sounding for depth about eight kilometres ahead of the ships, sometimes finding fifty metres of clear water and at other times coral reefs just a metre and half under the surface. At midmorning the whaleboat was heading back to Bligh with a report of a safe passage but the cutter, with Lieutenant Tobin, midshipman John Busby and seven crew on board, remained behind. Four large canoes, which Matthew said were 'about 50 feet in length [and] hollowed out of a single tree [with] an outrigger on each side', came racing towards the cutter from an island Bligh had just named for his distant relative, John Bligh, the Earl of Darnley.

The canoes were under sail, but as they came closer, the Torres Strait Islanders powering them began using paddles to propel the vessels even faster. The lead canoe had a small shelter in its middle, upon which sat their chief, a small, muscular and naked man, yelling out orders in the Meriam Mir language.

Tobin called on the whaleboat for assistance but its crew was too far away to hear him. He and his men readied their muskets, fearing that cannibals were approaching.[14] Paddling close across the cutter's bow, one of the Indigenous men held up a coconut. Tobin made signs for him to take it to the ships in the distance. But then all the men in the canoe – there were fifteen – armed themselves with bows and arrows, which had been concealed from view inside the vessel's shelter. Tobin's blood ran cold as he realised the other canoes were closing in fast.

'Fire!' he roared, and a volley from six muskets echoed across the water at the first canoe twenty metres away. All the warriors on it fell face down in their vessel, and Tobin presumed some of them had been hit. Their chief remained defiant, though, still perched on the roof of the shelter. After the smoke had cleared from the first volley and the men had reloaded, Tobin's coxswain fired at the chief and brought him down.

The Islanders who could still paddle beat a hasty retreat, and the other canoes quickly left. Soon all four canoes were at a distance, the local warriors making shouts and watching cautiously. As Tobin steered the cutter back towards the ships, the canoes made a second attempt to cut the British off from safety, but Bligh sent Lieutenant Guthrie and his marines in the ship's pinnace to the rescue. The sight of reinforcements caused the local warriors to hoist their sails and return to Darnley Island.

Matthew, who had watched the skirmish nervously from the deck of the *Providence*, recalled:

No boats could have been manoeuvred better in working to windward, than were these long canoes … Had the four been able to reach the cutter, it is difficult to say, whether the superiority of our arms would have been equal to the great difference of numbers; considering the ferocity of the people, and the skill with which they seemed to manage their weapons.[15]

Blood was shed when the British cutter was attacked by Torres Strait Islander warriors in canoes. *State Library of NSW FL1606802*

'These people, in short, appeared to be dextrous sailors and formidable warriours,' he wrote, 'and to be as much at ease in the water, as in their canoes.'[16]

Matthew was furious, though, at the coxswain for shooting the chief:

> [He] did not fail to brag about it when he got on board, thinking he had done a meritous action … It is apparent the Lieut. did not fire before it was absolutely necessary … but to fire after they saw the poor Fellows so frightened, when they had given up the Contest, was not much better than Murder – at least it was Cruelty.[17]

Even so, Matthew pondered how others on the voyage, including himself, might have acted if they genuinely feared they were about to be 'devoured'.[18]

If he had been worried that the shooting would ruin any chance of contact with these people, Matthew's fears were assuaged the next morning, when more canoes from Darnley Island approached the *Assistant* at anchor; the Torres Strait Islanders remained watchful about the gunfire, but were keen to trade. The warriors clapped their hands upon their heads and let out a 'whooping' noise 'repeatedly with much vehemence', while at the same time brandishing their arrows and clubs and asking for *tooree*, by which they meant iron.

After much difficulty, two of the local men were persuaded to come onto the ship. 'They had bushy hair,' Matthew wrote '... were rather stout made, and nearly answered the description given of the natives of New Guinea'.

The cartilage, between the nostrils, was cut away in both these people; and the lobes of their ears slit, and stretched to a great length, as had before been observed in a native of the Fejee Islands. They had no kind of clothing; but wore necklaces of cowrie shells, fastened to a braid of fibres; and some of their companions had pearl-oyster shells hung round their necks. In speaking to each other, their words seemed to be distinctly pronounced. Their arms were bows, arrows, and clubs, which they bartered for every kind of iron work with eagerness; but appeared to set little value on anything else. The bows are made of split bamboo, and so strong that no man in the ship could bend one of them ... The arrow is a cane of about four feet long, into which a pointed piece of the hard, heavy, casuarina wood is firmly and neatly fitted, and some of them were barbed. Their clubs are made of the casuarina, and are powerful weapons ... the heavy end is usually carved with some device: One had the form of a parrot's head, with a ruff round the neck, and was not ill done.[19]

The crews traded with more of the local people off Campbell Island. Bligh noted that some of the men carried spears fourteen feet (four metres) long, and that the strings for their bows were made from the outer skin of bamboo. Most of the warriors had thick hair and beards. Some had lost their teeth and some had their foreheads daubed with red: some had a few feathers stuck in their hair, and others had the skin on their shoulders raised in circular rims that together formed a kind of badge. Some were old and their rough beards were tinged with grey.[20] They had dealt with Europeans before. Bligh noted:

> A looking-glass did not surprise them, but they cared for nothing but iron. I bought but one yam and that they wanted to cut in half to make a better bargain. They had a strange way of showing their astonishment by whistling and making a noise like a ball whizzing through the air.

The ships pressed on slowly, because the glare from the afternoon sun obscured the reefs and shoals.

At Dalrymple Island, they found a small and empty village consisting of about fifteen huts with flat roofs. Each had a doorway but no door, and several of the huts were joined together and formed one front, like terrace houses. They were slightly built and covered with mattings or palm thatch. Canoes were hauled up on the beach and a few dogs wandered about. The sails of the canoes were made of matting in an oblong form, roughly stitched together. The mast to which they were hoisted consisted of two bamboo poles, and some canoes had two masts.

In the afternoon, when the small boats were returning to the ship after taking depth soundings, the crew saw the villagers on the beach. The Torres Strait Islanders waded into the water to meet them, waving green branches and clapping their hands with excitement. Some climbed into the boats and became 'frantic' at the noise of metal jingling. One of the villagers had what Matthew described as 'a moderate sized dog with him of a brown chestnut colour'.[21] He had seen the same sort on Tahiti.

There were forty-two local people on the beach, including seven children. One of the youngsters was carried on the shoulders of a woman, who like the other wives had a covering round the hips, while the men remained naked. In return for iron, the villagers gave the visitors fruit and shell ornaments. Sadly, the friendly relations would not last.

The two British ships were carefully making their way between Dungeness and Warrior islands, with the *Assistant* in front and its cutter alongside. More than 100 warriors came racing across the water to them in a fleet of nine canoes. Matthew did not believe they had hostile intentions, since the two groups had now traded amicably for several days. But this time the warriors had planned to trade deadly fire.

One canoe containing about twenty men launched a rainstorm of arrows. Two men in the *Assistant*'s cutter and one on the deck were struck, and one, Quartermaster William Terry, died from his wound two weeks after taking an arrow to the hip. Lieutenant Portlock ordered his men to fire their muskets.

Matthew's heart was in his mouth as more canoes charged toward the *Providence*. When a musket was fired at the first boat, it did not scare the warriors; instead they let out a war cry and paddled forward. Then Matthew and others began firing at their attackers as fast as they could manage.[22]

Bligh ordered his gunners to fire the big quarterdeck cannons, loaded with cannonballs and grapeshot. 'Destruction and horrible consternation' was the result of the first blast, the captain recorded. Survivors fled from their canoes into the sea 'and swam to windward like porpoises'.[23]

Matthew watched them 'plunging constantly, to avoid the musket balls which showered thickly about them'. The attacking squadron then made off as fast as the Torres Strait Islanders could paddle. Afterwards they rallied at a distance and appeared to be ready to charge again, until a shot, which passed over their heads, made them disperse and give up 'all idea of any further attack.'[24] They returned later to collect one of their wounded companions, who was still in the shreds of what had been his canoe. The rescuers made signals to their companions on Dungeness Island 'expressive, as was thought, of grief and consternation'.[25]

Matthew was astonished by the power of the weapons used by

the Torres Strait Islanders, and the depth to which their arrows penetrated into the decks and sides of the *Assistant.*

'They must have had many Men killed,' he wrote, 'but here they certainly deserved it.'[26] No doubt the victims of the gunfire would have disagreed.

A WEEK LATER, and with no further bloodshed, the ships arrived at a small uninhabited island, and Bligh sent Lieutenant Guthrie and a small party including Banks' botanists on shore to claim all the islands in the strait for King George III.[27]

The crew took some coconuts and plum-like fruit, and left a knife and other trifles to pay for them. There was one more sighting of islanders when three canoes, each with seven or eight men, approached the ships' boats. A musket shot fired over their heads did not deter them, but Bligh was taking no chances and had one of the small swivel guns fired across their path, which made them flee.

By noon on 22 September, no land was in sight. Matthew wrote: 'Thus was accomplished, in nineteen days, the passage from the Pacific, or Great Ocean, to the Indian Sea.' Few voyages presented more dangers than the Torres Strait, but 'with caution and perseverance, the captains Bligh and Portlock proved them to be surmountable; and within reasonable time'.[28]

It was grudging praise from Matthew, who had enormous respect for Bligh's seamanship even if he had grown to think less of his character. Matthew was stubborn, determined to get what he wanted and developing what one biographer called 'a vein of self-importance amounting at times to arrogance'.[29]

Bligh's shortcomings were legendary. Matthew's shipmate Lieutenant Bond, the son of Bligh's half-sister, complained about his relative's unrelenting 'insolence and arrogance' and 'the fury of an ungovernable temper'.[30] Certainly Bligh's sickness on the voyage had not cooled his reputation for volatility. His flaming headaches affected his mood and manner. His eyes became badly inflamed[31] too, and his iron rule soon became even more unbending as the ships steered through safer waters towards Timor and the Dutch settlement of Kupang.

Making the breadfruit his top priority, Bligh rationed drinking water for much of what was left of the voyage, stipulating an allowance of a pint per day for each man, 'exclusive of his grog'. He reduced the water used in pea soup and gruel by more than half.[32] The men were thirsty and irritable, and Matthew and some of his shipmates 'would lie on the steps, and lick the drops of the precious liquid from the buckets, as they were conveyed by the Gardener to the Plants'.[33]

They reached Kupang on 2 October, and to Matthew it 'had more the appearance of an Indian Village than an European Settlement and but for the Dutch fort and Colours … it would scarcely have been taken for one'.[34] In Kupang, Matthew and the crews learnt that the HMS *Pandora*, sent to the Pacific to capture the *Bounty* mutineers, had sunk on the Great Barrier Reef near the Torres Strait. Thirty-one men, including four of the mutineers, had drowned. The ninety-nine survivors – eighty-nine crew and ten from the *Bounty* – had reached Kupang a year earlier, going on a Dutch vessel to England, where three of the mutineers were eventually hanged while Bligh and the *Providence* were at sea.

The *Providence* and the *Assistant* spent eight days in Kupang taking on provisions, and though there were slim pickings in the Dutch port after the arrival of so many hungry *Pandora* survivors, two buffalo were loaded for fresh meat.[35] Many of the crew came down with fevers; one of the marines, Thomas Lickman, was buried at sea after losing his battle with dysentery.[36] Despite the gardeners' best efforts to nurture the breadfruit trees, 200 were dead.

From Kupang, the ships sailed into the Indian Ocean, on their way to a brief stay at the Cape of Good Hope.

MATTHEW WAS MAKING COPIOUS NOTES, tables and charts of all he experienced, and after filling his logbook he started a new one on 20 November 1792. His first entry reflected a young man determined to make a name for himself in an age of enlightenment, adventure and discovery.

The Discoveries we have made and Dangers we have passed tho' perhaps not of the greatest Consequence to us as a trading Nation, will yet add to our well established Name as Discoverers, increase Geographical Knowledge in general and to the Cause of Navigation they will be an Acquisition. Captain Bligh, as the immediate Agent, will no Doubt receive the Honour and recompence equal to the Task he has performed and I as an Actor tho' in an inferior Station, shall have the Satisfaction, of having served my King in a Cause he has so much at Heart, my Country by assisting to put in Execution its benevolent Intentions and myself by gaining some Knowledge of Navigation, the universal Diffusion of which is one of our best National Characteristics.[37]

A month later, on 17 December 1792, the *Providence* and the *Assistant* arrived at the British settlement of Jamestown, on the Atlantic island of Saint Helena, midway between Brazil and the west coast of Africa.

Bligh now calculated that more than 800 plants had perished.[38] A passenger on the ship *Ganges* from Bengal gave Wiles and Smith mango and guava trees to take back for Joseph Banks.

Saint Helena's governor, Robert Brooke, had Mydiddee and Pappo stay at his house and gave them each a suit of red clothes. They, in turn, taught the governor's household how to prepare sago pudding, having brought sago from Tahiti.

The following night, Mydiddee and Pappo were in their traditional matting skirts as Brooke's special guests for a military parade, and appeared to be fascinated by all the pomp of the marching, the cannon salutes and the stirring performance of Saint Helena's military band, 'particularly with the little Drummers, which they laughed heartily at'.[39] Mydiddee's health had declined over the last few months of ocean sailing, and he overindulged on the liquid refreshments and became 'much intoxicated'. He was so embarrassed in the aftermath that he was still apologising to Bligh several weeks later.[40]

A bullock was taken on board to provide fresh meat for the rest of the voyage, and ships left Jamestown two days after Christmas 1792, to a thirteen-gun salute.

On 23 January 1793, they arrived in Kingstown Bay, on the Caribbean island of St Vincent, and were welcomed with a civic reception as they delivered 544 breadfruit plants to the slave owners there. Some of the plants were now more than three metres high, with leaves almost a metre in length. The triumph of the voyage was muted after crewman Henry Smith fell overboard and drowned. John Thompson, a survivor from the *Matilda*, deserted.

Smallpox was rampant at St Vincent, and Bligh insisted that Pappo and Mydiddee be inoculated. They were both naturally apprehensive about this baffling idea of voluntarily inflicting the disease on themselves, but when told that it was commonly practised in Europe, they received the infection favourably. Still, both Tahitians suffered badly after the inoculation.

Sailing on for another 2000 kilometres, Matthew sighted Jamaica on 4 February, and the ships reached Port Royal the following day. A committee in charge of the imports divided 623 of the remaining trees among various plantation owners.

Bligh's orders were finally completed, even though hundreds of the plants had died and the enslaved men and women ultimately did not care much for breadfruit, preferring to eat plantains, a type of banana. The Jamaica House of Assembly awarded Bligh 1000 guineas and Portlock 500 guineas as a reward for their efforts.

Bligh was preparing to publish his own account of the voyage, and in accordance with Navy protocols confiscated the personal journals being compiled by Matthew and others on board, lest they reveal official secrets. Matthew grudgingly did as he was told.

James Wiles took a job on £200 a year to run the botanical gardens in the Jamaican settlement of Bath, with Pappo as his assistant to take care of the breadfruit trees. The Tahitians remained ill after their inoculations, though, and on 3 March Bligh noted that he had to borrow a 'Kitterreen' (or carriage) to carry Pappo to Bath, for both he and Mydiddee 'were but barely fit to remove out of the House ... [Pappo] is a very chearfull

Man, and I am confident will be happy and well taken care of. I really think he will be the means of the Breadfruit being brought early into use, & on that account his life is valuable to Jamaica.'[41]

Pappo didn't survive on Jamaica for long, though, dying after suffering a heavy chest infection a few months later. Amid its regular list of runaway slaves and 'Diseased Negroes', Jamaica's *Postscript to the Royal Gazette* reported that 'Pappo, the Otaheitean who was left on this island to assist in cultivating the Bread Fruit and other Otaheite plants', died at Bath. James Wiles gave little praise to his assistant, a man who had sailed thousands of miles from his home to see the world beyond Tahiti: 'He was an exceeding good natured harmless creature, [but] had no ambition, learnt very little English and appeared to be about 34 years old.'[42]

Matthew was hoping to journey back to his father in Donington as soon as the breadfruit mission was complete. Wiles gave him a letter for Joseph Banks dated 16 March 1793, which advised the Royal Society's president that the 'young Gentleman who delivers this to you is a messmate and most intimate acquaintance of mine, whom I have authorized to receive the Part of my Salery due to me from my leaving the Ship ...'[43]

The return would be delayed, though. On 29 March 1793 the packet boat *Duke of Cumberland* arrived in Jamaica with the February mail from London. The French revolutionaries had beheaded King Louis XVI and declared war on Great Britain and Holland.

The outbreak of hostilities meant Bligh's ships had to remain on war duty in the Caribbean, escorting convoys and captured French vessels and transporting French prisoners. With the breadfruit voyage complete, Matthew was temporarily disrated to able seaman, a disappointing result after so long at sea.

Naval reinforcements arrived in Jamaica, and on 15 June 1793 the *Providence* and the *Assistant* were finally able to set sail for the Thames. Mydiddee was 'extremely ill' and Bligh little better, though the irritable skipper at last returned Matthew's journal. The *Providence* and the *Assistant* arrived at the Downs on the Kent coastline on 2 August. Matthew, eager for action in the war and for promotion, dashed off a letter to his great supporter Captain Pasley, then at Plymouth, on England's south-west coast. Pasley's warship *Bellerophon* was undergoing repairs there after colliding with HMS *Majestic* in gale-force winds while on patrol off the French city of Brest.[44]

Matthew had written to Pasley at every opportunity on the voyage, and the captain was delighted that his young protégé had made it home safe. Pasley promptly invited him to join his crew.

> Dear Flinders
> … I do not know what are your plans now you are returned back to England. I expect to sail in a day or two to join my Lord Howe on board the *Bellerophon*. I shall receive you with pleasure – after so long an Absence you will no doubt wish to see your friends pay them a short visit and return to join me, by that time I shall probably be returned into Port – if not you will only find a conveyance from Portsmouth or Plymouth by applying to the respective Admls they are both my friends.[45]

On 7 August, Bligh berthed at Deptford to begin unloading 1283 plants from the West Indies bound for Joseph Banks and the Kew Gardens. All along the river, men were rushing about preparing ships and crews for the war with France, while executed criminals hung in chains from gibbets, a warning to London's populace about misbehaviour and to any sailor thinking of mutiny.

The ghastly display sickened and revolted the ailing Mydiddee. The victims had not been fortunate to have their deaths commuted to transportation to the new prison settlement in Sydney, and George Tobin thought the sad spectacle made his Tahitian shipmate 'wish to be again among his countrymen'.[46]

It was not to be. Two days later, Bligh reported that 'our Otaheite Friend' Mydiddee had become so ill that the skipper was 'obliged to send him to Lodgings & Sick Quarters at Deptford'. On 3 September he died there.

'I sent a surgeon from Town to see him opened,' Bligh wrote. 'His Lungs were found decayed.' Bligh organised Mydiddee's funeral and he was buried in St Paul's Churchyard at Deptford.

Matthew had returned to England a vastly experienced navigator and a changed young man. At seventeen he had never ventured far outside of Lincolnshire. At nineteen he had seen men killed in battle and had many sexual experiences in a tropical paradise. He had seen friends die in maritime accidents and from the diseases that inevitably followed long sea voyages in cramped conditions. He had sailed where few Europeans had ever ventured, and he was hungry to explore more of the world.

Matthew's father was overjoyed that his firstborn had returned home in one piece, though Matthew had come back to Britain with a fever and was thinner than when he left. Flinders Sr wrote in his diary:

> By the Mercy and Divine Providence of God – my Son hath safe returned from his Long & Perilous Voyage. [C]ompleating it in little more than 2 Years ... his Captain latterly was not on the best terms with him, which was an unpleasant Circumstance ...[47]

Matthew had experienced many run-ins with Bligh, but the esteemed *Naval Chronicle* later reported that the skipper had been impressed by Matthew's 'useful auxiliary ... for he was ever ready to assist in the construction of his charts, and in astronomical observations; indeed, although still but a very juvenile navigator, the latter branch of scientific service, and the care of the time-keepers, were principally entrusted to him.'[48]

But Matthew no longer needed Bligh's approval or patronage. He had enjoyed the support of Commodore Pasley, and would soon find an even more powerful and influential mentor. Matthew was about to win the favour of the king's confidant, the most influential man in the world of science. The man who presided over the vast, wide world of Britain's global exploration.

Chapter 6

'Never I believe was such a scene of ruin and devastation seen before ... several of the French ships that were dismasted continued with invincible obstinacy to fire upon our ships.'

LIEUTENANT HENRY WATERHOUSE, ON A GREAT NAVAL BATTLE HE FOUGHT ALONGSIDE MATTHEW FLINDERS.[1]

A S MATTHEW PACED DOWN the gangplank of the *Providence* and onto Deptford's naval dock, he found everyone around him agog at horror stories of what was happening just across the English Channel. A 'reign of terror' had begun in France and would see at least 40,000 public executions by guillotine, sanctioned by the revolutionary government's ironically named Committee of Public Safety. The French democrats had also turned their attention to other European monarchies, declaring war on Britain and Holland. They had already stormed into the Austrian Netherlands – today's Belgium – and taken Savoy, Nice and parts of the Rhineland.

In 1793, Britain already had 175 large war vessels – the 'ships of the line' – but forests of trees were now being cut down to build another 150 as quickly as possible, as the Channel Fleet, under Lord Howe, patrolled the waters between England and Europe. Matthew was eager to serve, but the nineteen-year-old first rushed home to Donington to see his family.

He also had important business to conduct. On 15 August 1793, he rode thirty kilometres north from Donington to the Revesby

Estate just outside his stepmother's home town of Spilsby. Sir Joseph Banks, his wife Dorothea, his sister and their servants were there, on the family's annual migration north for the summer months. At the time, Banks was working with hydrographer Joseph Huddart on a plan to change the course of the Welland and Witham rivers, to drain more marshland for farming.[2]

Just as he was the great sponsor of British science and exploration, Banks bankrolled the village of Revesby and its surrounds as a benevolent landlord and employer.

The Banks family had owned the Revesby estate for four generations, and when he inherited the property at twenty-one, along with other landholdings and mining interests, Joseph Banks became one of the wealthiest young men in Britain. In his mid-twenties, he sailed to the Arctic, wrote of new worlds with James Cook, climbed a volcano in Iceland and became the most important man in British science as president of the Royal Society.

When Matthew was ushered in to see the great man by one of Banks' smartly dressed servants, he met a portly fifty-year-old who now lived a relatively sedate life as a country squire, while remaining the 'great panjandrum of British science'.[3] He was still sponsoring voyages of exploration around the world by men such as William Bligh and George Vancouver, and was also, effectively, the man in charge of the new settlement in New South Wales.

Banks lived a life of opulence, but like Matthew he had also spent years on a voyage around the world on a small, crowded ship, living rough. He knew what it meant to be on the edge of infinity, ever unsure what lay over the horizon.

Matthew felt small looking up at King George's confidant. Everything about Banks was big. Even his home office occupied two rooms, such was the scale of his library and the cabinets containing all his reference materials and correspondence. Banks was almost a head taller than Matthew and, at well over 100 kilograms, twice as heavy.

Nevertheless, Matthew and Banks had much to talk about. Their voyages had covered much of the same territory, the coast

Joseph Banks at about the time Matthew first met his great benefactor. By William Ridley (engraver), 1802. *nla.obj-136001803*

of New Holland and the perilous Torres Strait. And both had taken advantage of sexual opportunities in Tahiti.

Banks told Matthew he was delighted at the success of the breadfruit expedition. He had written to congratulate Bligh and had other jobs for him, and he had recommended Portlock for promotion. Matthew told Banks of his plans to sail with Pasley on the *Bellerophon* in the war against France. Then he presented the letter that James Wiles had given him in Jamaica, authorising Matthew to receive the rest of the gardener's salary for his work with Bligh.[4] Banks assured Matthew that the money would be promptly paid to Wiles' father.

Matthew left for Donington, and Banks took a pencil and scrawled in the corner of the letter: 'Delivered by a Mr Flinders who is now on board the *Bellerophon* C. Pasley. Augt 15.93.'[5]

Banks had only just met Matthew, but there was something about the young man's confident manner and eagerness for the sea that suggested they would encounter one another again.

HAVING GROWN ACCUSTOMED to the warmth of the South Pacific and the Caribbean, Matthew now found his father's house cold and grey. The arrival of a chilly autumn brought Matthew a persistent cough. He had also sustained a cut to his leg that became infected, making his brief time with his family less pleasant than it should have been.[6] Matthew would soon learn of the death from fever of his cousin Lieutenant Jackey Flinders. Jackey had died, aged just twenty-six, after suffering with yellow fever for three weeks while sailing from the West Indies.[7]

Flinders Sr was still working constantly to support his large family, and was still burdened by the ceaseless tide of debt. His marriage was not always happy, and there was friction between Mrs Flinders and her stepchildren. Dancing lessons for eleven-year-old Samuel Flinders were burning a hole in the family budget, but not as much as the cost of board and education for two of Matthew's other siblings, Susanna, fourteen, who would soon be apprenticed to a milliner, and John, twelve, who seemed to attract trouble wherever he went, and whose family described his appearance as 'not agreeable' and his character as 'dull, perverse, obstinate, weak and untoward'.[8] John's erratic behaviour would soon hint at a mental illness. Matthew played with his two little half-sisters in his father's house, and friends and relatives, knowing he would soon be off to fight the French, flocked to see the young sailor and to hear his tales of the 'South Seas'.

Matthew socialised with his sister Betsey and her circle of fashionable young women, which included Mary Franklin, his stepmother's young relative, and another friend, Ann Chappelle.[9] Matthew would come to address these friends as his 'charming sisters'.[10] They were captivated by his tales of death-defying adventures on the high seas, and of peoples and places few Europeans had encountered.

Matthew's acquaintances and relatives were from a new generation of women in Britain, no longer seen as mere housekeepers or decorations for potential husbands. Among the yeoman class of genteel citizens, young women were being educated – not too much, mind you: university was out of the question, but reading at home was encouraged in order for them to make polite conversation with educated men. Young women and teenage girls, such as Ann Chappelle and Matthew's sisters, took lessons in dancing, deportment and sewing, and often had private tutors to teach them literature and languages, history and geography. Arithmetic was also seen as essential for managing household budgets when they wed.

Ann Chappelle was small, slight and delicate, with alabaster skin, and she had a gentle nature. Like Matthew, she had dark eyes and dark hair, wearing hers in thick ringlets that cascaded over her pale shoulders. She was barely five feet tall, and so even Matthew looked tall beside her. She was three years older than him, and had sufficient education to make her interesting, but not intimidating, to potential suitors of the time. Born in Hull, Ann was the daughter of John Chappelle and Anne Mallison, a descendant of a Norman baronet who was among the regicides who signed the death warrant of Charles I in 1649.[11]

John Chappelle, a small and kindly Yorkshireman, had been a skipper in Britain's merchant navy. Despite being plagued by frequent headaches, he had earned a reputation as 'the Gentleman Captain', a thoughtful man who loved books and never sailed without his dog-eared copy of *Paradise Lost* or Edward Young's contemplative *Night-Thoughts on Life, Death & Immortality*. The books gave John little comfort, though, as he died with fever, a sweat-soaked towel wrapped around his head, off Java[12] on a trading voyage when he was just forty. Ann was just four at the time and never forgot the heartbreak and hardships she and her mother endured after the news reached their cottage in York.

They managed to get by on the savings John had left them, and lived a quiet and sheltered life, centred on the church. Ann's mother was the youngest of nine children, a stout, practical, capable woman with a florid complexion and dignified bearing. By all reports, she was strict but kind, and instilled in her daughter Puritan ideas of God and good manners, encouraging her to emulate her father's love of prose and poetry.[13]

When Ann was ten, her mother remarried, to Reverend

William Tyler, in York. They moved to the Lincolnshire fens at Partney, and became friends with the Franklin family at nearby Spilsby. Ann's mother delivered another daughter, Ann's half-sister Isabella 'Belle' Tyler.[14] The two girls would remain close for the rest of their lives.

Not long after, though, Ann almost lost hers. Matthew's father was doing his best to combat smallpox in Donington but it tore through Partney with a vengeance when Ann was still a girl, and her small, pale body was covered in ugly, weeping sores as her life hung in the balance. Her eye became infected with a smallpox blister, and lancing the sore left her blind in one eye. Along with that, Ann inherited the constant headaches that had so plagued her father.

Despite this pain in her young life, Ann's family regarded her as 'clever, with a sweet and perfect temper, beloved by all who knew her, witty, generous, nervous, with aptitudes for poetry, literature, singing, verse, and painting flowers from nature'.[15] She was an impressive young woman and Matthew asked if he could see her again if he returned from the war.

Matthew was small and active, with dark hair and alert, dark eyes. She liked the cut of his jib. He reminded Ann of her father, or at least what she remembered of him through the fog of lost childhood. Ann was on guard, though. What she recalled clearly was how she and her mother had wailed at the loss of John Chappelle thousands of miles from home. From the time she was a little girl, Ann told Matthew, she had always sworn she would never marry a man who was already wedded to the sea.

THOMAS PASLEY HAD overseen major repairs to his great ship after its collision with the *Majestic* in July, and by 7 September 1793, after carpenters in Plymouth had worked around the clock, Pasley and his mighty vessel were ready to rejoin the Channel Fleet.

Admiral Howe had assigned the *Bellerophon* to a flying squadron made up of the fastest warships he had, and he put Pasley in command, with the temporary rank of commodore.

Pasley would have a crew of almost 600, including Matthew, who became Pasley's aide-de-camp. Also on board was Lieutenant Henry Waterhouse,[16] whose father, William, was a page to the King's brother Prince Henry, the Duke of Cumberland. Henry Waterhouse had been named after the duke, his godfather.

Henry had sailed as a midshipman on John Hunter's *Sirius* with the First Fleet to Sydney Cove in 1788, having been recommended to Governor Arthur Phillip by the duke. He had seen warfare of a kind before, standing beside the governor when an Aboriginal warrior named Willemering speared Phillip in the shoulder at a place near Sydney that the governor called Manly Cove because of the 'manly' bearing of the Eora warriors there. Waterhouse had also fathered a daughter, Maria,[17] with convict Elizabeth Barnes at a time when many British officers formed liaisons with convict women.

Matthew started another journal for his time on the *Bellerophon*, and on 11 September he noted proudly: 'Hoisted a broad pennant by order of Lord Howe, Capt Pasley being appointed a commodore of the fleet. Weighed and anchored in our station in Torbay.'[18]

The following month, as the French queen Marie Antoinette was about to face the guillotine in Paris, Admiral Howe sent the *Bellerophon* to lead a hunt for five French vessels that had chased the British frigate HMS *Circe* into Falmouth. On 18 November, Matthew and the *Bellerophon*'s crew prepared themselves for battle. As the *Bellerophon* made its way through the English Channel, Matthew saw nine or ten large ships in the faint distance.

At 9 the Admiral made the sign for the strange fleet being an enemy, and for our sternmost ships to make more sail. At 10 the signal to engage as the other ships came up was made. The enemy though escaped with as much sail as they could carry. Split one jib; got another bent as fast as possible. We were now the headmost line of battle ship and gaining fast upon the enemy; but the main part of our fleet seemed rather to drop from them.[19]

The next morning, Matthew could make out six of the enemy's ships, two frigates and two brigs among them. The *Bellerophon* tacked in a squall, as did the 36-gun HMS *Latona*, which brought the ship near the rear of the enemy's vessels. The *Latona* fired several cannon blasts, but the French ships were on the run and at sunset passed to windward of the *Bellerophon* ten kilometres away. The crew set the topgallant sails but had to take them in again for fear gusts would tear away the masts. One of the French ships sent cannonballs through the *Latona*'s breadroom and galley, but 'happily' no one was hurt, Matthew noted, 'and but little injury received'.[20]

Eight days later, Matthew saw an unusual ship to the south. It hoisted a Union Jack at the main topmast head and a red flag at the fore. HMS *Phoenix* sent up flags asking for the pass signal, but the imposter did not answer and Pasley immediately gave the order to chase. The *Phoenix* and the *Latona*, which were closer, fired a few cannonballs and the fleeing corvette then hoisted French colours and surrendered.

'She proved to be *La Blonde* of 28 guns and 190 men,' Matthew wrote.[21] The French captain, 'Citizen Gueria', came on board and surrendered his sword to Pasley. The prisoners were locked up, and four days later an officer of the *Phoenix* sailed the captured vessel to Falmouth, where it was sold and the proceeds divided between the crews of the *Bellerophon*, the *Vanguard*, the *Phoenix*, the *Latona* and the *Phaeton*.[22]

With Pasley now responsible for a squadron, Captain William Johnstone Hope[23] was given command of the *Bellerophon* in January 1794, although Pasley, about to be promoted to rear admiral, would still fly his pennant from her. Matthew remained on patrol through a cruel winter as the Channel Fleet sailed off the French island of Ushant and the Brittany coast.

ON 20 MARCH 1794, four days after his twentieth birthday, Matthew was at Spithead, off Portsmouth, where the *Bellerophon* was undergoing more repair work. Matthew took the opportunity to write to Joseph Banks, now back at his Soho mansion, on behalf of botanist James Wiles, who, because of a problem over his power of attorney, had not yet received his final pay for his work on Bligh's voyage. Matthew reminded Banks of their meeting at his country house at Revesby, signing off as 'your much obligd and most humble Servant Mattw. Flinders'.[24]

Banks replied promptly: 'Mr Wiles letter & the instrument came to my hand yesterday & I shall lose no time in demanding from the Treasury the balance due to him for wages which will be about but not quite one hundred Pounds.'[25]

Problem sorted, Matthew was soon back on the water and ready for battle again. British spies had alerted the Navy that France's new government, the National Convention, had organised a convoy of merchant vessels to bring grain and other food from America. The starving French were bleeding internally from a chaotic revolution, the failure of France's own harvest and the blockade of French ports by British and Dutch ships.

More than 100 merchant vessels[26] were leaving America's Chesapeake Bay for Brest in April, as the Americans repaid France for its support against the British during the American War of Independence. Lord Howe, now sixty-eight and determined to end his long career with a bang, made plans to intercept the cargo and bring France to its knees. On 2 May, his Channel Fleet put to sea from the anchorage of St Helens Road, off Spithead, with twenty-nine massive ships of the line and fifteen frigates. Together they created an awe-inspiring spectacle of great towering masts, white sails and flying pennants. Howe was aboard his flagship, the 100-gun HMS *Queen Charlotte*, along with John Hunter, who had farewelled Matthew and Bligh in Cape Town two years earlier on his way back from Sydney Cove.

Two weeks later, Rear Admiral Louis-Thomas Villaret de Joyeuse[27] sailed from Brest with twenty-six giant French ships of the line, along with five faster frigates and two corvettes.

Matthew's *Bellerophon* was 740 kilometres west of Ushant on 23 May when he observed the preliminary contests of a huge naval conflict:

The *Southampton* brought a strange brig into the fleet and destroyed her ... a.m. A fine little ship, called the *Albion*, of Bermuda, set on fire by the *Glory*. A galliot, with Dutch colours inverted, passed through the fleet, having been set on fire by the *Niger* ... A French man-of-war, captured and brought into the fleet by the frigates, was set on fire.[28]

Skirmishing continued for days as the boats manoeuvred around each other, firing broadsides, and the French corvettes tried to shield a huge merchantman flying American colours. Matthew was about to take part in what would become one of the greatest naval battles of all time.

On 28 May, he saw two vessels in the distance through his spyglass, and soon after signalled with a flurry of flags that there looked like a 'strange fleet south-south-west', four leagues (about twenty kilometres) away. At about 8 a.m. he counted thirty-three ships, twenty-four of which looked like large warships, 'and all standing down towards us. At 8.30 our signal was made to reconnoitre the enemy as we were now certain they were. A frigate of their's was likewise looking at us.'[29]

As gales whipped up the surging, grey waves, and with rain clouding vision, a swell from the west tossed the ships about. Howe ordered his fleet to give chase, and as the French ships began to tack away from the fight, he ordered an attack. Pasley, resplendent in his blue coat and white breeches, stood in command on the *Bellerophon*'s quarterdeck, as Matthew and the other aides-de-camp trembled with the fear and the excitement of battle.

Sailors rushed about, extinguishing fires in the galley and throwing sand across the decks to aid traction and stop the spread of flames from cannon fire. The gunports were opened and the cannons unchained, rumbling about on their heavy wheels as they were positioned to blast away. Ashen-faced crew hauled gunpowder from the magazine stores below, hoping they wouldn't meet a spark on their way. There was a cacophony of shouting and screaming, men sprinting about and the drums of war beating as the crew raced to position themselves on the decks and at the rigging.

At 3 p.m., HMS *Russell*, being a mile or two to windward of Matthew's *Bellerophon*, began to fire on the enemy, and was soon joined by HMS *Thunderer* and some frigates, getting into the wake of the French fleet.

Matthew was stunned by the skill of Pasley and the crew as they manoeuvred the enormous *Bellerophon* through the rough seas like it was a small racing skiff. One wrong move could have seen its sails torn from the masts, but eventually, with the crews working fanatically, the leviathan of the water was moved into a position from where it could open fire on the first ship within range.

The gunners aimed for a large French frigate which was bringing up the enemy's rear, but it soon went to windward of the next ship. Then, looming ominously in front of Matthew was the unnerving sight of a ship almost twice the size of his, the 2600-ton three-decker *Révolutionnaire*.[30] It was a black-hulled 110-gun monster that carried 1000 crew and had been the pride of the French Navy for almost thirty years.

Pasley moved the *Bellerophon* within about 1500 metres of the French vessel and ordered, 'Fire!' Black powder was ignited and the first volley from the *Bellerophon*'s seventy-four big guns exploded across the Atlantic like a thunderstorm, a deafening roar of explosions amid clouds of black smoke on the boiling sea. Cannonballs tore through the wooden hull of the *Révolutionnaire* as its crew held on for dear life. Chains, too, were fired from the British cannons: they ripped through the French sails and rigging, and through Frenchmen. Immediately, the *Révolutionnaire* began firing back with what Matthew called 'great spirit'.[31]

As the French cannonballs landed with terrifying force, the *Bellerophon*'s maincap was smashed and the main topsail had to be dragged in. A sheet block burst into splinters. Aboard the *Queen Charlotte*, Lord Howe and John Hunter saw Matthew's ship engaged, and the next French vessel ahead firing at the *Bellerophon* too. Howe signalled to the *Russell* and HMS *Marlborough* to come to its assistance.

About dusk, the HMS *Leviathan* and the HMS *Audacious* had joined the battle against the *Révolutionnaire*, and soon more British ships arrived to rescue *Bellerophon*, which was now so badly battered that Howe ordered Pasley to bring it into his wake. The ship stayed astern of the *Queen Charlotte* as Matthew helped the crew make urgent repairs.

The firing continued through the night, and when the sun came up and the crews could once again see what they were shooting at, a cannon blast killed the skipper of the *Révolutionnaire*, Daniel Vandongen. The huge French ship was separated from its support vessels and defenceless as the British cannon fire burst its wooden hulls. The pride of France had lost its mizen and main masts, and soon 160 crewmembers were killed or wounded, including most of the senior officers.

Normally, the British would have taken such a prize as a spoil of war, but the *Révolutionnaire* escaped because of Howe's order to regroup for a renewal of hostilities the next morning. During the night, the 74-gun *Audacieux* towed the *Révolutionnaire* to safety.

The enemy ships now gathered about five kilometres to windward, and at night both sides lit lanterns to prepare for another firefight. Matthew noted:

> At daybreak the enemy's line was formed about 2 miles distant, and our commander in chief made the signal to form the line of battle, and take stations as most convenient. We bore down and took ours astern of the *Queen Charlotte*, the *Marlborough* and *Royal Sovereign* following. About 8 our fleet tacked in succession, with a view to cut off the enemy's rear, the *Caesar* leading and my Lord Howe the 10th ship ... At 10 the firing commenced ...[32]

Admiral Howe signalled to cut through the enemy's line and the French were now 'well within point-blank shot, which began to fall very thick'.[33] Wood shattered on both the French and British ships, and men on both sides were blown to pieces.

My Lord Howe in the *Charlotte* ... cut through their line between the 4th and 5th ship in the rear. We followed, and passed between the 2nd and 3rd. The rest of the fleet passed to leeward. Their third ship gave us a severe broadside on the bow as we approached to pass under her stern, and which we took care to return by two on her quarter and stern. Before we had cleared her, her fore and maintop masts fell over the side, and she was silenced for a while, but it was only till we had passed her. Their rear ship received several broadsides even from our three-deckers, but kept her colours up.[34]

Lord Howe made the signal for a general chase, but few of the British ships were in a state to follow because of the damage sustained. The rigging on Matthew's ship was 'entirely cut to pieces forward and the foresail was rendered useless, and was cut away'. The crew spent the next day, 31 May, splicing and knotting the rigging, mending mastheads and repairing sails.

Then 1 June 1794 arrived – or, as it became known in naval history, 'the Glorious First of June'. This was the date of the largest naval conflict ever between Britain and France.

At 8.10 a.m. Howe made the signal from the masthead of the *Queen Charlotte* for close action and each ship was to engage his opponent with everything it had. Sink them or swim. Howe ordered his ships to sail on an oblique course down upon the French line. He wanted to break it near the centre, each British captain sailing round the stern of his opponent and concentrating the attack on the enemy's rear.[35]

The *Bellerophon* was the second ship in the British line, after the 80-gun HMS *Caesar*. Both dread and exhilaration clawed at the men as this close-quarters fight neared. Every man on every ship knew he might soon become a hero or a corpse as the huge vessels jockeyed for position, the better to destroy each other.

Henry Waterhouse dashed off a quick note for his father and put it in his pocket in case his body was found after the battle. His gravest concern was for the fate of his little daughter in Sydney.[36]

Matthew's spine was tingling as he prepared to show Pasley his true mettle. The eager youngster was on the quarterdeck as the *Bellerophon* steered through a narrow gap in the French line. The sterns of the two ships seemed so close that Matthew thought he could almost punch his enemies on the 74-gun enemy *Éole*.

He seized his moment to stand tall. There was a brief moment when the gunners were aloft trimming sails as Pasley manoeuvred for the attack. The cannons on the quarterdeck were loaded and primed ready for use, but Pasley didn't want to fire them until he could pour a broadside into the *Éole* with the most damaging effect. Matthew, though, already full of self-importance despite his youth, seized a lighted match and fired in succession as many of the guns as he could, right into the *Éole*. The French warship was hit again and again by the blasts, but Pasley was furious that someone was getting ahead of his orders. He ran down as fast as his legs could carry him to where Matthew was scurrying about setting off more cannon fire. Pasley grabbed his protégé and shook him violently by the collar. 'How dare you do this, youngster, without my orders?' Pasley roared.

As the *Éole* tried to blow Bellerophon out of the water in response, Matthew replied that he was sorry but 'he thought it a fine chance to have a good shot at 'em'.[37] At 8.45 a.m., Matthew's ship now began 'a severe fire' upon its opponent, which the *Éole* returned 'with great briskness'.[38]

Three of the French ships were now blasting away at the *Bellerophon*, with the *Éole* and the *Trajan* doing the most damage. Matthew's ship was erupting in splinters and blood. Then, ten minutes before 11 a.m.,[39] Matthew watched horrified as the *Éole* sent an eight-kilogram cannonball hurtling through the barricading of the quarterdeck. It ripped Pasley's left leg off below the knee, and Pasley was sent crashing across the deck, screaming in anguish as blood gushed out around him.

Pasley was in deep shock, as were the men who carried him below deck, where the ship's surgeon went to work immediately to save his life. Torn flesh would become infected quickly. Pasley was fed rum to dull the pain, at least a little, and bit down hard on a leather strap to stifle his screams. The doctor sawed through the mangled flesh and broken bone of Pasley's ruined leg as quickly as could.

Matthew remained in what he called 'the heat of the action' as Captain Hope took command of the ship. Admiral Howe's intricate battle plans descended into a chaotic free-for-all as damaged, helpless ships and wounded, dying crews on both sides shot muskets and cannons at anything they could hit. On HMS *Barfleur*, Rear Admiral George Bowyer also had a leg shot off by enemy fire. Henry Waterhouse on the *Bellerophon* said this scene of ruin and devastation was incomprehensible,[40] and wrote to his father again, giving his return address as 'At Sea & nearly a Wreck'.

'After one of the most obstinate Engagements ever known we are victorious & thank God I have not recd the smallest hurt,' Waterhouse wrote. 'Adml Bowyer & Pasley have both lost a Leg ... We have taken 7 Sail of the Line & one 74 [gun] was sunk dreadfull sight. Write to Govr Phillip as I have not time.'[41]

Matthew recalled:

Our shot was directed on three different ships as the guns could be got to bear. In ten or fifteen minutes we saw the foremast of the third ship go by the board, and the second ship's main-top-sail-yard down upon the cap. At 11¼ ... they both bore away and quitted the line, their Admiral being obliged to do the same some time before by the *Queen Charlotte* ... And now, being in no condition to follow, we ceased firing; the main and foretopmast being gone, every main shroud but one on the larboard side cut through.[42]

Matthew counted the dead and injured on the *Bellerophon*:

We had but three men killed outright (a fourth died of his wounds very soon after) and about 30 men wounded, amongst whom five lost their limbs, and the other leg of one man was so much shattered as to be taken off some time after. Our brave Admiral [Pasley] was unfortunately in this list ...

Fortunately no accident happened with the powder, or with guns bursting ... Most of our spars were destroyed, and the boats severely injured.[43]

Overall, the British had 1200 men killed or maimed. It was estimated that 4200 Frenchmen had been killed or wounded, and another 3300 taken prisoner. Six of the French ships were captured, and another, the *Vengeur du Peuple* – the *Avenger of the People* – sank due to the devastating broadsides of HMS *Brunswick*,[44] the ship on which the *Bounty* mutineers had been hanged. The *Avenger* went down with 600 men on board.

Matthew said the French prisoners were 'in a very bad state both with respect to discipline and knowledge of their profession'.

Out of our 198 prisoners there certainly cannot be above 15 or 20 seamen, and all together were the dirtiest, laziest set of beings conceivable. How an idea of liberty, and more so that of fighting for it, should enter into their heads, I know not; but by their own confession it is not their wish and pleasure, but that of those who sent them; and so little is it their own that in the *Brunswick* ... they could see the French officers cutting down the men for deserting their quarters.[45]

Admiral Louis-Thomas Villaret de Joyeuse limped his shattered vessels back to France as best he could. Admiral Howe, with a large portion of his fleet no longer battleworthy, was unable to resume his search for the American merchant boats in the Bay of Biscay, and the food began arriving in France under a convoy of French frigates on 12 June. Thus, both Britain and France claimed victory on the Glorious First of June – France because its desperately needed supplies arrived unharmed, and Britain because it had taken more than 3000 French prisoners and captured or sunk seven enemy warships without losing any of its own.

The Channel Fleet, torn, tattered but triumphant, returned to Spithead with its six captured ships, and nearby Portsmouth turned on a huge celebration for its conquering heroes. King George, still smarting a decade after the loss of his American colonies, arrived on his royal barge with Queen Charlotte to salute Howe's flagship, which bore Her Majesty's name, and to congratulate the crews for ensuring that Britannia still ruled the waves. Matthew was swept up in the pomp and pageantry.

The King presented Howe with a diamond-studded sword. Thomas Pasley would never sail into battle again, but the King made him a baronet, and his service was rewarded with promotion to admiral and more than £1500 worth of gifts. Pasley had lost a leg, but Matthew had lost an important mentor. Without Pasley's direct support, his prospects for advancement in the Royal Navy had taken a bad turn. As a midshipman, his share of the French ships which were sold as prizes was just £10.[46] Had Pasley not been wounded, there is little doubt Matthew would have sailed into battle with him again, to the music of cannon fire and screams.

Matthew's five days of bloody conflict off the coast of France would be his last experience of naval warfare, though his battles with the French were far from over. Matthew wrote to his father to relate everything that had happened over those five days of mayhem.

'Victory over the French fleet – June 1. 1794,' Matthew's father wrote in his well-worn diary.

> On Wed: June 18. 1794, we had letters from my dear Matthew, announcing his safety and arrival at Portsmouth (for which blessing I humbly offer up my gratitude to Divine Providence) after having been in the great Naval Engagement between the English grand Fleet and that of the French … the Slaughter amongst the French has been very great … How Merciful is Providence to us, that hath protected my Son thro' such a hard and most perilous Scene.[47]

Chapter 7

'In Mr. George Bass, surgeon of the Reliance, *I had the happiness to find a man whose ardour for discovery was not to be repressed by any obstacles, nor deterred by danger; and with this friend a determination was formed of completing the examination of the east coast of New South Wales.'*

MATTHEW FLINDERS, ON HIS GREAT PARTNER IN EXPLORATION.[1]

TWO MONTHS AFTER he had dodged the French cannonballs aboard the *Bellerophon*, Matthew was in his father's sitting room at Donington, announcing that he was leaving for the colonial outpost of Sydney, New South Wales, and so he would likely not see his father again for five years.

Matthew had travelled to Lincolnshire on three weeks' leave from the Navy, but his father lamented that he still did not get to see much of his son. Matthew always seemed to be off visiting his mother's people in Tydd St Mary – and, more frequently, his young female friends, including Mary Franklin and Ann Chappelle in Spilsby.

Still, Matthew Sr was grateful even to see a little of his son. More than 1000 British families were mired in grief after losing men in the great naval battle against the French.

'I thank God [Matthew] looks well, and is in good health and spirits,' his father wrote. 'He made this journey in order to take leave of us, previous to his going on another long voyage – [the]

government having fitted out two vessels for the new settlement of New Holland.'[2]

Matthew told Flinders Sr that he had been appointed to the *Reliance* as master's mate five days before he arrived in Donington on 15 August 1794. Later, Matthew would write that he was 'led by his passion for exploring new countries' to accept the opportunity Henry Waterhouse had offered him to go to New South Wales, because 'above all others' it 'presented the most ample field' for his 'favourite pursuit'.[3] He had already spent time at Adventure Bay in Van Diemen's Land, and he wanted to explore more of the vast southern land.

The *Reliance* would carry the second governor of New South Wales, Captain John Hunter, to Sydney, and after Captain Bligh's offsider Nathaniel Portlock was unavailable, Henry Waterhouse was made second in command behind Hunter on the *Reliance*. Hunter had recommended Waterhouse as 'well qualified' for the job after their time together on the *Sirius* at the head of the First Fleet.[4] Waterhouse would also be able to visit his three-year-old daughter, Maria. Hunter and Matthew both liked Waterhouse for his easy manner and straight talking.

Governor Arthur Phillip had returned to England in May 1793, worn out by the stress and privations of running a penal colony on the far side of the world. He had arrived back in London with two Indigenous Australians, Wangal men of the Eora Nation, Bennelong[5] and Yemmerrawanne.[6]

Phillip's soldiers had originally kidnapped Bennelong to ascertain all they could about the Indigenous people: their numbers, customs and language, as well as their attitude towards the European invaders. But in time Bennelong had learnt a little English and he and the Governor became friends – or as much as kidnapper and captive could be called that. At Bennelong's request, Phillip ordered a hut to be built for Bennelong, twelve feet (three and a half metres) square, on a point of land that had been used to unload cattle from the ships – and which, two centuries later, would become the site of the Sydney Opera House.

Bennelong visited England with another Eora man, Yemmerrawanne, in 1793. Yemmerrawanne died there but Bennelong returned to Sydney on the *Reliance* in 1795. *State Library of NSW FL214557*

Phillip had told Joseph Banks that he was bringing Bennelong to England because when he became proficient in the English language, 'much information' about Indigenous Australians 'may be attained, for he is very intelligent'.[7]

Bennelong was about 168 centimetres tall (five foot six) and was described by Captain Watkin Tench as being 'stoutly made' and with a 'bold intrepid countenance, which bespoke defiance and revenge'.[8] Yemmerrawanne was said to be a 'good-tempered lively lad' who had become 'a great favourite' waiting on the table at Government House.[9]

In London, the two Wangal men stayed with Henry Waterhouse's father, William,[10] at his grand home at 125 Mount Street, Mayfair. They were measured for clothes that were the height of London fashion – coats, silk stockings, blue and buff striped waistcoats, ruffled shirts and slate-coloured breeches. They toured the sights, including St Paul's Cathedral and the

Royal Zoo at the Tower of London, swam in the Serpentine and watched comic operas, pantomimes, plays and a circus.[11] They even attended the Theatre Royal at Covent Garden during a royal visit there.[12] But away from their own people and land, in a strange and often condescending world, they seemed 'constantly dejected', *The London Observer* reported, 'and every effort to make them laugh has been for many months ineffectual'.[13]

Yemmerrawanne fell ill in October 1793, and despite expensive treatments, including regular doses of laxatives, bark medicines, liniments, hot plasters and painkillers – possibly opium – he died from a lung ailment aged just nineteen on 18 May 1794. He was buried in the churchyard of St John's in Eltham, under a tombstone costing £6 6s, which Arthur Phillip insisted the Crown provide.[14] Bennelong ached to go home.

GOVERNOR PHILLIP HAD left Francis Grose in charge at Sydney as Lieutenant Governor of New South Wales. Grose had established military rule with his New South Wales Corps, while handing out generous land grants to his fellow officers. The New South Wales Corps gained a monopoly on buying and selling rum, and liquor gradually became the main currency in the colony. Five years after its founding, the colony was still lurching about trying to find its feet.

Grose had doubled the salary of a friend and brother officer, the ambitious and pugnacious John Macarthur, making him both paymaster of the New South Wales Corps and Inspector of Public Works. The second job gave Macarthur control of the colony's captive workforce, materials and machinery. Macarthur oversaw a trade that involved soldiers of 'the Rum Corps' holding a monopoly on imported goods and the liquor to pay for them. Much-needed wares arriving in the colony were sometimes sold at a 1000 per cent markup, while Macarthur and his confederates built huge landholdings, driving away the Indigenous inhabitants and often killing them.

The British government had decided that if John Hunter could stare down French cannon fire during the battle of the Glorious

First of June, he could certainly bring John Macarthur and his confederates into line.

Hunter had been instrumental in establishing the settlement of Sydney. He had met many of the Indigenous people there, and had helped feed the settlers when they were starving. There were now more than 3500 Europeans in the colony, made up of settlers, soldiers, convicts, emancipists, government officials and their respective families. Most were living in Sydney, with smaller settlements a few kilometres west at Parramatta and Toongabbie, and about sixty kilometres north on the Hawkesbury River.[15]

Matthew told his father that he had a better chance of promotion if he travelled to Sydney, rather 'than staying in the home Service'.[16] Two Royal Navy ships were making the voyage.

Losing his firstborn for so long was a bitter pill for Flinders Sr to swallow, but then Matthew told his father that eleven-year-old Samuel wanted to go to Sydney as well. Despite all the money their father had spent on little Samuel's dancing lessons, the boy had 'for some time expressed a desire for the Sea'.[17] Matthew had sought Henry Waterhouse's permission for Samuel to come and the deal was sealed.

Flinders Sr felt much heartache, given the personal problems he was already battling. 'My Family as they grow up,' he complained, 'encrease my cares and my expences, and I begin to find my Constitution not quite so adequate to the fatigues of Business.'[18] The relationship between daughters Betsey and Susanna and their stepmother was sometimes frosty, and Betsey had not been in good health. Matthew's brother John was back in Donington after two years boarding at a school in the village of Market Deeping, where Flinders Sr had sent him because of 'his rude Companions'[19] and 'unpolished' manners. John's behaviour was only getting worse.

Despite his economic worries, Matthew's father took £30 out of his well-worn purse to ensure Samuel was outfitted properly for the voyage.

'[Samuel] is very young,' Flinders Sr wrote of the boy, who would not turn twelve for another three months, 'but if we had missed this opportunity several Years must have passed before so good an opportunity might again occur, & Matthew has often lamented he did not go sooner, as he thinks he should now have been promoted to a lieutenancy.'

'The Ship's name,' he continued, 'is the *reliance* – 16 guns – 80 men – the other the *Supply* – a smaller one I suppose …'[20]

THE *RELIANCE* WAS A 27-metre-long, 394-ton merchant ship, and the *Supply* was 382 tons and American-built of black birch. The *Supply* had started its working life as the *New Brunswick*, but was rechristened in honour of the First Fleet transport ship *Supply*.[21] The Navy Board had purchased both ships late in 1793, and since then they had been undergoing refits at the Deptford and Woolwich shipyards, and occasionally doing surveying work. Both needed a lot of repairs.

The Navy gave Governor Hunter's nephew Lieutenant William Kent[22] command of the *Supply* and Kent's wife Eliza would act as the first lady of the colony given that Hunter was a bachelor.

As a master's mate or 'petty officer' reporting directly to Henry Waterhouse on the *Reliance*, Matthew was authorised to take command under Hunter if Waterhouse was incapacitated. As well as having navigational responsibilities, Matthew would be the chief inspector of the vessel, making sure all was shipshape, and reporting to Waterhouse any problems with sails, masts, spars or rigging.

So it was that on 3 September 1794, Matthew Flinders Sr, the apothecary and surgeon of Donington, a man 'Gratefull to Providence', threw his arms around his sons Matthew and Samuel, and bade them farewell from the house where he had helped deliver them. 'Pray Heaven bless them both,' he wrote in his diary, 'and grant them Prosperity and Success.'[23]

Samuel, listed on the ship's log as a 'volunteer', was carrying his clothes in a bag almost as big as himself as the brothers climbed aboard a carriage that would take them to the adventure of a lifetime.

MATTHEW AND SAMUEL humped their gear onto the *Reliance*, moored at Deptford, in the first week of September and stowed it below deck. Together, the *Reliance* and the *Supply* sailed on to Plymouth Sound, and Matthew wrote home to tell his father that they would be at Portsmouth for some time. Their ships needed their final refitting, and they were also awaiting Lord Howe's escort of naval and merchant vessels that were leaving England at the same time. The warships would protect them until they were beyond the reach of French raiders.

In Portsmouth, Matthew and Samuel went ashore with shooting parties and enjoyed themselves as much as a dreary, cold season would allow.[24] Days morphed into weeks and months, during a bitter English winter when there was ice on the Thames. Matthew wrote to Mary Franklin telling her that his fond memories of his dear friends induced 'a Sentiment which if possible now warms my Heart with greater ardor than ever'.[25] He was becoming lonely and had started 'to look upon Samuel and I as two poor orphans whom nobody knows'.[26] There was always the company of the ship's cats – vital in the constant battle against rats and other vermin – but Matthew was still forlorn.

As the ships bobbed about in the grey water, Matthew admitted that he had 'continually dreamd [sic] of Chappelle and Mary F.', and that in the cold loneliness of his surrounds he ached for the 'Comforts of a Spilsby fire side [and] the agreable Chat and lively [jest] … of my two lovely friends so pure and warm is my friendship'.[27]

Hunter told John King, the Under Secretary of State at the Home Office, that 'our tedious detention in this part of the world will, I fear, be much felt in that to which we are going'. The settlement in Sydney was always starving.

Though Howe's Channel Fleet had scored decisive victories on the Glorious First of June, Hunter feared the rapid progress of the French army after they had marched into Amsterdam. He warned the British government that if the French made a surprise attack on the Dutch settlement at the Cape of Good Hope, they being 'so very active a people' could swoop, and 'if the Republicans once get footing there, we shall probably find it difficult to dislodge them'.[28]

French rule at the Cape, he said, would throw the East India trade into chaos and imperil the new colony in Sydney, which, being so far away from reinforcements, was at the mercy of a French invasion. Worried about what might happen if the *Reliance* and the *Supply* touched at a French-controlled Cape, Hunter and Waterhouse plotted a course to avoid the south of Africa, instead making stops at Tenerife and Rio de Janeiro, and then across the South Atlantic and Indian Oceans directly to Sydney Cove. Hunter recommended sending British troops to the Cape as soon as possible, and sure enough, control of southern Africa fell to the British later that year.

Hunter also had grave concerns for the health of his famous passenger Bennelong, who was returning to his people three years after leaving them. Bennelong had boarded the *Reliance* at Chatham, in Kent, back on 30 July 1794, after the British government paid one guinea to transport him from London to the docks on a fancy post-chaise carriage. The ship's muster included 'A Chest for Mr Benalong', which contained bonnets, gowns and ribbons 'for the Native Women'. Now Hunter told John King that after months of waiting on board the vessel, Bennelong was homesick, frustrated at the long delay of his return voyage, and broken-hearted over the death of Yemmerrawanne. He was, Hunter said, 'in a precarious state of health'.

'I do all I can to keep him up,' Hunter said, 'but still am doubtful of his living.[29]

THE TASK OF KEEPING Bennelong alive was given to the young Lincolnshire surgeon George Bass[30] and thirteen-year-old William Martin, known as a 'loblolly boy', a slang term for surgeons' assistants on a ship, deriving from a porridge served to sick or injured crew. Bass was a handsome, strapping and gregarious youngster three years older than Matthew, and at 183 centimetres (six feet) almost a head taller. He often wore spectacles and had a penetrating, compelling gaze,[31] and he and

Matthew quickly became firm friends, although Bass could be harsh on his little mate, chopping away with criticisms.[32] Physically they were complete opposites and their personalities contrasted, but despite their occasional squabbles, Matthew said, they ran 'together like two drops of mercury',[33] both being enthusiastic with dynamic energy.

Matthew's new friend showed him a small wooden boat he had brought onto the *Reliance* and suggested they could do some exploring in it when they finally reached Sydney. It had an eight-foot keel[34] and was five feet across (2.4 by 1.5 metres), and Bass had named it the *Tom Thumb* after the English fairytale about a tiny little boy. The boat was really just a skiff: it had a small sail that could be used when winds were favourable, and it could be steered by a sweep oar, though its design was more for idling along in calm water, not fighting the open sea. It was small enough to be stored inside the cutter on the *Reliance*.

Bass was born on Grange Farm at Aswarby, a hamlet near Sleaford, sixteen kilometres from Donington. He was the only son of a tenant farmer, George Bass Sr, and his wife, Sarah, said to be a local beauty. Bass Sr died when the boy was just six, and Sarah took her boy to Skirbeck, on the outskirts of seaside Boston, where she sent him to the Boston Grammar School.

Bass wanted to go to sea from a young age, but his mother insisted he learn a trade. For five years from the age of thirteen he was apprenticed to Boston's surgeon-apothecary Patrick Francis. After an examination in London, he was accepted, at just eighteen, as a member of the Company of Surgeons – the forerunner of the modern Royal College of Surgeons. Two months later he was certified as a surgeon's mate for the Royal Navy, being far more qualified in experience and learning than many seagoing surgeons, who were reputed to be 'half-qualified butchers or shaky-handed candidates for delerium tremens'.[35]

Within a week Bass was at sea on HMS *Flirt*, and he spent three years with the Channel Fleet and made a voyage to Newfoundland. While still a youngster, he built up a library of ninety books, and he made the exploration of the Pacific one

Flinders formed a successful partnership in exploration
with big, burly George Bass, another daring young
adventurer from Lincolnshire. *State Library of NSW FL1782862*

of his passions. When he heard that the *Reliance* was sailing for
New South Wales, he obtained a transfer to the ship and applied to
take William Martin with him. Hunter regarded Bass as 'a young
man of a well-informed mind and an active disposition'[36] and
'of much ability in various ways out of the line of his profession'.[37]

Bass and Flinders together formed the idea of completing the
examination of the east coast of New South Wales.[38]

The crew of the *Reliance* also included Lieutenant John
Shortland,[39] who had served under Hunter as master's mate on the
Sirius in the First Fleet, and had been with the captain on Norfolk
Island after its shipwreck before returning to England with him.

The *Reliance* and the *Supply* finally left Plymouth Sound on
15 February 1795,[40] as part of a grand fleet of 500 ships, then
probably the greatest number of ships that had ever left English
shores together.[41] Two days later, Howe's Channel Fleet returned
to their bases off the French coast, and the huge convoy, like a

flock of white birds, broke up and headed off in their different directions around the globe.

It was 'a long and boisterous Passage' of three weeks and 2600 kilometres to Tenerife, which the *Reliance* and the *Supply* reached on 4 March, but Matthew was proud that 'my little Samuel ... stood the Gales of Wind exceedingly well, he is in high Spirits and has lost no Part of that enterprising Spirit which brought him on board with me, I hope he'll make a good Sailor, a good officer and a good Man which last is the Groundwork of the other two and the Foundation of all Happiness'.[42]

Matthew was back in Santa Cruz three and a half years after sailing there with Bligh, and he wrote to his 'charming Sisters', Ann Chappelle and Mary Franklin, asking them to correspond with him constantly while he was in New South Wales – to 'sit down and direct as many Sheets of Paper to me as your Pen can run over, if it cannot be sent that Month never mind, it may the next, the greater the Interval ere it comes, the greater Quantity I hope to receive and the greater will be my Happiness in receiving it, by every Ship which shall touch at the Settlement'.[43] Englishmen were not meant to be happy, he told them, but it would make him so joyful to see them both again.

Ann was hesitant in answering Matthew's letters. While young ladies of the late eighteenth century were encouraged to write regularly to relatives and friends, it was thought undignified for an unmarried woman to write to a gentleman unless she was engaged to him.[44]

CAPTAIN HUNTER ENCOURAGED his junior officers to take every opportunity to perfect their navigational skills, and Matthew used the sextant and chronometer to identify locations on the Canary Islands.

'Our next Passage is to Rio de Janeiro on the Coast of Brasil,' he wrote, 'and from thence to the Colony at Sidney Cove New South Wales – what a charming Climate are we now in, while you in England with your double fortified great Coats can scarcely keep out the cold, we with the slightest covering are almost too warm'.[45]

Matthew continued his navigational calculations when the *Reliance* reached the Cape Verde islands, and then on the way to Brazil. On 2 May 1795, he calculated the latitude and longitude for Cabo Frio, which translates as 'Cold Cape' – a perilous area of shipwrecks 100 kilometres east of Rio. Five days later, the ships moored in Rio's spectacular Guanabara Bay. Brazil was a colony of Portugal, and Hunter feared that the vice-regent there had French sympathies.

Twenty-five years earlier, the crew of Cook's *Endeavour* had been treated inhospitably by the authorities in Rio, and on this visit Hunter had to wait six days before the local viceroy deigned to meet him, and then not until 7 p.m., when it was dark, and at a time Hunter said was 'extraordinary and inconvenient'.[46] Eventually, though, the crews were able to take on water and provisions, including a few barrels of high-priced salted pork, bound for the hungry in Sydney.[47] While there had been abundant harvests of wheat and corn, there was always a scarcity of fresh meat in Sydney.

Hunter wrote again to the Duke of Portland to say that while he had hoped to load a cargo of live cattle at the Cape, he had decided that sailing into Dutch territory was too risky. 'If the French should be able to possess themselves of that settlement,' he wrote, 'it will be rather unfortunate for our distant colony.'[48]

Waterhouse deemed the *Reliance*'s bumbling second lieutenant, Nicholas Johnson, as 'unserviceable' and left him in Rio, and Matthew was given that rank temporarily.[49] The promotion put him at odds with a rival for the position, the ship's master, Henry Moore, who had deep republican sympathies. Hunter perhaps felt that such sympathies in a time of conflict with France should not be encouraged.[50]

Moore had befriended passenger Daniel Paine, a 25-year-old boatbuilder, who also had radical political leanings – a hankering for democracy that British loyalists had seen as the cause of carnage in America and France. Paine was a nonconformist in religious matters, too, shunning the Church of England for Congregational and Baptist services.

Matthew avoided both these potential troublemakers, but Moore resented the youngster's promotion.[51] Matthew later wrote of Moore's 'rank pride'.

In Donington, Flinders Sr reported that both Matthew and 'my little Samuel' wrote to him from Rio to assure him that they were in 'good Health & spirits' and would be staying in 'Brasil' for about three weeks.

In Rio, Matthew received permission to climb the rocky outcrop, Ilha das Enxadas, in the middle of Guanabara Bay, and after he reached the summit he carried out a series of navigational calculations.

From Rio, the ships then sailed towards southern Africa, bypassing the Cape of Good Hope to enter the Southern Ocean. After passing Van Diemen's Land, they turned north towards Port Jackson. At sunset on 6 September 1795, Matthew saw the distant sandstone cliffs along the coastline of New South Wales, near 'Bottany Bay'.

At 7 a.m. the next day, the ships rounded Port Jackson's South Head, where the Union Jack fluttered from a flagstaff, and at eight o'clock fifteen cannons boomed out a welcome. For the first time, Matthew gazed upon the settlement of Sydney Cove, a hardscrabble village struggling for survival against the backdrop of the magnificent harbour.[52] A heavy wind began blowing from the north-west.[53] Both the *Reliance* and the *Supply* needed a great number of repairs when they finally anchored, but at least they had delivered their crews safely.

There were khaki hills on either side of the harbour, grand sandstone cliffs giving way to golden beaches, and a myriad of coves and inlets. There were towering bottle-green eucalypts and some cleared land. Dirt paths bisected the few rows of small, low houses. Pigs and sheep wandered about the village here and there, and in places kangaroos bounded. The most impressive building was the only two-storey structure in the colony, a whitewashed brick and stone building that would be Hunter's new home, Government House.

Hunter also saw disease, disarray and crime – a starving colony at the mercy of the Rum Corps and the profiteering and corruption of its officers. The Scottish sea captain was rowed ashore, to be met by Captain Grose and other officers of the Rum Corps, unaware that Grose's military force would prove the greatest threat to Hunter's office. Four days later, with Matthew and Samuel watching on, the 58-year-old bachelor, grey-haired and grandfatherly, was sworn in as the second governor of New South Wales. Hunter wrote to Portland to say there was 'scarsely a pound of salt provision' in the government stores and the colony was 'destitute of every kind of tool' used in agriculture and construction.[54]

As Matthew and Samuel explored their new home on the far side of the world, Henry Waterhouse wrote to his father back in Mayfair to tell him sad news. The convict Elizabeth Barnes, the mother of Waterhouse's daughter Maria, had died giving birth to another daughter, Rebecca, who did not survive long either. That baby's father was Thomas Smyth, a local storekeeper, who had been a marine on the First Fleet before taking up land on Sydney's southern boundary, and he would soon become the colony's provost marshal. Smyth had been looking after Maria since her mother's death and was, according to Waterhouse, 'a very respectable man [and] so very fond & much attach'd to the child that he will scarce trust her out of his sight'. Waterhouse planned to send little Maria to his father in London so 'she might be kept at school, as much as possible without a knowledge of her birth till a proper time, as it will never do much credit to either.'[55]

Waterhouse wrote to the previous governor, Arthur Phillip, who was recuperating with his new wife in London's Marylebone. 'Benalong desirs me to send his best wishes to yourself & Mrs. Phillip,'[56] Waterhouse wrote, though Bennelong had endured more heartbreak on his return, finding that his wife had taken up with a young Gadigal man. Having experienced the stark contrast of London, with its bustle and grandeur, Bennelong was now living in and out of the European settlement in Sydney, visiting Hunter and other friends he had made, but spending most of his life among his own people.

Waterhouse told Phillip: 'A Mr Flinders is appointed our 2d Lieut ... with respect to the *Reliance*, I think I mentiond to you before, I never saild in so compleat a tub ...'[57]

DURING JOHN HUNTER'S first few weeks at Government House, he praised the colonists who displayed initiative and industry by raising goats, pigs, sheep and poultry, because they always had food on the table. The British population of New South Wales at this time was 3211: there were 1908 convicts, and the remainder were mostly military and administrative personnel, as well as emancipists who had stayed in the colony after their sentences ended. 'Some Settlements which have been lately formd on the Banks of the Hawkesbury River,' Hunter wrote, 'are [also] delightful & promising.'

He was impressed by his first sight of the locally grown wheat, which was 'as beautifully luxuriant' as 'in any part of the world'. During the voyage of the First Fleet, Hunter had brought a horned bull, one male calf and five cows to the new prison settlement, but the cattle had wandered away, into the thousands of miles of bush all around Sydney Cove. Now Hunter heard reports that his cattle had multiplied into a healthy and impressive herd of horned beasts that were living wild near the Nepean.[58]

But the new governor's glowing early reports of Sydney and its surrounding settlements were soon replaced by sharper realities. Macarthur and his confederates were mired in corruption, and before long Hunter was describing 'this town of Sydney' as 'a mere sink of every species of infamy'.[59] He complained to Portland that the New South Wales Corps was composed of characters considered disgraceful by every other regiment in His Majesty's service.[60] Many, Hunter wrote, were superior in wickedness to the worst of the convicts.[61]

Macarthur and his cohort had taken control of the courts, as well as the public stores, convict labour and much of the settled land. Hunter told Portland that 'the debts of the settlers in general are chiefly, altho not wholly, owing to a disposition to indulge in drunkenness', and that Macarthur was the villain fuelling their addiction.[62]

The new governor wanted the land around Sydney to be explored as much as possible, because he found it endlessly fascinating and considered its potential limitless. He sent wildlife specimens to Banks in England, while making sketches of kangaroos and parrots, as well as the platypus, which he called an 'amphibious animal of the mole kind'.[63]

Matthew and Samuel soon wrote home to their father, giving their letters to the captain of a ship sailing for England. The letters took ten months to reach Donington, but they made the country surgeon's heart leap with joy after what had been an unusual year with a freakishly cold winter and an earthquake shaking the Flinders family in their beds. Matthew's sister Susanna was now apprenticed to a milliner in Boston, while his stepmother was suffering terribly with rheumatism that resulted in partial paralysis for weeks. Matthew's brother John was 'doing no good', having had a month's trial as an apprentice with a Boston printer before being sent home. An apprenticeship with a 'Chemist and Druggist' in the same town seemed more promising. Matthew's oldest sibling, Betsey, had married a well-off young draper named John Harvey in Donington, with her father, sister Susanna and best friend Mary Franklin as witnesses.[64]

Matthew informed his father that he was now a second lieutenant, thanks to the recommendation of William Kent on the *Supply* and a merchant friend from London named Thomas Wilson. He wrote that he soon expected to sail to China or Bengal.[65]

But Matthew and his friend George Bass knew there was a world of adventure on their doorstep in New South Wales, waiting to be explored.

Chapter 8

*'The furor of discovery is perhaps as strong and can overlook obstacles,
as well as most other kinds of mania.'*
MATTHEW FLINDERS, ABOUT TO EMBARK ON HIS FIRST VOYAGE OF
EXPLORATION IN THE LAND THAT HE WOULD NAME AUSTRALIA.[1]

MATTHEW WAS ASTONISHED by everything he saw in
New South Wales. Not even the grass under his feet was
like 'whatever had before met his eye', and he and Bass were eager
to uncover all they could about this strange land.[2] Captain Cook
had mapped the east coast, while Governor Hunter had conducted
his own expeditions along Port Jackson and the Parramatta River,
with explorations of Botany Bay to the south and Broken Bay to
the north. But the investigation of the landmass known then as
New Holland had not gone far beyond that in seven years.

The settlement at Sydney Cove was no quaint English town
like Donington. The colony was still in its brutal infancy, growing
slowly around what Matthew regarded as 'one of the finest
harbours in the world'.[3] It was a spectacular setting, although it
offered many convicts the opportunity to eventually own land,
livestock and businesses, life was harsh for convicts, their masters
and the First Peoples. There was a stone arch bridge running over
the fenced-in Tank Stream, which provided the town with fresh
water, but the waterway was always in danger of being fouled by
animal waste and runoff.

The town stank like an abattoir, with blood and entrails left in the streets after butchers had slaughtered their beasts in public places. Another observer thought Sydney looked less like a town than a ragged camp, 'mixed with stumps and dead trees'.[4] At the western end of Sydney Cove, in an area called The Rocks, the homes were mostly dirty thatched huts made of wattle and daub – mud smeared over a framework of sticks. Few of the buildings had glass in their windows.

Everywhere Matthew looked there were men and women with bent bodies and hard, threatening faces tramping dusty streets. In different locations around the harbour, chained men in iron gangs laboured under a bright sun. There was a wattle and daub church on the outskirts of the settlement, and it was compulsory for all convicts to attend it every Sunday, Hunter believing that religion and a reverence for the Bible would aid in the reformation of the prisoners and build a harmonious society. Some of the prisoners seethed at this: to many of them Sydney was more hell than heaven.

Crime was rife, despite the horrific penalties inflicted on those who were caught.

Young Daniel Paine, now the settlement's official boatbuilder, noted that 'the Soldiers as guardians of the Public Stores' were the worst offenders when it came to stealing, 'probably from their not being able to receive any other pay from their Officers than Spirits, and other Commodities at very exorbitant rates'.[5]

But Matthew was more interested in the geography of the colony than its social ills. He noted that by the time he and Bass arrived in September 1795:

It appeared that the investigation of the coast had not been greatly extended beyond the three harbours; and even in these, some of the rivers were not altogether explored. Jervis Bay, indicated but not named by captain Cook, had been entered by lieutenant Richard Bowen [Commander of the Third Fleet]; and to the north, Port Stephens had lately been examined … but the intermediate portions of coast, both to the north and south were little further known than from captain Cook's general chart …'[6]

The knowledge of the coast, he said, was confined to ten or twelve miles south of Sydney's harbour and fifteen to twenty miles to the north.[7]

While skirmishes between European settlers and the Dharug people along the Hawkesbury had devolved into open warfare, relations between the settlers and the Gadigal people around Port Jackson were more cordial. Only a few months earlier, judge advocate David Collins had been a guest at a Kangaroo and Dog Dance corroboree and initiation ceremony at Farm Cove.[8] The ceremony lasted two days. Fifteen boys from the eastern harbour clans were initiated as warriors by having their front teeth knocked out by what Collins said were the 'Cam-mer-ray' (Cammeraygal) elders from the North Shore.[9]

Not long after Matthew arrived in Sydney, he heard from Gadigal hunters that there was a wide and deep river inland from Sydney Cove, suitable to carry large boats. He suspected it was a continuation of the Georges River, which Hunter had visited years earlier. He and Bass approached Hunter for permission to explore the river and the area around Botany Bay.

'Projects of this nature, when originating in the minds of young men, are usually termed romantic,' Matthew wrote, 'and so far from any good being anticipated, even prudence and friendship join in discouraging, if not in opposing them.'[10] As it happened, Sydney was about to open its first ever theatre, with productions staged by convicts, but Bass and Flinders wanted to star in their own show, and Hunter let them.

Matthew had recovered from a kidney complaint when he and Bass sailed the tiny *Tom Thumb* out of Sydney Cove on 26 October, with William Martin assisting them. They sailed past South Head and down a coastline of golden beaches, rounding Cape Banks and entering Botany Bay. They reached the Georges River and, winding along its serpentine course, travelled thirty kilometres beyond Hunter's explorations. They sailed the *Tom Thumb* for as far as they could, before traversing further on foot through the thick bush along the riverbank.

Nine days after they set sail from Sydney, they reported back favourably to the new governor, presenting him with a sketch of their journey and telling him that the soil seemed fertile along the river and that the area around it was heavily wooded. Hunter travelled south to see for himself, and soon established a community along the river, which he named Banks' Town in honour of Joseph Banks.[11]

Hunter was impressed by the work of his two young crewmembers, but although the pair planned more ambitious explorations along the coast of New South Wales, the governor first wanted to send them as part of the *Reliance*'s crew to Norfolk Island, which had grown into an agricultural community of 887 people, farming 11,000 acres (4400 hectares).[12]

Matthew had to stay aboard the *Reliance* as it was being prepared for the voyage, but Bass, being a surgeon, was not captive to the same restrictions, and nor was Henry Waterhouse. The two young officers joined the governor and David Collins as they went exploring on a mission close to Hunter's heart.

SYDNEY WAS STILL A NEW SETTLEMENT, but Henry Hacking,[13] the colony's game hunter tasked with supplementing the salted-meat rations for the hungry population, already had a long history of violence there. He had arrived in Sydney with the First Fleet as Hunter's heavy-drinking quartermaster on the *Sirius*. Hacking was a crack shot and may have been the first colonist to kill an Indigenous person.

While chasing game on the north shore of the harbour in September 1789, he claimed that he had been in thick woods when 'a stone was thrown at him from one of two [Eora] natives whom he perceived behind him, and that on looking about he found dispersed among the trees a number that could not be less than forty'.[14] According to David Collins:

Wishing to intimidate them, he several times only presented his [rifle] toward them; but, finding that they followed him, he at last gave them the contents, which happened to be small shot for birds. These he replaced with buckshot, and got rid of his troublesome and designing followers by discharging his piece a second time. They all made off; but some of them stumbling as they ran, he apprehended they had been wounded.[15]

Hacking had also tried to find a way across the seemingly impenetrable blue-coloured mountains west of Sydney, and though he failed, he claimed to have penetrated thirty kilometres further into their dense forests and deep ravines than any other European.

To Governor Hunter's delight, in 1795 Hacking heard stories from Indigenous people about great horned beasts roaming in a lush but remote area south-west of Parramatta. Hacking went to investigate and found the cattle that had wandered away from Sydney seven years earlier. He immediately reported the find to Hunter.

At daylight on 17 November 1795, the governor set off with Bass, Collins and others from Parramatta and headed into the wilderness beyond to see for himself what had become of his prized animals. Two horses carried their tent and provisions. By the second night of a trek through heat so oppressive that one of their terrier dogs died,[16] they had walked about fifty kilometres south-west and were pitching their tents in the area now known as Menangle. They heard the lowing of cows in the distance and climbed a hill for a better view. They saw forty head of cattle grazing, untroubled, in beautiful pasture.

Trying to recover or domesticate the cattle would be 'both difficult & dangerous', Hunter concluded, so it would be better to leave them alone for a dozen years or so to multiply. Within a decade it was estimated there were 5000 head of cattle roaming free in the area now known as the Cowpastures. On the way back from finding the herd, the party also found more lush pasture country in an area behind Prospect Hill.[17]

BASS AND WATERHOUSE rejoined Matthew and Samuel Flinders soon after on the *Reliance* and the ship left Sydney on 21 January 1796, delivering supplies to Norfolk Island. The ship also carried Captain George Johnston,[18] who was to take over as lieutenant governor on the island, replacing Philip Gidley King, who was ending his second stint in charge.

Johnston became ill on the *Reliance* and returned to Sydney in March 1796 along with some of the cannons salvaged from Hunter's *Sirius*. The *Reliance* was starting to look decidedly unwell too, as the wear and tear from its voyages began to erode its timbers.

Matthew assisted in the refit of the ship back in Sydney, while Bass was involved in trying unsuccessfully to save the lives of political exiles William Skirving and Joseph Gerrald, who had been transported for sedition as part of the group known as the Scottish Martyrs. Both were seriously ill despite kind treatment from Governor Hunter, who allowed them to live as free men in cottages.

HENRY HACKING KNEW the country around Sydney better than just about anyone in the infant colony. He knew the Georges River well, but he told Matthew about shooting kangaroos beside another big waterway south of Botany Bay. He said it was somewhere a little south of Point Solander, though it was not marked on any of Captain Cook's charts. Bass and Flinders sought a new vessel to investigate this uncharted river. The new boat was most likely built by Daniel Paine, who noted that it was not 'above twelve feet [3.7 metres] long'.[19]

'As *Tom Thumb* had performed so well before,' Matthew wrote, 'the same boat's crew had little hesitation in embarking in another boat of nearly the same size, which had been since built at Port Jackson.'[20]

Tom Thumb II was almost certainly government-owned, as Hunter, vexed by the number of convicts attempting to escape by sea, soon announced a government order 'strictly to forbid hereafter the building of any boats whatever for the use of private persons'.[21]

The new boat had a mast, sail and sweep oar for steering, and just before sunset on 24 March 1796, Bass, Flinders and Martin sailed it down towards the heads to begin their search for Hacking's river. They carried two muskets, two pocket compasses, a watch and food for ten days, including five watermelons. They realised too late, though, that their water had been placed in a thirty-litre wine barrel called a 'barica' and it was soon 'exceedingly bad'.

At nightfall they bedded down upon a rock on 'Sharks Bay', in modern-day Vaucluse, and at 3 a.m. the next day they began preparations to sail into the open sea beyond South Head. Matthew expected a breeze from the north-east in the late morning and they rowed their little vessel out to sea to catch it, passing a golden arc of sand the Gadigal people called 'Boondi', because of the noise that the crashing waves made on the rocks.[22]

On this hot and humid autumn day, they drifted south even faster than expected on a strong current, the same current that had forced the *Endeavour* back for miles as it had approached Botany Bay twenty-six years before. That first evening, as *Tom Thumb II* sailed in towards the coastline and its high sandstone cliffs, Matthew thought they were off Cape Solander at the southern end of Botany Bay, but soon realised they were more than thirty kilometres beyond it. Rising swells and unfavourable winds prevented them turning back.

Soon, 'an eminence in the high land ... presented itself under the semblance of Hat Hill'.[23] The three explorers could not imagine how they could have drifted so far; Hat Hill, now known as Mount Kembla, eight kilometres inland from the city of Wollongong, had been named by James Cook in 1770 and was, by Matthew's estimate, fifteen leagues or eighty-three kilometres south of Port Jackson. Coming closer, they realised that it was indeed the hat-shaped mountain Cook had named, and they recognised the land to the north of it as the coastline they had seen just weeks ago on the *Reliance* during their journey back from Norfolk Island.

As the conditions grew rougher, Bass, Flinders and Martin pulled in from about ten kilometres offshore to 'a bending in the coast'[24] – probably Bulli Beach – and Matthew tossed out the stone anchor. It was now 8 p.m. and in the moonlight they saw ominous white surf in the distance and heard the crashing of waves. They knew trying to land would be too dangerous, so 'after making a miserable supper and drinking a melon',[25] all three slept in the tiny boat, or 'as well as three people may be supposed to in so small a space as the bottom of *Tom Thumb*'.[26]

The next morning, 26 March 1796, the three explorers found they still could not land in an area Matthew called Barn Cove, but the sun was baking and they were in desperate need of fresh water.

To the north was a seemingly endless wall of high cliffs, but the coast appeared lower to the south, with slopes leading to the shore. With the breeze blowing from the north-east again, they let the wind carry them until 10 a.m., when they 'passed a reef which projected farther than usual, and to leeward of which was less surf'.

Just south of what is now known as Towradgi Point, and with the surf still too strong for the little boat, Matthew cast the anchor just outside the breakers as Bass, the strongest swimmer of the three, decided to search for fresh water. Bass stripped off his shirt and knee-length breeches, exposing his lily-white Lincolnshire skin to the harsh southern sun, and dived naked into the waves, carrying the wine barica with him. The strapping surgeon made his way up the beach to a creek, where he filled the barrel, then he swam back through the pounding surf with it in tow.

Matthew and Martin hauled Bass aboard, but as they began to heave on the anchor line to pull the boat further out to sea, the anchor came up too quickly just as a huge wave swept in, filling the *Tom Thumb II* with water and then lifting it and its occupants high in the air and flinging them onto the beach. The three young explorers were dumped, dumbfounded, on the sand, their guns and ammunition soaked and their provisions full of seawater.

If that wasn't bad enough, they could see smoke within five kilometres, and they became anxious to get back to sea as quickly as possible, since they had heard a rumour in Sydney – false – that

Indigenous people south of Botany Bay 'were generally believed to be cannibals'.[27] They bailed out their vessel as quickly as they could, loaded the most spoilable of the provisions back into the craft and pushed it back into the sea. Matthew and Martin leapt back into the boat, working the oars until they were through the breakers.

Bass remained on the shore, then began swimming out with the boat's lead line, ferrying the rest of the food and equipment back on board. 'The musquets were near being lost from the breaking of the lead line,' Matthew wrote, 'but by lashing the heaviest articles to the oars and masts, everything else came on board safe.'

By 3.30 p.m. they were underway again, but were in a 'miserable plight'. They had managed to salvage five days' worth of flour, a few potatoes and a little salt beef, pork, rice, sago and some soup – but three days' worth of bread had been ruined, along with their tea and coffee and half their sugar. Three horns full of gunpowder were wet. The guns were full of sand and saltwater, and it turned out that the water Bass had collected was brackish. One of their oars was also broken.

They made for two small islands, hoping for shelter late in the afternoon, but the surf was smashing against those 'small rocky barren spots' too, and a landing was out of the question. So they headed for 'two larger isles lying near a projecting point of the main, which has four hillocks upon it presenting the form of a double saddle'.[28] This was near the area of reddish earth which Cook had named Red Point, modern-day Port Kembla.

Again there was nowhere for them to land on the small islands, so they spent another night cramped, damp and uncomfortable, all three squeezed into their little wooden craft as it bobbed about in an angry sea. Bass had spent five hours naked in the scorching sun as he ferried goods to the boat, and by nightfall he was unable to sleep, his once pale skin now 'one continued blister'.[29]

At dawn on 27 March, the young explorers were woken by loud cries from the shore; they recognised the language of the Indigenous people they had met in Sydney. Two local people were calling out, offering these Europeans fresh water and fish. Matthew and his companions were apprehensive, as their muskets were still wet and useless, but since there were only two warriors with fish spears, they rowed over to them and swapped some potatoes and two handkerchiefs for water and two fish. One of the men said his name was Dilba. The visitors communicated with the local people as best they could, and in a display of friendship Matthew showed them a pair of scissors with which, he said, white men trimmed their beards. As a sign of friendship he gave the two warriors a quick trim, which apparently delighted them both.

Dilba and his friend said they were not local but visiting from Botany Bay, but they offered to take them to a freshwater river a short distance to the south. Matthew and the others were desperate for good water, so they invited the two Indigenous men into the boat to act as guides. On the way to this water supply, Matthew's new friends amused him and the others with 'some stories of white men and two women' who, they said, were living with the Indigenous people and 'who had Indian corn and potatoes growing'. 'The [white] women, they said, they would bring to us, as well as plenty of black ones; and that we should get quantities of fish and ducks in the river.'[30]

With the new guides pointing the way, Matthew steered the little boat for about six kilometres south-west, to the entrance of a small stream descending from a lagoon. The stream was the Canoe Rivulet; on the map he sketched, Matthew called the large body of water Tom Thumb's Lagoon. It is now known as Lake Illawarra.[31]

The visitors rowed up the stream for about a kilometre and a half until the water became too shallow to continue against a strong current. Dilba and his friend went onto the shore and were soon joined by ten or so of their companions. Matthew felt nervous about the strength in numbers of this party. The water was now just knee height, and sometimes the boat touched the riverbed. The guns and powder were still wet and useless. Matthew began to fear that he and his companions had been lured into a trap. Although Matthew saw these Aboriginal men as menacing, they may have only been trying to help him. Still, he was taking no chances.

Bass and Flinders decided to land and dry their guns and ammunition in the sun. On the opposite riverbank, the number of Indigenous onlookers had quickly grown to twenty. Now all three of the visitors were in a state of anxiety, made worse when some of the Dharawal warriors waded into the water and came across to them. Matthew's heart rate rose alarmingly, but he was relieved to find that they only wanted to assist Bass in repairing the oar that had been broken in the surf.

The sight of the muskets drying in the sun provoked alarm among the Indigenous group, who evidently knew the damage these weapons could inflict. Bass was wearing an old waistcoat of the same colour as the British soldiers' red jackets, which added to the intimidating appearance of a man who towered over everyone else in the gathering.

To Matthew's delight, Dilba and his companion were proudly showing off their trimmed beards to their friends 'and persuading them to follow their example'. Matthew suddenly found himself in a makeshift barber's store with a long queue of customers:

> I began, with a pair of scissors, to execute my new office upon the eldest of four or five chins presented up to me ... Some of the more timid, were alarmed at a double-jawed instrument coming so near to their noses, and could scarcely be persuaded by their shaven friends to allow the operation to be finished but when their chins were held up a second time, their fear of the instrument, the wild stare of their eyes, – the smile which they forced: – formed a compound upon the rough, savage countenance, not unworthy the pencil of a Hogarth. I was almost tempted to try the effect of a snip on the nose; but our situation was too critical to admit of such experiments.[32]

While Matthew snipped away, Bass and Martin were loading the drying gunpowder and fresh water onto the boat. Matthew was still worried by the large numbers Dilba seemed to have gathered around him, and by the fact that they kept pointing upstream, 'insisting' they travel further into the lagoon. Dilba was the most 'violent' of them with his insistence, and Matthew became certain that the promises of women and food were just dangerous bait.

Tom Thumb II was readied for departure in the opposite direction to the lagoon, and when Dilba and the others continued to protest that these British newcomers were not following their entreaties, Matthew explained that they were only going to store the vessel in a safe spot for the night on a green bank beside the river's mouth. There would be no turning back, though. He checked the gunpowder and still doubted it would be much use.

As the visitors rowed down the narrow river toward the ocean, Dilba and the others followed close behind on the banks of the rivulet. Four warriors jumped into the boat and others jumped into the shallow water and dragged the *Tom Thumb II* along towards the green bank. All the way down the narrow waterway, the Dharawal men were singing and shouting, and nervously their three guests played along, singing and shouting too, though Matthew admitted the experience 'was far from being pleasant'.

Then they all neared the mouth of the river, as Matthew explained:

On coming on to the green bank, they brought us to the shore, and those in the boat leaped out; one of them with a hat on, but which he returned on being asked. Some of them still kept hold to prevent us from going further; but ... with a menacing countenance, we resolutely pushed away from them: one observing to the rest that we were angry, let go his hold; and the others immediately followed his example. Whilst we got down to the entrance, they stood looking at each other, as if doubtful whether to detain us by force; and there is much reason to think, that they suffered us to get away, only because they had not agreed upon any plan of action: assisted, perhaps, by the extreme fear they seemed to be under of our harmless fire-arms.[33]

The escape from the rivulet was blocked by strong winds and high surf, so Matthew anchored the boat just within the breakers. The water was 'tolerably deep' and the stream from the lagoon ran rapidly, meaning Dilba and his friends would not venture in. But Dilba and two or three others kept hovering along the shore, 'constantly importuning us to return and go up to the lagoon'.

Matthew was not sure if it had all been a trap or just a bad misunderstanding, but as the sun was going down beyond a row of steep hills, five or six warriors entered the surf to swim towards the vessel. Matthew checked his powder and decided the guns were in order. He fired a shot in the air, which created panic among the Aboriginal men, who quickly swam back to land. The three explorers were then left alone, though for a time Dilba continued shouting at them from the shore to come back.

'We slept by turns til 10'oclock,' Matthew noted, 'and the moon being then risen, the weather calm and water smooth, we pulled out towards Saddle Point; not a little pleased to have escaped so well.'[34]

Matthew and his companions rowed until 1 a.m., when they anchored by the most northerly of the rocky islets they had passed a day earlier. Matthew named them 'Martin's Isles after our young companion in the boat'.[35] The group, including Martin Islet, is now known as Five Islands. The adventurers spent a third cramped and uncomfortable night in their little craft.

The next afternoon they landed at a small beach 'fronted by a reef of rocks' near present-day Bellambi Point, and they were able to haul their boat out of the water unmolested, find fresh water and cook their dinner. Bass tried to soothe the badly sunburnt skin on his back.

Matthew wrote with a flourish:

For the first time, we slept on shore, and perhaps the softest bed of down was never more enjoyed. The liberty of lying in any posture and stretching out our limbs, was an indulgence, which our little bark could not afford: but I ought to have had a back covered with one continued blister, to describe the sensations of my companion.[36]

The next morning they began their return journey north, but a heavy breeze coming from that direction meant they made little headway despite constant rowing. There were sheer cliffs and no beach offering respite, so that afternoon they anchored as close to shore as they could, near 'a high projecting head with a small reef running off it' in the area now known as Coalcliff.

The sky grew overcast as night fell, and by 10 p.m. they were threatened by the approach of a brutal 'southerly buster'. 'Heavy black clouds full of electric fluid were flying about in all directions,' Matthew wrote.[37] He and Bass thought it best to haul up the anchor stone and run before the wind, rather than founder where they were. Matthew later recalled:

> The wind kept increasing, and swept along the high steep cliffs with great violence. The waves became seas and began to break … The night was extremely dark, the moon having not yet risen. We could see the shade of the dark, grim-looking cliffs over our heads; and the thundering noise of the surf at their feet told us a tale that forbid all idea of approaching them. We were thus running in the dark, with the sail flying away before the mast like a flag, Mr. Bass keeping the end of the sheet in his hand, and hauling aft a few inches occasionally, to keep the boat ahead of those seas, which appeared eagerly following after, to overwhelm us with destruction. I was steering with an oar, and it required the utmost exertion and care to keep her directly before the sea. Breakers appeared right ahead.[38]

William Martin was bailing out the water as fast as he could, but despite his determination, the sea was filling the vessel even faster. After running for nearly an hour in this critical manner, Matthew feared the seas were becoming so high that within ten minutes they would break over the little boat and its hapless occupants. A single wrong movement with the oar or a moment's inattention would have sent them to the bottom.[39]

> As no situation could be worse than the present, we determined to make a push at all hazards. On coming to what appeared to be the extremity of the breakers, watching a favourable moment, we brought the boat's head to the sea, had the mast and sail down in a trice, and got upon our oars. We cautiously pulled towards the reef during the intervals of the surges, and ... got in under the lee of it; and in three minutes, were in smooth water, and out of danger ... We thought Providential Cove a well-adapted name for this place, but by the natives it is called Watta Mowlee [now known as Wattamolla].[40]

They landed next morning, 30 March 1796, and found water, but after noting that the country around was sandy and barren, they raised their sail and coasted northward for about eight kilometres, past cliff walls that gradually decreased in height. Finally, just before noon, they found the entrance to their original destination, which they named Port Hacking, though Flinders added a rider that the First People there called it 'Deeban'.

They entered the wide expanse of Hacking's river, and noted many inlets and small creeks. They made their camp in a small cove on the north side, where many cabbage trees were growing. That area is now the southern end of Cronulla. They cooked a meal and then decided to rest from their days of struggle with some idle fishing, but the sharks were so numerous that no fish dared to make an appearance.

'These sea monsters appeared to have a great inclination for us,' Flinders wrote, 'and were sufficiently daring to come to the surface of the water, eyeing us at the same time with voracious keenness. The size of our vessel did not place us at a great distance from them.'[41]

The three explorers then retreated to their cove and tore out grass to make beds for the night. Two of the local Gweagal people visited them in the afternoon 'and behaved very civilly'. 'We understood them tolerably well,' Matthew wrote, 'and were not under the least apprehension considering ourselves as almost at home; and had it not been for the numberless mosquitos that inhabit here, should have passed a comfortable night.'

They spent the next day examining Port Hacking and its silver river, the glorious azure ocean behind them and all around a mass of green and autumnal tones over the thickly wooded hillsides flowing down to the cool, refreshing water. But they declared the wide expanse of the river 'very shoal' and not really fit for shipping. They camped again further upstream, but despite their eager expectation could find no ducks or any other game to shoot.

On the morning of 1 April[42] they sailed out of Port Hacking, aiming for Cape Solander, on the southern end of Botany Bay, about twelve kilometres to the north-east. Matthew was busy preparing a journal with maps and charts for Hunter, and reports of fertile soil and what looked like coal on the green slopes along the coast south of Sydney. They passed Botany Bay at 10 a.m. and entered Port Jackson in the afternoon, mooring their vessel beside the decaying *Reliance* at Sydney Cove.

Just after sunset, all three were back in their bunks, enjoying a well-earned rest. It would not be for long.

Chapter 9

'The signs of superior intelligence which marked his infancy, procured for him an education beyond what is usually bestowed upon the individuals of his tribe …'
MATTHEW FLINDERS, ON ADOPTING A BLACK AND WHITE KITTEN NAMED TRIM.[1]

BY THE TIME MATTHEW arrived back from Port Hacking, news had reached Sydney that the British were firmly in control of Cape Town, having overrun the Dutch garrison there. Hunter was well pleased. He needed the *Reliance* and the *Supply* to sail for the Cape, because New South Wales and its 3000 European souls were desperate for livestock. But both his ships were looking less and less seaworthy. Despite Daniel Paine's arrival as the colony's official boatbuilder, and sixteen shipwrights, caulkers, labourers and watchmen in the dockyard at Sydney Cove,[2] the colony still lacked the capacity of an English shipyard.

Hunter confessed to Under Secretary Evan Nepean that while William Kent would not complain about the *Supply* 'whilst he can make her swim', the ship was beyond repair.[3]

With limited resources, Hunter decided that the *Supply* would have to make do with basic repairs. But Matthew wrote later that 'with great repairs required by the *Reliance*', he and Samuel were confined to the ship for months while '[m]y friend Bass, less confined by his duty [as a surgeon], made several excursions, principally into the interior parts behind Port Jackson; with a view to pass over the back mountains, and ascertain the nature of the country beyond them'.[4]

It must have been frustrating for Matthew to be compelled to curb his zest for adventure in order to help with the overhaul of his ship while Bass geared up for an assault in June 1796 on that steep, rocky wall west of Parramatta – the Blue Mountains.

It was rumoured among many convicts that China was at the other side of the steep precipices, but Hunter was hoping there would be fertile plains, suitable for producing food for the growing number of colonists. The Dharug and Gundungurra people of the mountains had pathways through the forests and ridges, but even with scaling hooks and ropes, Bass and his two companions gave up after two frustrating weeks of thirst, hunger and fatigue.

'His success,' Flinders noted, 'was not commensurate to the perseverance and labour employed: the mountains were impassable, but the course of the river Grose resulted from one of these excursions.'[5]

Bass returned to the *Reliance* to learn that not only was the list of repairs extensive, but the ship had lost its longboat and half a dozen crew when a fishing expedition to Botany Bay failed to return.[6]

Home Secretary Portland wrote to Hunter to say, 'In the hope of enabling you to employ the ships under your command to the best advantage in stocking the colony with live cattle, it is with particular pleasure I inform you that the Cape of Good Hope, the Island of Ceylon, Cochin, and the Dutch Settlements in Malacca, now make a part of his Majesty's dominions.'[7]

Getting to any of those places would be a problem for Hunter's ships, but he resolved to send the *Reliance* and the *Supply* to the Cape. On 20 September 1796, the *Supply* sailed for Norfolk Island and it was joined there by the *Reliance*, the little colonial schooner *Francis* – named after Francis Grose's son – and the hired transport *Britannia*, bound for London. *Britannia* was carrying home the ailing David Collins, and Philip Gidley King, who complained of a multitude of ills. Lieutenant Governor William Paterson was on the *Britannia* too, heading back to England because he needed specialist treatment for a severe eye infection. Henry Waterhouse was taking his daughter Maria as far as Cape Town on the *Reliance*, and Paterson and his wife, Elizabeth, would have charge of her for the rest of the journey to Mayfair and the home of Maria's grandfather.

Waterhouse thought Paterson 'a very sensible pleasant Gentleman, & more so his wife', though he warned his father that Paterson and his wife were 'both Scottch people' and not to believe them if they said they had incurred expenses looking after Maria. As well as the little girl, Paterson also had charge of Matthew's sketches of Port Hacking and the area he and Bass had surveyed south of there. He planned to have them published alongside topographical surveys of Broken Bay by the colonial surveyor, Charles Grimes.

On 25 October 1796, the *Reliance*, the *Supply* and the *Britannia* set off from Norfolk Island for London. The ships headed into the Southern Ocean, catching the roaring westerly winds that revealed ice in their sharp teeth. They took on the dangers of Cape Horn at the southern tip of South America and powered on.

Matthew's friends Elizabeth and William Paterson. Captain Henry Waterhouse warned his father that they were 'both Scottch people' and careful with money. *State Library of NSW FL553505, FL1056084*

The *Reliance* anchored in Table Bay on 16 January 1797. Three days later, Waterhouse ordered the firing of his cannons in 'a royal salute in honour of the Queen's birthday' – a salute that was repeated by the 'Ships and the almost innumerable Forts' now under British rule.[8]

Matthew then sat for an oral exam to qualify as a permanent lieutenant, having now spent more than six years in the navy and two as a midshipman. Waterhouse recommended his promotion to Rear Admiral Thomas Pringle, who was in command of the Cape. Matthew appeared before a board of officers, who grilled him on navigation and seamanship. He passed with flying colours. Despite the conflicts he had experienced with Captain Bligh, Matthew also presented the examining officers with certificates from all his former commanders stating that he possessed the skills necessary in an officer.[9]

On 24 January 1797[10] he was awarded a certificate listing the ships on which he had served, and acknowledging his 'diligence, sobriety and obedience to Command'.

> … He can splice, Knot, Reef a sail, work a ship in sailing, shift his Tides … observe by the Sun or Star, find the Variation of the Compass, and is qualified to do his duty as an Able Seaman and a Midshipman.[11]

Matthew sent letters to both his father and stepmother informing them of his success, and it was an added celebration for them after Betsey gave birth to Flinders Sr's first grandchild, James Harvey. Although aged just forty-six, the apothecary of Donington was feeling positively ancient after twenty-six years of dispensing potions and being on call for women in labour night and day. He wrote in his diary that if he had 'any prospect of being freed from the fatigues of Business selling my land begs fairest for it'.[12] Soon, burdened further by increased taxes to fund a Provincial Cavalry to fight Napoleon,[13] he would write that his workload and responsibilities weighed 'heavier upon me every year'.[14]

MATTHEW PREPARED FOR the return voyage to Sydney as the ships were refitted, and fresh water, firewood, fresh vegetables and salted meats were taken on board. Pens were erected for the livestock.

Southern Africa was one of the few places in the world with homegrown merino sheep, a breed thought to have originated in North Africa, and which became so prized in Spain that, until the early 1700s, exporting them was a crime punishable by death. The South African merinos had come from a flock originally given by King Carlos III of Spain to the Dutch Prince William V of Orange. In 1789, Prince William had sent two rams and four ewes to the warmer Dutch colony at the Cape, under the care of Robert Gordon, a soldier of Scottish descent who had been in charge of the Dutch garrison at the Cape.

Three months before Matthew arrived in Table Bay, Gordon had shot himself after surrendering his garrison to the British. Now Gordon's widow was selling his merinos. She first offered them to the darkly handsome commissary of New South Wales, John Palmer. He was more interested in buying beef cattle than the widow Gordon's woolly flock, but Waterhouse and Kent bought thirteen sheep each at £4 a head.[15] Six more were given to William Paterson and Philip Gidley King aboard the *Britannia*.

The *Reliance* and the *Supply* were turned into livestock transports as hundreds of bales of hay and a variety of animals were loaded onto the vessels, including the merinos and other breeds of sheep. Waterhouse counted forty-nine head of black cattle, three mares, 107 sheep[16] and a few goats on the *Reliance*. On the *Supply*, Lieutenant Kent loaded about forty head of cattle, almost fifty sheep and five mares. The *Supply* was judged to be weak, decayed and in such a leaky state as to be unfit for the voyage, but Kent was eager to push it to the limit, and he told Matthew and the other officers that the importance of meat and beasts of burden in the colony was so great that he was 'determin'd to run every risk'.[17]

John Palmer had bought most of the livestock for the government but all the goats, two-thirds of the sheep, three horses and a few of the cattle were bought by officers for use on their own farms. Waterhouse was carrying twenty-six sheep, eight wethers, two cows, two mares and a colt, while George Bass had a cow and nineteen sheep.[18]

Matthew declined to buy livestock as he had no real interest in farming, but soon he would procure an animal that he regarded as more valuable than the whole menagerie on the *Reliance*. It would take up far less room, too.

The *Reliance* and the *Supply* were now dangerously overcrowded with so many animals constantly lowing and bleating, and on 9 April 1797 the *Supply* was torn from her anchor by a gale. Despite this, two days later the ships left Table Bay, heading into the Indian Ocean for what they expected would be a voyage of no more than forty days to Sydney. For Matthew, though, the journey would be an ordeal much longer than that.

On both ships the crowding of men and livestock together was uncomfortable and unhealthy. Most of the bigger animals were packed into stalls filled with straw on forward decks, but the cows and sheep belonging to the officers were herded into their cabins.[19] Waterhouse shared his cabin with enough animals to stock a farm. The smell and droppings made it a nauseating voyage; he told his father, 'I believe no ship ever went to sea so much lumber'd'.[20]

Six days into the voyage the *Supply* was now 'a complete mass of rotten timber',[21] and it began to fall apart. The stern worked loose and every part of the sides opened up as the ship took on 'considerable quantities of water'. Kent was dismayed that there hardly appeared 'an atom of sound wood'.[22] The frantic crew were constantly at the pumps and making repairs, to a chorus of noises from the terrified animals. As Kent surveyed a menacing sky and 'the vast extent of ocean' that lay between the two ships and Port Jackson, he thought his best protection was to pray to 'the Great Author of the universe'.[23]

The storms came like the wrath of God, and Matthew was in constant fear of his life on what Waterhouse described as 'one of the longest & most disagreeable passages I ever made'. 'We met with one Gale of Wind the most terrible I ever saw or heard of,'

Waterhouse recalled, 'expecting to go to the bottom every moment, something more than I can account for preserv'd us.'[24]

Huge seas slammed into the increasingly fragile ships, sending men and beasts crashing around the decks. Vicious squalls stretched the sails and rigging to breaking point, and white foam spewed across the decks as the rolling, mountainous waves threatened to sink the ships without trace. Still Matthew and the others remained afloat, though the water smashing through the widening gaps in the wood knocked over many of the animals. Many perished. The saltwater raced through the crew's quarters, spoiling bread, biscuits, flour and hay. The hungry animals that survived 'liv'd upon air part of the time'.[25]

In spite of this chaos, Matthew found a rainbow. For him the perilous voyage was memorable not for the violent storms but for the birth of a creature that became one of his greatest friends. While the *Reliance* was still in the Indian Ocean, one of the ship's original rat-catching cats delivered a litter of kittens.

A tiny bundle of black and white fur stole Matthew's heart, and he called his kitten Trim, after a character in one of his favourite novels, *Tristram Shandy*, a character who was irritating, loyal, funny and brave. It was, however, Matthew wrote, from the 'gentleness and the innate goodness' of the cat's heart that he gave him the name of this 'honest, kind-hearted, humble' character.[26]

The kitten, he said, showed signs of superior intelligence from infancy, 'and being brought up amongst sailors, his manners acquired a peculiarity of cant which rendered them as different from those of other cats, as the actions of a fearless seaman are from those of a lounging, shame-faced ploughboy'.

In playing with his little brothers and sisters upon deck by moonlight ... the energy and elasticity of his movements sometimes carried him so far beyond his mark, that he fell overboard; but this was far from being a misfortune; he learned to swim and to have no dread of the water; and when a rope was thrown over to him, he took hold of it like a man, and ran up it like a cat.[27]

Sydney Cove about the time Matthew began his explorations from there in the closing years of the 1700s. *State Library of NSW FL1146307*

William Kent pushed the *Supply* hard, wanting to get to Port Jackson as quickly as possible, lest his ship disintegrate in the middle of nowhere. The vessel flopped over the finish line at Sydney Cove on 16 May 1797, just thirty-five days after leaving Table Bay. As they approached home, the crew could see the colony's first windmill, newly built and standing proudly on Flagstaff Hill – now Observatory Hill. As they came nearer to their mooring, they passed the rotting corpse of Francis Morgan, still hanging in rusty chains from a gibbet on Pinchgut Island (now Fort Denison) as a warning to the colony six months after being executed for murder. Governor Hunter thought Morgan's putrid body was only in marginally worse shape than Kent's ship.

Hunter noted that the *Supply* had landed twenty-seven cows and thirty-five sheep. Eight of its cows, two bulls and thirteen sheep had died on the voyage.[28] The *Supply* was in such 'distressing and dangerous condition',[29] Hunter said, that it should never again set sail. Indeed the ship was soon dismasted and anchored in the harbour for use as a storage hulk.[30]

Matthew and Trim took a much slower route home. Waterhouse sailed the *Reliance* in a way that would cause the least amount of stress to its fragile timbers, so it wasn't until 26 June, fully forty-one days after the *Supply*, that Matthew saw Sydney again. Although two-thirds of the livestock arrived safe, half of the twenty-six merinos bought by Waterhouse and Kent had died.[31] John Macarthur offered Waterhouse fifteen guineas a head for the surviving merinos on condition that he could buy them all, but Waterhouse refused. Instead he grazed them at 'Maria's Farm', also known as the Vineyard, his new 55-hectare property on the Parramatta River, before eventually distributing a few between Macarthur, the Reverend Samuel Marsden, Lieutenant Kent and Captain Thomas Rowley, a prominent landowner. The merinos would become the foundation of one of the world's biggest wool industries.

Waterhouse told his father the *Reliance* was 'so leaky' when he sailed her into Sydney that even though the harbour was as smooth as a 'Mill pond, She keeps our whole Ships Company pumping from three to four hours a day ... How she will be patch'd up I do not know. It will take near a twelve Month, & then I shall dread going to sea in her.'[32]

The overhauling of the *Reliance* would once again thwart Matthew's ambitions to investigate more of the vast land all around him. He took comfort in Trim, who was constantly by his side or nuzzling at his feet over the next few months.

Matthew wrote that from the care that was taken of Trim, 'and the force of his own constitution', the little cat eventually 'grew to be one of the finest animals I ever saw ...

his weight being from ten to twelve pounds ... His tail was long, large and bushy; and when he was animated by the presence of a stranger of the anti-catean race, it bristled out to a fearful size, whilst vivid flashes darted from his fiery eyes, though at other times he was candour and good nature itself. His head was small and round, – his physionomy bespoke intelligence and confidence, – wiskers long and graceful, – and his ears were cropped in a beautiful curve. Trim's robe was a clear jet black, with the exception of his four feet, which seemed to have been dipped in snow; and his under lip, which rivalled them in whiteness; he had also a white star on his breast, and it seemed as if nature had designed him for the prince and model of his race.[33]

THE DAY AFTER WHAT was left of the *Supply* lurched into Sydney Cove, a small fishing boat anchored near it, with the crew on board shouting for help for three shipwreck survivors they had rescued – sun-blistered, bedraggled men so emaciated and exhausted they could barely speak, let alone walk. The three had survived the loss of their ship hundreds of miles away, and after months tramping and staggering through the wilds of New South Wales they were skeletal remnants of the men who had set out on a trading voyage from Calcutta six months earlier.

The survivors were carried a few hundred metres uphill to Hunter's Government House. One of them, 27-year-old Scottish highlander William Clark, had been speared through both hands by Indigenous warriors. He carried with him a tattered letter addressed to the Governor, written by another Scotsman, Gavin 'Guy' Hamilton, explaining that the ship the *Sydney Cove* had been beached off Van Diemen's Land. Hunter's eyes widened at an astonishing tale of human fortitude. The three men in his parlour had somehow survived fourteen weeks of hell, stumbling for hundreds of miles through wild bush.

As the survivors slowly recovered, they explained to Hunter that on 10 November 1796, Captain Hamilton, a 37-year-old who had been in India for ten years, sailed the 250-ton *Sydney Cove* through the shoals of Kolkata's Hooghly River and into the Bay of Bengal.[34] On board was a crew of fifty-five men, mostly young lascars – Indian sailors – and all manner of livestock and goods for sale in Sydney – meat, sugar, rice, tea, fine porcelain, clothing, and gunpowder – even a carriage and an organ. It also carried what the owners thought would be the big seller – 7000 gallons of Indian liquor. The *Sydney Cove* had started her life as a rice transport named the *Begum Shaw*, but hearing reports from other ships about the dearth of goods in Sydney, the Calcutta-based trading company Campbell and Clark sent their new ship on a speculative voyage to New South Wales.

By early December, the ship had crossed the equator on benign seas, but by 13 December black clouds covered the sun and high winds whipped up huge seas. On 25 January 1797 the ship was hit by what Captain Hamilton called a 'perfect Hurricane' in the Southern Ocean. The second mate was sent flying from the yardarm like a slingshot into the briny deep and a horrible death.[35]

Many of the lascars 'were so benumbed by the severity of the weather' that for a long time they abandoned the pumps and huddled below deck. The next day the Antarctic winds were so horrific that two Lascars, cold, weak and malnourished, dropped dead; a third died a few hours later.[36] The leaking vessel reached the southern tip of Van Diemen's Land on 1 February and bore north, twisting and rocking in the demonic weather. On 7 February winds with hurricane force[37] burst open more timbers and tore the sails into strips. On 8 February Captain Hamilton beached his stricken vessel on a sheltered area on what is now called Preservation Island, in the Furneaux Group, north of Van Diemen's Land.

Most of the crew had somehow survived, and the cargo, as well a distressed mare and cow, chickens and pigeons, were carried off to safety. Hamilton sent men to find fresh water, and as a way of safeguarding the ship's fortune in precious liquor, he had the barrels rowed across to a smaller island 200 metres away. He called it Rum Island.

With the *Sydney Cove* beyond repair, Hamilton realised the only way to save the lives of his men was to fix the longboat damaged in the storms and sail it for 850 kilometres north to Sydney, the only European settlement within 5000 kilometres. He chose his best seventeen men, chief mate Hugh Thompson, the supercargo William Clark, three other Europeans, eleven lascars and Clark's Indian manservant.[38] He gave Clark a letter he had written for Governor Hunter explaining their location, predicament and the fact they were carrying valuable cargo, unaware that Hunter was trying to stamp out the rivers of booze in his colony. On 28 February, the seventeen lifesavers left the thirty or so other members of the crew on the beach beside the wreck of the ship, hoisted the sail on their longboat, and to shouts of 'Godspeed!' and 'Good luck!' began what they hoped would be a death-defying odyssey.[39]

Before long, they were lurching about in heavy seas again, their small craft filling with saltwater from the foaming waves. After three days they glimpsed through the rain and spray the longest beach any of them had ever seen but as they neared its protection, violent breakers threw all of them out of their craft and smashed their vessel to pieces.

The seventeen men collapsed exhausted on the wet sand, having survived their second wreck in less than a month and now with little more than the sodden clothes on their back. They were still alive – but for how long? They had arrived at the northern end of what is now known as Ninety Mile Beach, in the Victorian region of the Gippsland Lakes, but they no longer had a boat and Sydney was still 800 kilometres away across land they had never experienced before. They were cut off from any hope of rejoining their companions; without food, without most of their weapons, and with little hope of subsistence or defence. In this vast, alien landscape, they seemed doomed now 'to all the horrors of a lingering death, with all their misfortunes unknown and unpitied'.[40]

The sad seventeen spent three days collecting anything that had been thrown on shore from the wreck, including a musket, two damaged pistols, two small swords, some rice and rolls of cloth, hammers, nails and axes. They began their weary march north on 15 March 1797, like a procession of dead men walking.

They tried to stick as near to the coastline as they could. They covered about thirty kilometres along the beach and beside the great lakes on their second day, but on the third spent much of their time lashing together a rough raft to cross a large river.[41]

The survivors ploughed on, catching fish when they could, tramping through rain, fording rivers, including the Snowy, and, when the rain cleared, marvelling at the vast country before them. On 2 April some of the Thaua people, who had been observing them from a distance, gave the weary travellers a gift of shellfish –

which was most welcome, since their rice, rescued from the surf at Ninety Mile Beach, was almost exhausted.

After six more days of hard walking, and having crossed what is now the Bega River, they came face to face with fifty Yuin men armed with long spears and 'throwing sticks'. The visitors were terrified, but Clark implored his men to 'betray no symptoms of fear'. Finally, they were allowed to pass with the payment of a 'few yards of calico'.[42] The warriors threatened them again over the next two days, but on 10 April Clark convinced them to take some cloth in return for a large kangaroo tail, with which his men made soup.

On 11 April, the day Matthew left Cape Town on the *Reliance*, they came to a river where a party of Walbanga men conducted them 'to their miserable abodes in the wood adjoining to a large lagoon [Wallaga Lake] and kindly treated us with mussels, for which unexpected civility we made them some presents'.[43]

For the next few days, Aboriginal people would help them cross more rivers, including the spectacular Wagonga, near present-day Narooma, using their 'rude little' bark canoes, about two metres long and barely half a metre wide, 'tied at both ends with twigs'. The Europeans and lascars found the canoes impossible to steer, though Clark noted that the Aboriginal people could ride in them with three or four people and paddle about in them 'with the greatest facility and security'.[44] Assistance with river crossings would ensure these strangers kept moving north and away from sacred places.

On and on the men stumbled and staggered north, barely surviving now on local berries and the nectar of banksia plants. The meagre fare was not enough to sustain more than half the men. Some of the barefoot lascars had been limping for days, their bloodied feet now infected with stinking sores. Scurvy was swelling their joints, loosening their teeth and bringing on paralysis.

On 16 April, Clark took a pencil and scratched into his little diary: 'At this place we were under the painful necessity of leaving nine of our fellow-sufferers behind, they being totally unable to proceed further.'[45]

He likely knew they had no chance of survival, but he was not about to die with them or leave his skipper stranded on Preservation Island. The nine lascars slumped to await their fate as eight men continued limping towards Governor Hunter in Sydney.

Soon two more lascars were left behind to fend for themselves. On 20 April, the remaining six strugglers reached Batemans Bay, named by Cook twenty-seven years before. The Walbanga people guided the men safely through thick forests around the area. But on 26 April, when they were near modern-day Sussex Inlet, a hundred of the Wandandian people came at the travellers waving spears and clubs. They tossed some of the spears at the visitors' feet as a warning, and when Clark held up his right hand, one of the men threw a spear straight through it. Defensively, he held up his left hand for protection and soon it had a long spear sticking through it as well. Thompson and Clark's manservant were wounded too. The Wandandian could have killed all six of the intruders but retreated after what was most likely a payback punishment for the disrespect on the Wandandian land. The spears were removed carefully and the wounds washed with salt water and wrapped in calico. On they staggered to the large bay named for Admiral John Jervis.

On 30 April, they came to the wide expanse of what George Bass would later call the Shoalhaven River. Six Dharawal men helped them across with their canoes. Barely able to move now, they struggled on for two more weeks until 15 May, when they collapsed more dead than alive on a beach covered in black rocks. Clark recognised it as coal, and spotted a coal seam in the steep cliffs above them. He used the coal to warm them all against the cool winds coming off the ocean, but Thompson, the ship's carpenter, and one of the two remaining lascars finally surrendered, saying they could go no further. Clark knew he would likely never see his friend Thompson again, but there was no use in all of them dying, and Clark, his Indian manservant and crewman John Bennet pressed on.

The next day, 16 May, after sixty-three days of torture, the three last men standing reached the small, sheltered cove that Matthew had visited with Bass the previous year – 'Wattamowlee'.

To their joy, the men saw a small rowboat with fishermen casting nets bobbing up and down just off the coast. Although he was knocking on death's door, Clark's spirit was instantly resurrected. He screamed and leapt about like a madman. There was no acknowledgement from the men on the boat, but on blistered, bloodied feet Clark ran towards the water, yelling like he had never done in his life. At first the fishermen thought the travellers in the distance were Dharawal men, but then Clark removed his tattered shirt and waved it frantically above his head, all the while crying for help in English.

Clark wrote that when they were finally carried into the Governor's home in Sydney, Hunter treated them with 'such kindness and humanity as it were impossible to describe'.[46] Within a few hours, the Governor sent a crew in his personal whaleboat on a rescue mission for Thompson and the two men with him, but they were lost forever.

'Some articles they had were pick'd up cover'd with blood,' Hunter wrote, 'so that we have reason to believe they have been murder'd in this helpless state.'[47] Matthew would later hear that their killer was Dilba, the same man who had tried to lure him deep into Tom Thumb's Lagoon a year earlier.[48] Dilba's involvement was only a rumour, but it was well known that some of the Sydney Cove crew had been cruel to the First Peoples and perhaps had brought their fate upon themselves.

Matthew was, as always, busy with repairs to the *Reliance*. So Hunter sent Bass south to investigate the coal Clark had warmed himself with on the day before his rescue.

The cliff with the huge seam of coal was predictably named Coalcliff, and the mine that was eventually established there would produce coal for more than a century.

Hunter now turned his attention to the other survivors still on Preservation Island. He had few vessels at his disposal for such a long voyage. The *Supply* was a wreck and the *Reliance* still unseaworthy, but on 27 May, as soon as it became possible, the

Governor dispatched the colonial schooner *Francis*, of just forty-one tons, under the command of William Reed, together with the *Supply*'s ten-ton sloop-rigged longboat the *Eliza*,[49] skippered by that ship's master, Archibald Armstrong.

When they reached the emaciated men off the coast of Van Diemen's Land, the two small craft could only take part of the cargo and crew. John Bennet, who had guided the rescue boats south, elected to stay with five of the lascars to keep watch over the rest of the supplies. The *Francis* and *Eliza* then sailed for Port Jackson on 21 June.[50] But in bitter winter winds they became separated just north of the Furneaux Islands. The *Francis* arrived in Sydney 'after a stormy passage of fifteen days'[51] with Hamilton and some of the lascars, but the *Eliza*, which was carrying eight of the rescued Indian crewmen, was never seen again.

Matthew had spent months cooling his heels, eager to investigate more of the vast continent beyond Port Jackson. The chance to collect the last of the survivors on Preservation Island would also provide him with the opportunity to solve one of the great riddles about the coastline of New Holland – a riddle that had intrigued mariners for more than a hundred years.

Chapter 10

'I sent in the schooner Lieut. Flinders, of the Reliance *(a young man well qualifyed), in order to give him an opportunity of making what observations he could amongst those islands ...'*

GOVERNOR HUNTER, TELLING THE HOME SECRETARY ABOUT A NEW
VOYAGE OF DISCOVERY BY MATTHEW FLINDERS.[1]

HIS LONG ABSENCE FROM Ann Chappelle and Mary Franklin made Matthew's heart grow fonder. Naturally, he remained coy in letters home about any romantic attachments he had made at his new home on the far side of the world, although relationships out of wedlock were not uncommon in New South Wales. His friends Waterhouse and John Shortland had both fathered children in the colony, and even Lieutenant Governor Philip Gidley King had sired two sons with his convict mistress on Norfolk Island. It could get lonely on the *Reliance*, but Matthew was cheered somewhat by the promotion of his brother Samuel, now fifteen, to the rank of midshipman,[2] and by Hunter's grant to Matthew, Bass and Shortland of forty hectares each in the new settlement of Banks Town.[3]

He could always get a laugh when needed from Trim, who rarely left his side:

Notwithstanding my great partiality to my friend Trim, strict justice obliges me to cite ... a trait in his character which by many will be thought a blemish: He was, I am sorry to say it, excessively vain of his person, particularly of his snow-white feet. He would frequently place himself on the quarter deck before the officers, in the middle of their walk; and spreading out his two white hands in the posture of a lion couchant, oblige them to stop and admire him. They would indeed say low to each other, 'See the vanity of that cat!' But they could not help admiring his graceful form and beautiful white feet …[4]

The cat's capers also helped to take Matthew's mind off a painful kidney inflammation that had troubled him for months, and which he had managed to alleviate with lashings of cold water. He assumed that he was flushing out the cause of his complaint, but kidney problems would continue to plague him.[5] So too did months of frustration at being kept at work as the *Reliance* was all but rebuilt,[6] a confinement to his adventurous spirit made worse by the explorations and discoveries being made by his friends on long, important expeditions.

On 9 September 1797, while Matthew was watching the *Reliance* being re-rigged and fitted with new timbers along its hull, John Shortland was on the hunt for runaway convicts who had stolen the colonial vessel *Cumberland*. The theft of the ship was a severe blow to a colony always under threat of starvation, and since the only other colonial vessel, the *Francis*, a 41-ton schooner, was at Norfolk Island, Hunter dispatched the only crafts he could, rowboats fitted with sails.

One journeyed south past Botany Bay while Shortland and his crew sailed north, eventually as far as Port Stephens, 200 kilometres away, where earlier escapees had once made their home. Hunter encountered an uncharted inlet that was in fact the mouth of a wide river. The expedition was 'fruitless as to the proposed object', Matthew wrote many years later, as the convicts were never found.

[B]ut in returning along the shore from Port Stephens Mr. Shortland discovered a port … capable of receiving small ships; and what materially added to the importance of the discovery was a stratum of coal, found to run through the south head of the port. These coals were not only accessible to shipping, but of a superior quality to those in the cliffs near Hat Hill. The port was named after His Excellency governor Hunter and a settlement, called New Castle, has lately been there established.[7]

Soon after, Shortland wrote to his father in England to describe 'a very fine coal river' just south of Port Stephens, which 'in a little time, will be a great acquisition to this settlement'.[8] To add to Matthew's exasperation, Hunter authorised George Bass and six volunteer naval seamen for a long voyage in an 8.5-metre-long whaleboat, equipped with oars and a sail, and built from native banksia and cedar at Daniel Paine's shipyard. They were to make a voyage towards Van Diemen's Land in the hopes they could map

Returning from Port Stephens, north of Sydney, Matthew's friend John Shortland explored a port where a settlement called 'New Castle' was established. *State Library of NSW FL1765233*

it more accurately and reveal whether it was an island or part of the mainland of New Holland.

Hunter instructed them to follow the coast of New South Wales for as far south as safety would allow, and they were given provisions for six weeks. One of the volunteers was John Thistle, who was always at the ready to sail with Bass and Flinders.

The Governor had written to Joseph Banks with news that the survivors of the *Sydney Cove* had told him that the tides around Preservation Island were 'Shory, running sometimes 4 Knots, its direction nearly East & West', and that the 'account seems to indicate what I have long Suspected, that there is a Strait thro this part of the Coast & that Vandeimans land is an Island'.[9]

Hunter implored Joseph Banks:

I earnestly Wish Government woud send a Maratime Surveyor here with fit Vessels & have this Coast Examined; I am much inclind to think many useful discoverys woud be made: We have much Ore in the land beside Iron at least I think so this woud be good amusement for a Mineoroligist but we are such an Abandond set, that My time is wholly taken up in looking after the Public Concerns & in endeavouring to establish some decency & order ...[10]

If Van Diemen's Land were an island and a passage existed between it and the rest of New Holland, it would cut valuable time for voyages from India or the Cape, and it would allow captains to avoid sailing so far south into the towering, icy waves of the Southern Ocean. Hunter was determined to discover the truth, and in the absence of a 'Maratime Surveyor', Bass and his team would have to do.

Hunter told Portland that because of 'the tedious repairs which his Majesty's ship *Reliance* necessarily required before she could be put in a condition for going again to sea', he had given the opportunity of exploration to the ship's young surgeon. When Hunter had asked Bass 'in what way he was desirous of exerting himself' on behalf of the colony, the strapping young officer had replied that 'nothing could gratify him more effectually than my allowing him the use of a good boat and permitting him to man her with volunteers from the King's ships'.[11]

So, to the cheers of his shipmates on the *Reliance*, including a somewhat downcast Matthew, Bass and his crew left Sydney on 3 December 1797, and headed south in the whaleboat that Bass had named the *Elizabeth* after Henry Waterhouse's sister, with whom he was in regular correspondence. The crew ran into storms and sheltered the first night in what is now called Little Bay. The bad weather continued and they spent their second night just a few kilometres south in Port Hacking. On day three Bass noted 'a deep ragged hole of about 25 or 30 feet in diameter, and on one side of it the sea washed in through a subterraneous passage with a most tremendous noise.' This remarkable feature is now known as the Kiama Blowhole.

Soon after, Bass named the 'Shoals Haven river',[12] and further along sailed into Twofold Bay. On 20 December, the *Elizabeth* rounded Cape Howe and, on great rolling blue waves capped with white foam, sailed west along a largely uncharted coast of what is now the state of Victoria. Savage weather forced the team to shelter for ten days in an area now called Wingan Inlet, their Christmas spent dining on fish and rationed provisions, and explorations of a countryside rarely seen before by Europeans.

The expedition passed a 'lofty hummocky' promontory[13] which Bass called Furneaux's Land, and on 3 January 1798[14] Bass saw smoke coming from an island in what is now known as the Glennie Group. He supposed the smoke was from fires made by Indigenous people, and he and his men rowed cautiously towards it to see if he might communicate with them, since the few he had seen since leaving Sydney had always appeared timid and had run away.

Instead, Bass was shocked to find seven escaped convicts, in various states of exhaustion and ill health, their clothes in tatters. They had been part of a prison gang of fourteen, mostly Irishmen, who had staged an escape from Sydney in October 1797, not long after the hijacking of the *Cumberland*.[15] They told Bass they had been 'treacherously left on this desolate island' five weeks earlier by the other convicts because the boat was too small for all of them. They had remained alive by eating seals and birds.

Bass told Hunter that 'these poor distressed wretches, who were chiefly Irish, would have endeavoured to travel northward and thrown themselves upon his Majesty's mercy, but were not able to get from this miserable island to the mainland'.[16] Bass's whaleboat was far too small to carry all of them, but he gave them what sustenance he could and promised to call at the island on his return from the expedition.

Bass had sailed with only six weeks of provisions, but Matthew later wrote that with the assistance of occasional supplies of 'petrels, fish, seal's flesh, and a few geese and black swans, and by abstinence', he was able to prolong his voyage.[17]

The *Elizabeth* continued to sail west, and on 5 January 1798 reached 'a large sheet of water branching out into two arms which end in wide flats of several miles in extent, and … [with] two outlets. He named the place Western Port,[18] and he and his crew spent twelve days there exploring the area and what is now called Phillip Island.

The mission was hampered by bad weather and damage to the leaking vessel caused by the heavy seas. Reluctantly, Bass decided to return to Sydney, having failed to prove there was a strait between Van Diemen's Land and New Holland, although he remained convinced that one existed. He was only about thirty kilometres from reaching the vast expanse of Port Phillip Bay and what would become the site for the settlement of Melbourne.

Poor weather blighted much of the return voyage, but Bass was a man of his word and stopped to see what he could do for the seven stranded convicts. He could only fit two into his small, damaged craft, so he elected to take the two who were faring the worst – 'one of whom was old and the other diseased'.[19]

Then he ferried the other five to the mainland, where he gave them a musket, half his ammunition, a pocket compass, fishing line and hooks, and what clothes and food he and his crew could spare. Bass also gave them detailed directions of how to walk the 800 kilometres back to Sydney, telling them to stick to the coast as best they could, where they might find food and where he thought the local Indigenous people seemed especially hostile. He told them to keep going no matter what.

When the five parted from Bass and his crew, tears were shed on both sides.[20] None of the five was ever heard of again.

MATTHEW FINALLY RECEIVED his own chance to investigate the questions surrounding Van Diemen's Land when Hunter mounted a mission to rescue the last six crewmen of the *Sydney Cove* and the remaining cargo on Preservation Island.

Hunter ordered that the schooner *Francis* be deployed, with William Reed again in command. Matthew requested a place on the mission, and Hunter, already well acquainted with the young man's enthusiasm and capabilities in mapping, agreed. Matthew defined his role on the voyage as being 'for the purpose of making such observations, serviceable to geography and navigation, as circumstances might permit'.[21] Hunter told the Home Office that he sent 'Lieut. Flinders, of the Reliance (a young man well qualifyed)', and that he would combine the charts of Bass and Flinders for a clearer picture of Van Diemen's Land.[22] It was Matthew's first major hydrographic survey, and the chance he'd been waiting for.

Bass and his men were still at Furneaux's Land, on their way home, when Matthew and the *Sydney Cove*'s skipper, Gavin Hamilton, left Sydney on the *Francis* on 1 February 1798. Matthew asked Reed to stay as close to the coast as possible as he plotted places already known, and others that were new. On the third day, near Batemans Bay, Matthew noted:

Soon after noon, land was in sight to the S.S.E., supposed to be the Point Dromedary of captain Cook's chart; but, to my surprise, it proved to be an island not laid down, though lying near two leagues from the coast ... This little island, I was afterwards informed, had been seen in the ship *Surprise*, and honoured with the name of Montague. When captain

Cook passed this part of the coast his distance from it was five leagues, and too great for its form to be accurately distinguished.[23]

Matthew named Green Cape, near present-day Eden, writing:

[T]he shore abreast of the schooner was between one and two miles distant; it was mostly beach, lying at the feet of sandy hillocks which extend from behind Green Cape to the pitch of Cape Howe. There were several fires upon the shore; and near one of them, upon an eminence, stood seven natives, silently contemplating the schooner as she passed.[24]

With Captain Reed staying close to the coast, they came upon a group of unmapped islands which Flinders called the Kent Group, after his friend William Kent. The *Francis* finally reached the *Sydney Cove*'s survivors eleven days after leaving Port Jackson.[25] The wrecked vessel had been run on shore between Preservation and Rum islands, and while the hull was still upright, 'the sea thrown in by westerly gales had, in great measure, broken her up, and scattered beams, timbers, and parts of the cargo upon all the neighbouring shores'.[26]

Of the six men left to guard the cargo, one had died, but the rest remained in reasonable health thanks partly to the bird life that nested on the islands—geese, 'pinguins' and 'sooty petrels', also known as muttonbirds because of their taste.

The sooty petrel ... frequents the tufted, grassy parts of all the islands in astonishing numbers. It is known that these birds make burrows in the ground, like rabbits; that they lay one or two enormous eggs in these holes, and bring up their young there ... These birds are about the size of a pigeon, and when skinned and dried in smoke we thought them passable food. Any quantity could be procured, by sending people on shore in the evening. The sole process was to thrust in the arm up to the shoulder, and seize them briskly;

but there was some danger of grasping a snake at the bottom of the burrow, instead of a petrel.[27]

Rough seas and damage to the *Francis* prevented Matthew's survey work for a few days, but as soon as it was safe, he and four others set off on 16 February on a five-day investigation of the rocky, seal-covered islands surrounding him. He sailed through Armstrong Channel, named after the *Eliza*'s ill-fated skipper, and he climbed hills where he could, mapping and measuring all that he could see and giving them European names. He saw a 'smooth round hill' from Preservation Island and named it Mount Chappelle,[28] dedicating it to the girl he longed to see again. The Chappelle Isles were nearby.

Matthew had done his best to stay in contact with his young female friends, sisters and half-sisters, and in one letter told the girls: 'never will there be a more happy soul than when I return. O, may the Almighty spare me all those dear friends without whom my joy would be turned into sorrow and mourning.'[29] He thought of Ann often: she was witty and generous, he remembered her small, petite figure, her long, raven-black curly hair cascading over her shoulders, her rich red-brown eyes. She was slight and graceful and pale, and everyone thought her so damned clever. Despite the sorrows of her childhood she had what Matthew thought 'a sweet, perfect temper'. He remembered that she wrote lovely poems and painted beautiful flowers.[30]

Matthew made extensive notes about the geography, weather and wildlife he saw around him. He could find no Indigenous people, but there were kangaroos, what he called ant-eaters (echidna) and a slow, 'bear-like quadruped' that burrowed 'like a badger' and was 'known in New South Wales, and called by the natives *womat, wombat,* or *womback'*.[31]

There were also two kinds of seal: one, which took six musket shots to kill, was as big as an ox. All were fair game for the hungry sailors.

At Cone Point on Cape Barren Island, where the multitude of seals exceeded anything Matthew had ever seen, he made surveys

Wood engraver Thomas Bewick published this image of a 'wombach'. From *A General History of Quadrupeds*, 1800.

and charts as several thousand timid animals cowered while sailors went among them, clubbing as many as they could for meat, oil and fur.

> The young cubs huddled together in the holes of the rocks, and moaned piteously; those more advanced scampered and rolled down to the water with their mothers; whilst some of the old males stood up in defence of their families until the terror of the sailor's bludgeons became too strong to be resisted ... The sailors killed as many of these harmless, and not unamiable creatures as they were able to skin ... and we then left the poor affrighted multitude to recover from the effect of our inauspicious visit.[32]

The echidnas were 'exceedingly fat' – 'the flesh [had] a somewhat aromatic taste, and was thought delicious'[33] – while wombat meat resembled lean mutton. Matthew didn't try eating the local tiger snakes, but he said these 'speckled yellow snakes, of three or four feet in length, were found upon Preservation Island, and exist, no doubt, upon the larger isles.'

They sometimes get into the burrows of the sooty petrel, and probably destroy the young. I saw one dragged out by a sailor who expected to have taken a bird; but, being quick in his movements, he was not bitten. These snakes possess the venomous fangs; but no person experienced the degree of virulence in their poison.[34]

By 25 February 1798, the *Francis* was ready to return with the *Sydney Cove* survivors, and although Matthew had not yet been able to find conclusive proof that Van Diemen's Land was an island, 'the great strength of the tides setting westward, past the islands, could only be caused by some exceedingly deep inlet, or by a passage through to the southern Indian Ocean ... and the schooner not being placed at my disposal, I was obliged, to my great regret, to leave this important geographical question undecided'.[35]

Matthew returned to Sydney on 9 March 1798, to find that Bass had made it back on 25 February after twelve weeks of exploration. Perhaps forgetting about Bligh's great voyage in an open boat to Timor, Matthew wrote that Bass's journey, 'expressly undertaken for discovery in an open boat, and in which six hundred miles of coast, mostly in a boisterous climate, was explored, has not perhaps its equal in the annals of maritime history'.[36] Bass provided all of his notes and observations to be added to the chart Matthew was completing, and the two now concurred that 'there seemed to want no other proof of the existence of a passage between New South Wales and Van Diemen's Land, than that of sailing positively through it'.

From his desk at Government House, Hunter wrote to the Duke of Portland to explain that Bass's 'perseverance against adverse winds and almost incessant bad weather led him as far south as the latitude of 40-00 S., or a distance from this port, taking the bendings of the coast, more than of six hundred miles'.[37]

Hunter wrote to Banks again, informing him of the work Bass had done and also reporting:

I sent in the Schooner the Second Lt: of the Reliance (Mr. Flinders) with directions to make such Observations amogst the Islands as he coud, Such Sketches as he had an Opportunity of Making Assisted by those of Mr. Hamilton the Master of the Wreckd Ship, a Sensible well informd Man, who I am Concernd to say died here lately.[38]

Banks was already familiar with the work Matthew was doing. The recently promoted William Paterson had brought home Matthew's charts of the coastline around Sydney on the *Britannia* for publication in London by the cartographer Aaron Arrowsmith.[39]

Banks had written to Evan Nepean at the Home Office to tell him that he had been in contact with the esteemed Scottish explorer Mungo Park, who had just returned from Africa and now offered himself as a 'volunteer to be employ'd in exploring the interior of New Holland'.[40] In Africa, Park had discovered the source of the Niger River, a feat which Banks said now offered Britain 'the means of penetrating into the centre of that vast continent, exploring the nations that inhabit it, and monopolising their trade'.

Banks said Park would 'want a deck'd vessel of about 30 tons, under the command of a lieutenant, with orders to follow his advice in all matters of exploring ... Lieutenant Flinders – a countryman of mine, a man of activity and information, who is already there – will, I am sure, be happy if he is intrusted with the command, and will enter into the spirit of his orders, and agree perfectly with Park.'[41]

In the end, Park, who was about to marry, decided to remain in Scotland and practise as a physician. Banks and Hunter would have to find two new explorers for the vast land that contained New South Wales.

Chapter 11

'We made much of him, and gave him some biscuit; and he in turn
presented us with a piece of gristly fat, probably of whale. This I
tasted; but watching an opportunity to spit it out when he should
not be looking, I perceived him doing precisely the same
thing with our biscuit.'

MATTHEW FLINDERS, ON SHARING FOOD WITH A MAN OF THE YUIN
PEOPLE AT TWOFOLD BAY.[1]

A FTER ALMOST A YEAR of repairs, the *Reliance* was
ready to sail again in May 1798, bound for Norfolk Island
with much-needed supplies and personnel. Hunter was sending
the dashing Irishman D'Arcy Wentworth[2] to relieve Thomas
Jamison,[3] the assistant surgeon there.

Wentworth – tall, handsome, blue-eyed and charming – had
been appointed a superintendent of convicts and then an assistant
surgeon despite a dubious past. In 1789 he was acquitted of four
armed robberies in England, despite having been caught carrying
a pistol, a black silk mask and a wig near the scene of a hold-up in
the wilds of London's Hounslow Heath.[4]

Matthew and Bass set sail on the *Reliance* for its two-month
voyage with a complement of military and convicts, 1200 bushels
of wheat and 100 casks of salt meat.[5] Also on board, as guests
were three adventurous Indigenous men wanting to see the world
beyond theirs – Nanbarry, Wingal and Bungaree,[6] an intelligent
and solidly built young man of about Matthew's height, weight

and age, who had arrived in Sydney with others of the Kuring-gai people escaping the deadly conflicts with settlers along the Hawkesbury.

Bungaree had become a popular figure among the colonists and was seen as a bridge to promote friendship between them and the First Peoples around Sydney. He learnt English quickly and was a brilliant mimic, though apparently – despite a lofty opinion of his own voice – a lousy singer. He was comfortable walking about naked except for a small belt, but at times he also wore the discarded uniforms and cocked bicorne hats donated to him by Governor Hunter and other officials, and would entertain audiences by affecting the walk and mannerisms of prominent Europeans in Sydney. Matthew liked Bungaree immediately and was soon praising his 'good disposition and manly conduct'.[7]

As the *Reliance* moved out of Sydney Cove, it passed not only the still rotting corpse of Francis Morgan, clanging in his chains, but also the small 65-ton trading ship *Nautilus*,[8] which had just arrived from Tahiti after months of battling Pacific typhoons and 'the most terrible Sea the oldest Seaman on board ever knew'.[9] The little vessel had lost her way to the north-west coast of America when trying to sail from China.

It had docked at Formosa, at the Kamchatka Peninsula (on Russia's eastern seaboard) and in the Hawaiian islands before reaching Tahiti in a battered state. There a group of English missionaries saw the ship's arrival as their saving grace. Not only had their message from the Bible been rejected by the Tahitians,[10] but the evangelists claimed they had been forced to barricade themselves within a fortress when the local men threatened to make off with the missionaries' wives. By the time the *Nautilus* reached Sydney, the three-masted Calcutta-built 'snow' was leaky, worn out and 'in very great distress' – though not quite as distressed as its passengers.

The English skipper and part-owner Charles Bishop[11] and his supercargo Roger Simpson sold their goods in Sydney, and with Hunter's permission the ship was repaired by Daniel Paine and reprovisioned from the government stores. Bishop looked to set sail again quickly, but changed his mind after speaking with Bass and Flinders about the enormous seal populations in the islands off Van Diemen's Land, and how the animal skins might pay for his ship's repairs.

Matthew was eager to do more mapping for Hunter, but on his return from Norfolk Island he was pressed into service with the Vice Admiralty Court of New South Wales, as a member of a panel of naval and military officers that also included his friends Waterhouse, Kent and Shortland.

The court had been convened to hear a case of alleged mutiny by members of the New South Wales Corps aboard the convict ship *Barwell* after it had left Cape Town bound for Sydney. The soldiers were accused of conspiring with prisoners in a plot to murder the ship's officers, take over the vessel and sail it to Mauritius. Ensign George Bond was said to be one of the ringleaders, and Matthew heard damning testimony from a convict witness that some of the soldiers had called for a bottle of brandy and then drunk a toast: 'Damnation to the King and Country.'[12]

The prosecution case fell apart, though, after six days, due to a lack of corroborating evidence, and by September 1798 Matthew was readying himself to gather evidence of a different kind about what he and Bass were sure was a strait north of Van Diemen's Land.

On 3 September 1798, Hunter wrote to Evan Nepean, now Secretary to the Board of Admiralty, to relate how Flinders and Bass had both made recent voyages along the south coast of New South Wales. Bass had encountered seas which 'rose to so mountainous a height that he had every reason to believe he was not covered by any land to the westward ... a circumstance which [corroborated] an opinion which I ventur'd ... that there was a probability of an open strait thro[ugh]'.[13] Hunter sent the Admiralty the chart that combined the findings of Bass and Flinders, and said he was endeavouring to 'fit out a deck'd boat of about fifteen tons burthen' for the two young officers to sail south again and finally confirm their shared opinion.[14]

THE GOVERNOR DID NOT have to look far for a vessel. The *Norfolk* was a 25-ton sloop most likely made from Norfolk Island pine. It was the first seagoing ship built on the remote island, in disregard of Hunter's edict curtailing boatbuilding in a place with a large convict population.

The island's new lieutenant governor, Captain John Townson, had initiated construction of the single-masted boat as a way of safeguarding the regular supply of communications and provisions to his small settlement. As the island had no proper harbour, the ship, thirty-five feet long and eleven wide (eleven metres by 3.5 metres), was launched from the shore without proper testing of timbers that later proved far too flexible. After its first voyage across the Tasman Sea, the *Norfolk* had arrived in Sydney in June 1798 leaking like a colander. Only the fact that it carried two pumps that were worked constantly had prevented serious problems.

Confirmation of Matthew's rank of lieutenant had finally been received. He was now authorised to command a vessel, and Hunter gave him:

> ... authority to penetrate behind Furneaux's Islands; and should a strait be found, to pass through it and return by the south end of Van Diemen's Land; making such examinations and surveys on the way as circumstances might permit. Twelve weeks were allowed for the performance of this service, and provisions for that time were put on board.[15]

The rest of the equipment was supplied 'by the friendly care of captain Waterhouse of the *Reliance*'. Matthew wrote: 'I had the happiness to associate my friend Bass in this new expedition, and to form an excellent crew of eight volunteers from the king's ships; but a time keeper [a marine chronometer], that essential instrument to accuracy in nautical surveys, it was still impossible to obtain ...'[16]

On 1 October, as final preparations were being made and equipment was being loaded onto the *Norfolk*, Matthew's gaze shifted from the harbour to a spot up the hill on what is now the intersection of Hunter, Bligh and Castlereagh streets. Between seven and eight o'clock in the evening, Sydney's first and at that time only church had become a towering inferno, lighting up the night sky on the east side of the cove as orange flames snaked heavenward. Convicts and soldiers combined to fight the flames, but the wattle and daub building was covered with thatch, which at this time was 'exceedingly dry and combustible',[17] and the church was consumed in an hour.

Governor Hunter said the fire was no accident, but the work of 'some worthless and infamous person or persons',[18] and a result of the enforced attendance of divine service. Hunter thwarted the arsonist's 'wicked design', though, by directing that a large storehouse, newly finished, could be used as a temporary church, and that the convicts would not miss a single Sabbath service. To David Collins, the fire illustrated the 'dreadful state of profligacy' in Sydney, where 'a more wicked, abandoned, and irreligious set of people had never been brought together before in any part of the colony'.[19]

Hunter posted a £30 reward and 'absolute emancipation' to any convict informer if he revealed the culprit, and a recommendation to the master of any ship to take that person from the colony. But to Hunter's chagrin, 'rewards and punishments alike' failed to find the arsonist.[20]

MATTHEW HAD HIS OWN investigation to pursue, and on Sunday, 7 October 1798, the same day as the first service in the new temporary church, he and Bass and their crew on the *Norfolk* – once again including the reliable John Thistle – sailed out of Sydney Heads. Matthew left Trim the cat in safekeeping with friends, but he did take a greyhound on board to help in the chase for kangaroos for the dinner table.[21] The *Nautilus*, more than twice the size of Matthew's vessel, and carrying a crew of twenty-five, accompanied them on the voyage with a mission to hunt the huge seal populations around Van Diemen's Land.

The south-flowing current pushed the two ships along and they covered about ten kilometres an hour. They passed Hat Hill at 4 p.m. and the next morning sighted Mount Dromedary. By 10 p.m. they had reached Cape Howe, where Matthew decided to rest for the night. The weather turned in the morning and they had to shelter for six days in Twofold Bay, which Bass had recently visited in the whale boat.

Bass took the opportunity to further examine the countryside, while Matthew, aided by Roger Simpson, made a survey of the bay.[22] On the long northern beach, the approach of the two men startled a group of Yuin women and children, who fled screaming. Soon afterward, though, a man made his appearance. 'He was of middle age, unarmed, except for a *whaddie*, or wooden scimitar,' Matthew said, 'and came up to us seemingly with careless confidence.'

Matthew gave him some hard biscuit, and he in turn presented Matthew with a piece of gristly fat, which Matthew guessed was probably from a whale. Matthew tasted it but, hoping he wasn't being watched, spat it out as quickly as he could – only to see his new friend doing the same with the biscuit.

'The commencement of our trigonometrical operations was seen by him with indifference, if not contempt,' Matthew wrote, 'and he quitted us, apparently satisfied that, from people who could thus occupy themselves seriously, there was nothing to be apprehended.'[23]

The next day, Matthew and Simpson surveyed the west side of the bay. The sun was visible, but squalls blotted out the horizon, so to calculate latitude Matthew used his artificial horizon, a shallow pan of mercury that would reflect the sun and allow him to make the necessary measurements of degrees. As he was pouring the mercury from a small flask, seven or eight Yuin men cried out from a bank above him, holding up their open hands, presumably to show they were unarmed.

We were three in number, and, beside a pocket pistol, had two muskets. These they made no objection to our bringing, and we sat down in the midst of the party. It consisted entirely of young men ... and their countenances bespoke both good will and curiosity, though mixed with some degree of apprehension. Their curiosity was mostly directed to our persons and dress, and constantly drew off their attention from our little presents, which seemed to give but momentary pleasure. The approach of the sun to the meridian calling me down to the beach, our visitors returned to the woods, seemingly well satisfied with what they had seen.[24]

After more foul weather, the two ships finally left Twofold Bay on the morning of 14 October 1798, and three days later sighted the islands Matthew had named for William Kent. Here they fell into difficulties. Matthew was taking depth soundings to thirty fathoms – around fifty-five metres – when the crew lost the lead. It was their only deep-sea line, but they realised too late it was totally rotten. This meant they could not record the depth of deeper waters for the rest of the voyage, and would have to sail far more cautiously when approaching land.

On 18 October, Matthew passed the 'smooth round hill' of Mount Chappelle, but more bad weather forced him to spend the night at sea under storm sails. The *Nautilus* anchored in Kent's Bay, off Cape Barren Island, and almost immediately its crew started a seal-killing frenzy that would last five months.

The following day, at noon, Matthew anchored the *Norfolk* at the east end of Preservation Island, and he and the others went ashore to see what remained of the shack Captain Hamilton had built, and of the henhouse where he had kept his fowls and pigeons before leaving the island nine months earlier. 'The house remained in nearly the same state,' Matthew noted, 'but its tenants were not to be found, having probably fallen prey to the hawks.'[25]

Strong winds prevented the *Norfolk* from sailing west, so it remained at anchor for eleven days at the mercy of the driving rain and howling gales. Finally, on 1 November, the little sloop left the *Nautilus* and the bloody work of its sailors behind, and headed west to investigate the possibility of a strait between Van Diemen's Land and New Holland.

Matthew surveyed more of the channels, inlets and islands of the Furneaux Group, adding further detail to the chart he had made aboard the *Francis*, while Bass made detailed notes in his log about the flora, fauna and topography of the land they saw. The *Norfolk* then sailed toward the north-eastern-most point of Van Diemen's Land, which Matthew named Cape Portland.

THE *NORFOLK* WAS NOW in uncharted waters, following a coastline that no European had mapped. Matthew sailed cautiously, but despite his careful attention, when the little breeze abated almost to nothing in the islets around Cape Portland, a strong current pushed the sloop towards a shallow reef in water not even four metres deep. The *Norfolk* became trapped in a swirling eddy – 'a curved line of rippling water', Matthew called it – and the ten men on board could do nothing except pray that the boat's flimsy timbers would survive the rocky ledge.[26]

Catastrophe was avoided, and when back in deep waters again they sailed on past golden sands and forests and a row of 'pleasant looking hills' on the mainland, which Matthew named Point Waterhouse. He also named a seven-kilometre long island for his friend. It was almost covered with seabirds and seals, and Matthew deduced that the Indigenous people of Van Diemen's Land had not been able to get across to it, and that, consequently, 'they had no canoes upon this part of the coast'.[27]

On 3 November, Matthew rounded a point he called Low Head to find a wondrous sight. He entered a wide river mouth that showed immense potential. There appeared to be three arms, or rivers, discharging themselves into this extensive basin, and Bass took the *Norfolk*'s small rowboat to explore as much as he could, though his attention quickly shifted to what he estimated were 300 black swans swimming within a small space. He gave chase. From experience he knew that only two-thirds of the flock could fly, and the rest would do no more than flap along upon the surface of the water, 'being either moulting, or not yet come to their full feather and growth'. It was a long chase, but the boat tired them out,[28] and Matthew and his crew rejoiced when Bass arrived with four birds to pluck.[29]

The next morning, 4 November, Matthew landed Bass and two men to examine the countryside on foot and began a survey of the wide expanse of water, which Governor Hunter would call Port Dalrymple. This was a mark of respect to Alexander Dalrymple, the hydrographer to the Admiralty, and the man who, long before Cook had sailed on the *Endeavour*, had promoted the theory that there was an undiscovered continent in the South Pacific, *Terra Australis Incognita*. William Paterson would name the river as the Tamar in 1804.

At low tide, Matthew walked across the wet sand to what he called Green Island. Seeing the shoals 'so numerous and extensive' around it, he was surprised that his little sloop could have reached so far without striking them. That afternoon he returned to the *Norfolk* to find Bass back from his expedition and carrying a forty-kilogram kangaroo he had shot 'from a considerable flock'.[30]

Surrounded by pleasant forests and distant blue mountains, this river system branched off into many inlets. The *Norfolk* could be manoeuvred by oars when necessary, and Matthew and his team stayed in the waterway for sixteen more days, examining everything they could about this place that was new to Europeans.

There were many traces of Indigenous people on the shore, and on 6 November the crew saw a man setting fire to the grass in different places. He did not stay to receive the newcomers. Matthew concentrated on mapping the course of the wide river, steering the *Norfolk* south as his surveys demanded, and sometimes taking the boat for a closer inspection of the riverbanks, creeks and shoals.

Bass often explored the countryside on foot, using his knowledge of medicine and science to study the wonders of this new world. Over the next two weeks he noted that there were probably as many Indigenous people in this part of the world as there were in New South Wales, judging by the number of huts which he saw, though he regretted what he took to be the 'extreme shyness' of the local people because it prevented any communication.

They had made fires abreast of where the sloop was at anchor; but as soon as the boat approached the shore they ran off into the woods. Their huts, of which seven or eight were frequently found together like a little encampment, were constructed of bark torn in long stripes from some neighbouring tree ... It is then broken into convenient lengths, and placed, slopingwise, against the elbowing part of some dead branch that has fallen off from the distorted limbs of the gum tree; and a little grass is sometimes thrown over the top ... The single utensil that was observed lying near their huts was a kind of basket made of long wiry grass, that grows along the shores of the river ... Their apparent use is, to bring shell fish from the mud banks where they are to be collected ... The most scrupulous examination of their fire places discovered nothing, except a few bones of the opossum, a squirrel, and here and there those of a small kangaroo. No remains of fish were ever seen.[31]

Bass also wrote about two species of eucalypts, previously unknown to Europeans, and he made detailed observations about black swans on the rivers. He cut open some to learn more about their diet. He had heard that they fed upon fish, frogs and water-slugs, but in their gizzards Bass only ever found the remains of small water plants and a little sand.[32]

He was likewise fascinated by the wombat, writing that its 'pace is hobbling or shuffling, something like the awkward gait of a bear. In disposition it is mild and gentle, as becomes a grass-eater; but it bites hard, and is furious when provoked.'[33] He dissected one and wrote the first anatomical description of the animal.

Bass wrote that large grey kangaroos abounded in the open forest, but there was a smaller, darker kind that he thought the Eora people of Port Jackson, called the 'Wal-li-bah' [or wallaby].[34]

The explorers finally left the Tamar on 20 November, but the fresh breezes of the morning turned into violent north-westerly winds and driving rain by the afternoon. The sloop was driven so far to the leeward during the night that Matthew admitted he was 'not without apprehensions' about being dashed against the shore.[35] The *Norfolk* reared and thrashed about in the wild sea, and late on the next day the storm-jib split. Although the *Norfolk* was drenched by towering waves, Matthew declared that 'upon the whole' the little sloop 'performed wonderfully ... Seas that were apparently determined to swallow her up she rode over with all the ease and majesty of an old experienced petterel.'[36]

The winds prevented Matthew travelling any further west, so he changed course and back-tracked to the north-east for a hundred kilometres or so, sailing for a safe anchorage at what he called Hamilton's Road, on the east end of Preservation Island. The next day, 23 November 1798, the crew spent the daylight hours drying and repairing the sails. At night Matthew watched an eclipse through the telescope of his sextant and calculated the altitudes of the stars Rigel and Sirius to log his position.

The next day he sailed the *Norfolk* into Kent's Bay. The butchering work of the seals was well underway, and the crew of the *Nautilus* were busy gathering 9000 skins and several tons of seal oil, which Bishop planned to sell in China.[37] Matthew gave Bishop his report on the voyage so far, to be passed on to Governor Hunter.

THE CLOCK WAS TICKING on the voyage: eight of the twelve weeks were nearly up, bad weather had already cost Matthew dearly, and the question of the mysterious strait remained.

Despite the dangers of sailing at night, Matthew took advantage of favourable winds on the evening of 24 November and set sail towards Port Dalrymple again. Before long the *Norfolk* faced another gale, 'more violent and of longer continuance than any of the preceding', but they reached the mouth of the Tamar and rode out the storm in the port. It took almost a week for sunny skies to emerge, and this long succession of adverse conditions caused Matthew to despair at ever accomplishing the principal object of his voyage.[38]

Finally, though, on 3 December 1798, they were back on course, gliding west under a light breeze. Two days and a hundred kilometres later, they reached the striking rock formations and stark cliffs of what Matthew called Rocky Cape.

On the following morning of 6 December, at about ten o'clock, they passed a hill Matthew called Circular Head, 'a cliffy, round lump', he wrote, 'in form much resembling a Christmas cake … joined to the main by a low, sandy isthmus'.[39] Soon after, to the north-west, 'three hummocks of land' came into sight. Matthew thought the southernmost and highest of the hummocks looked like a sugarloaf.

By 8 December, as they continued to struggle west against unfavourable winds, another great swell began to rise, and there was every appearance of an approaching gale. Matthew considered sailing to the north for a safe harbour in Western Port, where Bass had been the year before. It seemed evident that the *Norfolk* could make no further progress westward, and might even be driven back, yet Matthew knew that, with the area around the western side of Western Port also being uncharted, it would be dangerous trying to reach there as well.[40]

At 6 p.m. he anchored the sloop in a small sandy bight under the northern hummock. 'Mr. Bass and myself landed immediately to examine the country and the coast,' Matthew wrote, 'and to see what food could be procured; for the long detention by foul winds had obliged me to make a reduction in the provisions.'[41]

The countryside was hilly, and Bass found it impenetrable because of the thick brushwood, although it had been partially burnt not long before. He also found several deserted fireplaces, strewn with abalone shells. At dusk, he returned to the ship with nothing to eat but with much food for thought. He and Matthew observed that the tide had been running from the east all afternoon, and 'contrary to expectation' they 'found it to be near low water by the shore'. The floodtide, therefore, must have come from the west.

'This we considered to be a strong proof,' Matthew wrote, 'not only of the real existence of a passage betwixt this land and New South Wales, but also that the entrance into the Southern Indian Ocean could not be far distant.'[42] It wasn't.

They pressed on – only to be astounded by another revelation. At daylight the next morning, 9 December, the men on the *Norfolk* were staggered to first see a great flock of gannets coming out of a bight to the south. It was a remarkable spectacle, but only the opening act for one of the greatest shows Matthew ever saw. The gannets were followed by sooty petrels – muttonbirds – in a show that left the ten men on the little ship slack-jawed. A stream of birds 'from fifty to eighty yards in depth, and of three hundred yards, or more, in breadth'[43] flew past for a full hour and a half.

'On the lowest computation, I think the number could not have been less than a hundred millions,' Matthew wrote, 'and we were thence led to believe, that there must be, in the large bight, one or more uninhabited islands of considerable size.'[44]

Later Matthew did his sums more carefully, taking the width and depth of the stream of birds moving at the rate of thirty miles an hour, and allowing nine cubic yards of space for each bird. He thus calculated that, during the course of the morning, he and the *Norfolk*'s crew had seen 151.5 million petrels fly by. 'The burrows required to lodge this quantity of birds would be 75,750,000,' he explained, 'and allowing a square yard to each burrow, they would cover something more than 181 geographic square miles of ground.'[45]

A few hours after the men witnessed this stunning event, the *Norfolk* presented Matthew with another spectacle he would never forget.

The sloop rounded the northern end of an island Matthew named for Governor Hunter, and suddenly there was no more land to be seen to the west – just a long swell crashing heavily with foaming surf upon a small reef. Matthew, Bass and the crew cheered with joy and mutual congratulations: this grand sight announced their longed-for discovery.

They had found the strait into the Southern Ocean. They now knew that Van Diemen's Land was an island.

ALL THAT EXCITEMENT made the crew hungry, and Bass took the boat to a small, rocky island that was almost white from bird droppings. He returned from the newly named Albatross Island at 2.30 p.m., with his boat full of dead seals and birds.[46]

A few miles further, as the *Norfolk* at last turned south, they saw the dark, foreboding headland that Matthew called Cape Grim. For the next four days, under grey skies, the sloop raced south in brisk winds, and Matthew looked upon the stark mountains along the coast with 'astonishment and horror'. They were, he said, among 'the most stupendous works of Nature I have ever beheld, and, at the same time, are the most dismal and barren that can be imagined'. He could find no place of shelter and the shore was dangerous to approach.[47]

On 11 December, he named a 250-metre-high pyramid-shaped mountain 'after my little vessel', and he then named Mount Heemskirk and Mount Zeehan, for the two ships of Abel Tasman, who he believed saw this same coast in 1642.[48]

The *Norfolk* crew saw only two 'smokes' on shore, suggesting that the land there was thinly populated, and he chose not to explore the large body of water that became known as Macquarie Harbour because of the many rocks around its entrance. The coastline would soon be notorious as one of the most treacherous in the world, with the Roaring Forties winds gathering speed across thousands of kilometres of empty ocean to assail the vessels that would follow the *Norfolk* in future decades.

They continued on, fighting the savage winds and high seas, and on the morning of 13 December they steered around the south-west cape of Van Diemen's Land, which Furneaux had seen from the *Adventure* in 1773, and which had been noted by Cook and Bligh. Matthew now knew exactly where he was: on the homeward leg of his great adventure.

Whatever comfort that gave him and his crew after weeks spent battling the elements in uncharted territory, it evaporated quickly after they passed De Witt Island, where the grass had recently been burnt. Twilight approached with heavy clouds of black and red, and the little sloop was not much more than two kilometres from the rocky shore, under the South Cape.

The crew had taken in all the sails except the foresail when the storm burst upon them with thunder, lightning and hammering rain. The direction of the wind was west by south; had it been from the south, or had the squall come on an hour sooner, Matthew said, it was probable the *Norfolk* would have been destroyed on the rocks. The crew would have been left to bleach under the lonely high cliffs, 'and the separation of Van Diemen's Land from New Holland would still have been only supported by conjecture'.[49]

The danger passed within an hour, but still the weather played havoc with their homeward travels. The *Norfolk* passed the Friars – bald rocks named by Furneaux because they looked like the bare heads of monks – but Matthew's hopes of further exploring Adventure Bay, where he had gone ashore from the *Providence* seven years earlier, had to be abandoned. So too did his immediate plans to explore the waterway named by the Frenchman Antoine Bruni d'Entrecasteaux[50] as the Rivière du Nord in 1793, but which was renamed as the Derwent soon after by John Hayes[51] of the British East India Company.

With Bass often steering, Matthew continued to map the coastline, exploring Norfolk Bay and making it back to Adventure Bay before, on 21 December 1798, finally sailing into the Derwent. He and Bass went ashore at many places, including Sullivan's Cove, the site that would become the settlement of Hobart, and on Christmas Day Bass battled his way through the steep ridges and deep forests to reach the summit of the 1270-metre 'mighty Mount Table'[52] – kunyani, later renamed Mount Wellington – which towered over the river and the sea beyond. Matthew manoeuvred the sloop into Risdon Cove and then onto Herdsman's Cove, about twenty-five kilometres from the river mouth, before it became too shallow to travel further.

Three days after Christmas, Matthew and Bass made an excursion further up the river in the *Norfolk*'s boat. Matthew was taking notes about the width and depth of the water and the nature of the surrounding foliage when their attention was captured by a voice coming from the hills. Bass and Flinders were about to have their first encounter with the Palawa, the Indigenous people of Tasmania.

Matthew described the moment:

There were three people; and as they would not comply with our signs to come down, we landed and went up to them, taking with us a black swan. Two women ran off, but a man, who had two or three spears in his hand, staid to receive us, and accepted the swan with rapture. He seemed entirely ignorant of muskets, nor did anything excite his attention or desire except the swan and the red kerchiefs about our necks; he knew, however, that we came from the sloop, and where it was lying. A little knowledge of the Port Jackson, and of the South Sea Island languages was of no use in making ourselves understood by this man; but the quickness with which he comprehended our signs spoke in favour of his intelligence. His appearance much resembled that of the inhabitants of New South Wales; he had also marks raised upon the skin, and his face was blackened and hair ruddled as is sometimes practised by them. The hair was either close cropped, or naturally short; but it had not the appearance of being woolly. He acceded to our proposition of going to his hut; but finding from his devious route and frequent stoppages that he sought to tire our patience, we left him delighted with the certain possession of his swan, and returned to the boat. This was the sole opportunity we had of communicating with any of the natives of Van Diemen's Land.[53]

THE TWO YOUNG OFFICERS returned to the sloop to begin the run home to Sydney. They carried with them fourteen swans they had shot, which would feed the crew for a few days. They filled their water casks at Risdon Cove and made all the necessary repairs to the *Norfolk*. By the evening of 30 December all was in readiness for their departure, but dangerous weather meant it wasn't until 4 January 1799 that they left Storm Bay, at the mouth of the Derwent, and took a left turn for Sydney.

Matthew had seen a body of water between two capes that he thought might make a good harbour, and it did in later years as the convict settlement of Port Arthur.[54]

There were more gales and shifting winds, but by 9 January they passed the north-east corner of Van Diemen's Land and the following day were off Cape Howe. The foul weather followed them up the east coast of New South Wales, but Matthew noted joyfully that on 11 January 1799, with the gale still continuing:

> ... we anchored within the heads of Port Jackson at ten o'clock the same evening, having exceeded, by no more than eleven days, the time which had been fixed for our return. To the strait which had been the great object of research, and whose discovery was now completed, governor Hunter gave, at my recommendation, the name of BASS STRAIT. This was no more than a just tribute to my worthy friend and companion, for the extreme dangers and fatigues he had undergone in first entering it in the whale boat, and to the correct judgment he had formed from various indications, of the existence of a wide opening between Van Diemen's Land and New South Wales.[55]

Hunter alerted Evan Nepean that 'Lt. Flinders and Mr. George Bass, late surgeon of the Reliance [had] completely circumnavigated Van Diemen's Land, formerly consider'd a part of this country.'[56] Later, the largest island in the Furneaux Group would be named Flinders Island.

The mapping of Bass Strait shortened the route from the Cape of Good Hope to Sydney by a week and 1000 kilometres. And while the weather could be wild in the newly named Bass Strait, the wind and waves were not usually as damaging as those in the freezing Southern Ocean, meaning less harm to vessels and their cargo.

In Donington, Matthew's father was overjoyed to receive letters from his two sons, and he had much to tell them in his dispatches back to Sydney. Their sister Betsey Harvey was now the mother of two girls, though the fatigues of childbirth had taken their toll and her 'connexion' with her husband was not as happy as she had hoped. He was an older man, domineering, 'unkind and grasping'.[57] Their younger sister Susanna had found work in a husband-and-wife drapery and millinery business at Dartford, in Kent. Brother John's temper tantrums seemed to be subsiding and his father was trying to find him suitable work.[58]

For months, Flinders Sr and much of Britain had lived with their hearts in their mouths at the prospect of the French military leader Napoleon Bonaparte leading an invasion across the few narrow miles of the English Channel. But Matthew's father and his many friends had gathered at Donington's Red Cow Inn on the evening of 3 October 1798 to celebrate what he called 'the Great Naval Victory obtained by the blessing of God, by Admiral nelson over the French Fleett at the bottom of the Middeteranean Sea, even at the Mouth of the Nile'.[59]

In Sydney, Governor Hunter was facing the continuation of his own bitter warfare against the machinations of John Macarthur and his minions, who were constantly trying to undermine the Governor's authority, with a letter-writing campaign to Portland in London.

Hunter's days were numbered, but his fascination for the country around him would never wane. While the adventurous partnership of Bass and Flinders had come to an end, Hunter had another vital mission for his reliable young lieutenant of the *Reliance*.

Chapter 12

*'The vast interior of this new country was wrapped in total obscurity;
and excited, perhaps on that very account, full as much curiosity as
did the forms of its shores.'*

MATTHEW FLINDERS, ON THE QUEST FOR DISCOVERY TO FIND INLAND
RIVERS IN NEW HOLLAND.[1]

MATTHEW HAD LED A SMALL expedition that had
recorded one of the most important discoveries since the
mapping work of Captain Cook on the *Endeavour* almost thirty
years earlier. Proof that Van Diemen's Land was an island was a
revolutionary development in this new world.

Yet, as Joseph Banks pointed out to his government, almost
all of the new territory claimed by the British in what they called
New South Wales – more than 3000 kilometres of coastline
and 2000 kilometres into the interior – remained a mystery to
Europeans a decade after settlement. 'So much has the discovery
of the interior been neglected,' Banks opined, 'that no one article
has hitherto been discover'd by the importation of which the
mother country can receive any degree of return for the cost of
founding and hitherto maintaining the colony.'[2]

It is impossible to conceive that such a body of land, as large
as all Europe, does not produce vast rivers, capable of being
navigated, that such a country, situate in a most fruitful

climate, should not produce some native raw material of importance to a manufacturing country as England is.[3]

The Blue Mountains remained an impassable barrier for Europeans, but Governor Hunter and Matthew believed that the size of the river mouths they had seen on their voyages indicated a huge inland network of waterways just waiting to be revealed.

Hunter, though, had more on his mind than geography. He had the weight of the world on his shoulders, with political pressure from opponents in both Sydney and London. To him, New South Wales was full of loafers, corrupt officials and backstabbers. He railed against military officers such as Macarthur, who had taken over the colony's commerce with no thought for its general benefit, or that of their neighbours.

Hunter had no doubt that Macarthur was trying to break him – in the hope that Hunter would desert his post by 'heaping one vexation upon another untill the weight shou'd be felt too heavy for me singly to bear'.[4]

For now the governor was staying put, though, and so too was Matthew Flinders. Before making his next voyage, Matthew was again pressed into serving on a judicial panel, this time in a criminal trial that started on 12 March 1799. The young emancipated convict Isaac Nichols[5] was charged with having received stolen goods in the form of 'One Basket of Brazil Tobacco'.[6]

Nichols had sided with Hunter against Macarthur and his men, and Hunter believed that the charges brought against Nichols had been fabricated as revenge. The court consisted of three officers of the New South Wales Corps, who supported Macarthur and his faction, and the three navy men: Matthew, Waterhouse and Kent. The presiding officer was the new deputy judge advocate, Richard Dore, whom Hunter regarded as a dishonest, scheming drunk.

The trial lasted four days. Matthew, Kent and Waterhouse said that the evidence was either false or inconclusive, and voted to acquit, while the military men voted to imprison. Dore's deciding

The belligerent John Macarthur caused trouble for the early governors of NSW.
State Library of NSW 446894

vote convicted Nichols, who was sentenced to fourteen years' confinement on Norfolk Island.

Governor Hunter was livid, decrying the case as a mix of perjury, prejudice and political sabotage. He suspended Nichols' sentence before sending all the documents concerning the case to London. Hunter allowed Nichols to return to work, and three years later he received a pardon from the king. Nichols eventually became the first postmaster of New South Wales.

Dore's guilty verdict was totally at odds with Matthew's detailed assessment of the case. Matthew compiled a thorough report of the four days' of proceedings and gave his opinion of the evidence to Hunter, analysing the witness statements of the early colonists Macarthur, William Balmain and Henry Kable. The evidence of the character of the prisoner 'spoke highly of him', Matthew told Hunter, and considering how trifling the profit in retailing this stolen tobacco would have been, the likelihood of Nichols committing the crime seemed 'inconsistent'.[7]

MATTHEW TURNED TWENTY-FIVE the day after the Nichols trial ended, and he found himself beset by a restlessness for personal advancement. While his discovery of Bass Strait had gained local acclaim, he was agitated by the fact he remained a junior officer with little to show for his talent and courage other than praise from the governor of a remote prison outpost, who himself was barely clinging to his job.

Perhaps needing a shoulder on which to cry, Matthew wrote home to Ann Chappelle, who was now almost twenty-nine and at an age in eighteenth-century England when a life of spinsterhood seemed probable. Matthew had recently read Ann Radcliffe's Gothic novel *The Mysteries of Udolpho*, about a trapped heroine in a lonely Italian mountain castle. The heroine's tearful farewell from the dashing hero reminded Matthew of his separation from Ann, and he told her of his 'unveiled heart', reminding the young woman who had lost her father to the sea of their own parting in Lincolnshire four years earlier.

'Fatal, enervating moment!' he wrote. 'I have never since been satisfied with my profession … to be cooped up in a wooden box; year after year; one decade after another; and the ultimate object not a bit more forward! … forgotten.'[8]

He explained to Ann why he had chosen her, 'of all people in the world', for his 'recital'. It was no use pouring his heart out to another man, such as his father or his friend Thomas Franklin. He needed a woman's ear for his unfettered emotions. His sister Betsey was now married with children, and her family had 'superior' claims to her affection, and Matthew had fallen out with Mary Franklin, who he now felt would be 'too much embittered to hear my tale with patience'.

So Ann Chappelle copped all his frustrations, and his purple prose. He was a servant of the sea, he wrote, but complained to it in a rhetorical flourish:

[T]hy wages must afford me more than a bare subsistence …
Thou art but a rough master, hast little mercy upon the lives
and limbs of thy followers … Half my life I would dedicate
to thee, but the whole I cannot if thou keepest me in penury
all the morning and noon of life.[9]

The young lieutenant wanted money and status for his efforts
as he imagined a career beyond sleeping in a small ship's cabin
in places literally off the map, and with his life often at risk. He
was not prepared to devote himself to a life at sea for penury, he
told Ann – but for the right money? Well, maybe. Hearing such
sentiments could only have reawakened the pain in Ann's heart of
losing her father to the ocean all those years ago.

Matthew considered a business proposal from Bass, who had
decided to leave the navy to enter a trading partnership with
Charles Bishop on the *Nautilus*. Matthew wrote to his botanist
friend 'Paddy' Smith, now at the Botanic Gardens in Calcutta,
for advice. The reply took months to arrive, but in it Smith told
Matthew he should '[c]ontinue with the Navy', and that if he
did, he would become 'a great man yet'.[10] Bishop had accepted a
lucrative assignment to deliver a cargo of English goods to Norfolk
Island, and he then planned to take the 9000 seal skins he'd taken
from the Furneaux Islands to China, where he also planned to
sell the *Nautilus*. Bass would sail with him and then travel on to
England, where Henry Waterhouse's sister, Elizabeth, was waiting
with open arms.

On Sunday, 17 May 1799, Waterhouse gathered his crew on
the deck of the *Reliance* to farewell the strapping young surgeon,
who went ashore together with his servant William Martin, now
eighteen. Two days later the pair sailed from Port Jackson on the
Nautilus. Matthew had given his big mate a letter of introduction
to Paddy Smith to help him with his trading ventures when he
reached Calcutta. Although the partnership of Bass and Flinders
had sometimes been shaken by squabbling, the pair had formed
a strong bond during their death-defying adventures together,

when they'd made astonishing discoveries of a kind only few intrepid mariners could know.

Matthew later wrote to Bass to tell him that, despite their differences, 'Franklin – Wiles – Smith – Bass, are names which will be ever dear to my heart'.[11] Matthew reminded Bass of the deep connection that they shared, and chided him in a light-hearted way, though some modern-day scholars see the letter as revealing another side to their relationship.

> There was a time, when I was so completely wrapped up in you, that no conversation but yours could give me any degree of pleasure; your footsteps upon the quarter deck over my head, took me from my book, and brought me upon deck to walk with you; often, I fear, to your great annoyance; but your apparent coolness towards me, and the unpleasant manner you took to point out my failings, roused my pride and cooled my ardor. I will do myself the justice to say, that it was not the being told of a fault that hurt me, for I am well satisfied that to be told of almost any failing, by you, in such a way as should bespeak a friendly intention of correcting it, would be received by me with thanks.[12]

He told Bass – perhaps jokingly, perhaps not – that his affection for James Wiles reached further into his heart than his love for Bass, because 'I would take [Wiles] into the same skin with me!'

'There is one circumstance that will always keep you from me,' Matthew told Bass, 'your thirst after knowledge and information will not permit you to have the necessary consideration for one, who not only cannot afford you these; but has a far less stock than yourself.'[13] But Matthew left Bass in no doubt about his admiration when he ended his letter to the trusted companion he knew he might never see again: 'I am, my dear friend, with the highest considerations of respect and esteem, yours most sincerely, Mattw. Flinders'.[14]

The deeply personal nature of the letter has since led to speculation that Bass and Flinders were more than friends,[15] although Matthew had a habit of writing in a florid style to both male and female friends, amusing himself with hyperbole. Waterhouse's sister, Elizabeth, failed to see the joke. When the letter finally arrived in England months later, she wrote across it: 'This George is written by a Man that bears a bad Character no one has seen this letter but I could tell you many things that makes me dislike him rest ashured he is no friend of yours or any ones farther than his own interest is concerned.'[16] Perhaps she was jealous of the bond between Bass and his close friend, or perhaps she was suspicious about their relationship.

MATTHEW HAD SAID HIS goodbyes to Bass during one of the most tumultuous times in the eleven years of European settlement in New South Wales. In 1799, a long dry spell was broken by a heavy downpour that took the new settlers along the Hawkesbury River by surprise with widespread flooding, though some of the Indigenous Dharug people had warned farmers friendly to them of the looming disaster.[17] The river swelled more than fifty feet (sixteen metres) above its banks, and the powerful torrent carried all before it. The government store was washed away, and most of the livestock, pigs and poultry were drowned. The previous season's harvest was ruined.

Hunter's bosses in London were now receiving anonymous letters that had all the marks of John Macarthur's hand, and that claimed the governor was guilty of the very corruption over the liquor trade that he was trying to suppress.[18] Hunter called the attacks on his character the work of a 'dark and infamous assassin; guilty of a diabolical departure from truth', and so wicked he was capable even of 'vilifying the immaculate character of his God'.[19] 'My Lord,' Hunter wrote to Portland, 'by his attack upon my character he shews his cloven foot.'[20]

Hunter remained committed to the colony, and the discovery of Bass Strait strengthened Matthew's resolve to make further discoveries as a way of advancing knowledge about a prized British asset. He also saw this as a way for him to gain rapid promotion in the Navy. With the young lieutenant's service on the *Reliance* not immediately needed, Hunter agreed to Matthew's request to head another voyage of exploration. This time he proposed sailing north from Sydney, to explore what he called 'Glasshouse and Hervey's Bays – two large openings to the northward, of which the entrances only were known ... I had some hope of finding a considerable river discharging itself at one of these openings, and of being able by its means to penetrate further into the interior of the country than had hitherto been effected.'[21]

Hunter gave Matthew use of the *Norfolk* again, with nearly the same crew as had sailed on her through Bass Strait, including John Thistle. Matthew wrote with regret that 'of the assistance of my able friend Bass I was, however, deprived'. This time, he would take his brother Samuel, and Matthew's Indigenous friend Bungaree, whose company on the *Reliance* to Norfolk Island 'had attracted [his] esteem'. Bungaree would act as guide, translator and, if necessary, peacemaker. This time, Trim the cat would also be part of the adventure, and so would the *Reliance*'s 'timekeeper', the chronometer so important for accurate geographical calculations. As the *Reliance* might be needed back in service quickly, Matthew and his crew – and the chronometer – were required to return within six weeks. To guard against accidents, they carried provisions for eleven.[22]

MATTHEW SET SAIL ON THE cold winter's day of 8 July, and the next morning the *Norfolk* passed the entrance to Port Stephens, which Cook had noted on his chart. Neither skipper, though, explored the body of water, which was more than double the size of Sydney's harbour. Matthew was racing the clock to chart as much of the coast as he could.

Fifty kilometres on, he named Sugarloaf Point, near what is now called Seal Rocks, but an 'extraordinary current' had pushed the *Norfolk* thirty kilometres offshore and its timbers had started to crack.[23] One of the little sloop's two pumps was constantly at work. Matthew needed to find a place to make repairs, but three days after setting sail he was still looking for a suitable landing spot. In the meantime, he added five more of the Solitary Isles to Cook's 1770 chart of the coastline, off the coast of a heavily forested area around what is now Coffs Harbor. About sixty kilometres north of the rocky outcrop Matthew called North Solitary Island, he found a shady cove that he named Shoal Bay, 'an appellation it but too well merited',[24] because of the rocks and reefs that formed a jagged carpet just underneath the *Norfolk's* keel.

The area was unsuitable to repair the sloop, but Matthew went ashore to explore for a while, as his mission had been to find inland waterways. He had actually anchored in the estuary of what is now called the Clarence River. He climbed a vantage point to inspect the countryside, but vegetation hid what is often called 'The Big River', and he missed out on exploring a waterway that extended 400 kilometres into the interior. It would not be the last time a golden opportunity for discovery went begging on the trip north.

Having returned to the sloop about noon, Matthew sailed it to the south head of this 'bay', where the town of Yamba is now located. He wanted to make observations of the sun there for his calculations of latitude. There were white cockatoos and parakeets about, pelicans as well, with some gulls and redbills, and a crow 'whose note was remarkably short and hasty'. Then he spotted three Indigenous huts nearby, deserted.

They were of a circular form, and about eight feet in diameter. The frame was composed of the stronger tendrils of the vine, crossing each other in all directions, and bound together by strong wiry grass at the principal intersections. The covering was of bark of a soft texture, resembling the bark of what is called the Tea-tree at Port Jackson, and so compactly laid on as to keep out the wind and rain. The entrance was by a small avenue projecting from the periphery of the circle, not leading directly into the hut, but turning sufficiently to prevent the rain from beating in. The height of the under part of the roof is about four and a half, or five feet, and those that were entered had collected a coat of soot, from the fires which had been made in the middle of the huts.[25]

One of them was a double hut, comprising two recesses under one entrance, 'intended most probably for kindred families, being large enough to contain ten or fifteen people'. Bungaree told Matthew that these huts were 'much superior' to any he had seen before. The local people also used small hand baskets, 'made of some kind of leaf, capable of containing five or six pints of water, and very nearly resembling those used at [Kupang] in the island of Timor'.[26]

Around the deserted fireplaces beside the huts the visitors saw the fruits of local 'palm-nut' trees, the lower end of the nuts having been 'chewed and sucked in the manner that artichokes are eaten'.[27] Matthew thought this fruit might have been related to the oil-rich pandanus plants of the Nicobar Islands in the East Indies.

'The taste was rather pleasant at first,' Matthew noted, 'but left an astringency behind that scarcely tempted one to try a second time.' He said Bungaree, 'who was tolerably well acquainted with the country as far as Port Stephens', had never seen or heard of the plant before, and this area around 'Shoal Bay' was probably one of the most southern situations in which it would be found.[28]

Matthew still could not find a suitable place to beach the *Norfolk* safely in order to make the repairs it needed. He continued sailing north, the pumps working relentlessly.

By this time, Bungaree had formed 'an intimate acquaintance' with Trim, Matthew having declared that his little cat was to be put in charge of the bread bags and that he was to have special kitchen privileges. Matthew noted:

If [Trim] had occasion to drink, he mewed to [Bungaree] and leaped up to the water cask; if to eat, he called him down below and went strait to his [kit], where there was generally a remnant of black swan. In short, [Bungaree] was his great resource, and his kindness was repaid with caresses.[29]

On the morning of 13 July 1799, they passed Cape Byron, the most easterly part of the coast, named by Cook in honour of 'Foul Weather Jack' Byron, who finished his circumnavigation of the globe in 1766. The peak of Mount Warning loomed above as they avoided the reefs of Point Danger.

At ten the next morning, they 'steered West for a large space where no land was visible; and seeing breakers off the south point of the opening' between two sandy islands, were satisfied that this was Moreton Bay. Cook had named the bay in 1770 for the then president of the Royal Society, James Douglas, Earl of Morton, and a later misspelling[30] remained.

Matthew found the area around Point Lookout too treacherous to navigate, so he sailed further north and rounded what Cook had called Cape Morton. At dusk he saw the Glass House Mountains in the distance, volcanic plugs that had reminded Cook of the glass-making furnaces in his native Yorkshire.

Matthew anchored for the night, and the next morning busied himself negotiating the ever-present shoals at the entrance to the bay. When they were within a mile or so of the shore of what is now called Moreton Island, he saw five Indigenous men at the water's edge, calling to him and making friendly gestures. One was waving a green branch from side to side, while others ran into the water, hitting the gentle waves with sticks. Behind them there were five more people watching – from a distance Matthew thought they were women. The men were calling out using a word similar to one used as a greeting by the Eora around Port Jackson.[31]

Matthew could not anchor until much later, and then spent that evening taking navigational measurements. But on the morning of 16 July, eight days after leaving Sydney, he anchored off a point he would sadly name Point Skirmish, 'because of an unfortunate occurrence'.[32] In a bloody confrontation, Matthew was about to show the unsuspecting Indigenous people of Moreton Bay the horrific firepower of British guns.

Chapter 13

'[T]hey seemed to recognise Mr Flinders as the person who had fired
upon them before, and were more desirous that he should keep at a
distance than any other person. Three of the sailors, who were Scotch,
were desired to dance a reel, but, for want of music, they made a very
bad performance, which was contemplated by the natives
without much amusement or curiosity.'

AN ACCOUNT OF MATTHEW FLINDERS TRYING TO MAKE PEACE WITH
THE KABI KABI PEOPLE OF BRIBIE ISLAND.[1]

SURROUNDED BY COOL, blue waters under the winter
sun, and with low eucalypt woodlands all around, Matthew
wanted to lay the *Norfolk* on the shore to repair the leak, but shoals
still blocked his path. He decided instead to eat breakfast and
then take the boat with some crew to row so he could examine
an opening he'd seen about eight kilometres from the sloop.
It looked like it led to the peaks of the Glass House Mountains
in the distance.

After breakfast, he and Bungaree, together with their oarsmen,
climbed into the boat and slowly approached the sandy point on
the east side of the opening. Some dogs came down upon the
beach, and soon after several Indigenous men appeared, most of
them carrying fishing nets over their shoulders. Matthew ordered
his men to stay in the boat as Bungaree conversed with the Kabi
Kabi people by signs, and repeating the welcoming words from
the previous day. Seeing nothing in their hands but firewood,

Bungaree went ashore, almost as naked and as unarmed as the local people. The muskets were kept in the boat, to be prepared against what Matthew called 'any treachery'.

Bungaree quickly made an exchange for his only piece of apparel. He swapped a belt made from bark fibres for one of the Kabi Kabi headbands made of kangaroo hair. Matthew jumped out of the boat and joined Bungaree on the little beach, but took his gun with him. Matthew made some friendly signs, put down the gun and proffered a woollen cap as a gift. One of the men approached him and took the cap – but when Matthew made signs that he wanted the net bag in return, the man made signs that he also wanted the white, wide-brimmed cabbage-tree hat on Matthew's head.[2] Matthew refused. Both sides had difficulty trying to understand the other and their signs.

Bungaree was not from this area and didn't know the cultural protocols of these people.

The man came forward, threw the woollen cap behind him for safekeeping and gestured that he wanted the hat or the gun, or both. The negotiations carried on 'very amicably', but when Matthew and Bungaree began retreating slowly towards the boat, things turned ugly.

We only have Matthew's version of events. He said one of the locals, while talking with Matthew, tried to take the hat off his head with a long hooked stick as the others laughed. Another made a snatch at the hat with an outstretched arm, but remained wary of coming too close.

Matthew and Bungaree jumped into the boat as the Kabi Kabi men made gestures for them to come back. Then, angrily, one of them threw a piece of firewood at the boat. It missed. Another ran into the water and threw his firewood, but it too fell short. Then one of the Kabi Kabi ran into the water and hurled a spear, using the hooded stick as a launching device – a woomera, or throwing stick. The spear passed over the centre of the boat, about a foot and a half above the gunwale, but touched no one.

Guns were at the core of British diplomacy, and Matthew decided he had to answer what he called an 'impudent and unprovoked attack'. He said throwing the spear 'unfortunately obliged' his men to fire upon these people, 'in order to maintain that superiority which [the British] meant upon all occasions to assert'.[3]

Matthew took aim with his musket and pulled the trigger, but the flint had become wet on the beach and it misfired. His gun was loaded with buckshot, and while he momentarily thought about firing a blast into the midst of the entire group, he instead aimed once more at the man who had flung the spear and who, now standing in the water, had turned to call to his companions to follow him. Again the gun misfired.

He tried for a third time – and *boom*.

The man who'd called to his companions fell into the water at the shock of the noise, as did the others on the shore. Those on the beach sprang up almost instantaneously and retreated quickly towards the bank, some upright, some on their hands and knees. One of Matthew's crewmen fired his musket too, and all the local men fell again to the ground. Then they got up and ran behind a sandbank and into the woods. Even the man in the water rose up and made off, but his progress was slower than that of the others, and he stooped a great deal, carrying one hand behind him.

'From hence it was conjectured that he was wounded,' Matthew wrote, 'and he looked every now and then over his shoulder, as if expecting to see the spear that he supposed must be sticking in his back.' Bungaree heard later that another of the men had suffered a broken arm from the second shot fired. Matthew noted, unhappily, that these people were the first victims of gunfire in this part of the country.[4]

Even so, he wasn't particularly contrite. He was a British naval officer after all, who believed the British and their guns must be obeyed. He thought the shooting 'might be the means of preventing much future mischief, to give them a more extensive idea' of his firepower, and 'thereby deter them from any future attempt' in his dealings with them.[5] As this bay needed to be examined, and the leak in the sloop had to be fixed here, it was probable that the two cultures would encounter each other again.

Matthew fired twice more, but said it was only to scare one of the Kabi Kabi men who was watching the intruders from among the trees, and who, being several hundred metres away, thought himself secure. He ran away when the musket balls smashed into the tree trunks near him.

Furious over the spear throwing, and with all the conceit of an invading admiral, Matthew went back onto the beach, intending to punish the locals by taking the fishing nets they had left behind. He couldn't find any, but with an unknown number of adversaries still among the trees, he ordered Bungaree and a crewman back to the boat.

From this newly named Point Skirmish, on what is now Bribie Island, the crew proceeded up through the opening of what Matthew thought was the river leading towards 'the Glass-House peaks', which, 'as far as could be judged, had every appearance of being volcanic'. The large quantity of pumice stone lying at the high-water mark upon the eastern shore of this 'river'[6] seemed to

The sight of the Glasshouse Mountains that would have greeted Matthew from across what he called the Pumice Stone River at what is now called Banksia Beach, Bribie Island. *Grantlee Kieza*

confirm his theory, and he called the waterway the Pumice Stone River.[7]

There were a few huts standing near each other, each four to five metres in length. They resembled a covered archway, rounded at the far end. Their roofs and construction were nearly the same as those Matthew had seen at Shoal Bay, but these did not have curved entrances to keep out the weather. In one he found a small, light shield. In another was an old net that had a string bag attached to it, which was made in the same way as one created by a skilled European seine maker.

Back on the *Norfolk*, the crew had been patching the leak with oakum – old rope and other fibres soaked in tar. Matthew decided that, rather than further investigate the shallow Pumice Stone River by boat, he would explore more of Moreton Bay in the *Norfolk*.

The next morning, 17 July 1799, broke bright and cold on the wide expanse of the bay, and Matthew sailed south on the rising tide. At about half past ten, the crew landed near some steep iron-rich cliffs, which Matthew called Red Cliff Point.[8] They then pulled over to a green headland, about three kilometres to the west, 'round which the bight is contracted into a river like form, but the greatest part of it is dry at low water.'[9] This area is now called Clontarf Point and the bight beyond it was Bramble Bay flowing into the wetland and estuarine area of Hays Inlet.[10]

In a deserted hut on the western side of the headland, Matthew found another remarkable seine net, this one not more than a metre wide but, astonishingly, fourteen fathoms long – twenty-six metres – and seemingly much stronger than European nets. He guessed it was used to trap fish in a semicircular enclosure of woven branches that the visitors had seen on one of the nearby shoals. He believed it was a method of fishing entirely different from those used by the Eora around Port Jackson, where the men used barbed spears or lines with shell hooks. Matthew speculated that the method of fishing here meant that the Kabi Kabi people would have to carry heavy bundles of nets in order to provide food, and so were more likely to live permanently in one place – hence the superior construction of their huts. He was fascinated by the skill involved in making the nets and decided to take it for study, leaving behind a hatchet, which he felt represented a fair exchange whether the Kabi Kabi liked it or not. He chopped down some branches with the hatchet and placed the little axe in front of the hut to indicate its effectiveness.

The wood, which the Indigenous people collected for their fires, proved to be cedar, and of a fine grain. The remains of a canoe made of the stringybark was lying upon the shore, near the house. Near the hut, there were traces of dingoes, kangaroos and emus upon the beach. Two hawks of a moderate size were shot, but their plumage was unlike that of any hawks known at Port Jackson. Matthew then sailed south-east in a light northerly breeze and anchored the *Norfolk* that night at a point between what is now called Mud Island and the mainland.

The next day, the *Norfolk* continued on, sailing between Mud Island and what is now called St Helena Island, as Matthew mapped, numbered and made notes on the bay's small wooded islands, which were ringed by mudflats and mangroves leading into a calm, flat blueness. His method of map-making involved giving important geographical features numbers and assigning names much later. Cook had marked a peninsula guarding Moreton Bay, but Matthew now saw that it was not connected to the mainland and called it Moreton Island. He was mapping the little outcrop he marked as Island No. 3 – now called Green Island – when fear gripped everyone on board the sloop.

Their hearts began racing as they saw about twenty Indigenous men standing up in their canoes and pulling towards them with all their strength, 'in very regular order'. They seemed to have long poles or spears in their hands, with which they appeared to be paddling, all of them shifting their hands at the same instant 'after the manner of the South Sea islanders'. Matthew had heard the stories of the New Zealand war canoes seen by Cook and Banks on the voyage of the *Endeavour*. It looked as if these men were working their paddles 'with much resolution', and Matthew ordered his crew to be 'prepared for whatever might be the event'. The decks were cleared and each man was provided with musket balls, pistol balls and buckshot. 'It was intended that not a man should escape,' Matthew said, 'if the [locals] commenced an attack.'[11]

After some minutes, though, Matthew saw that, despite their exertions, the men did not approach much nearer to the vessel. Before long, to the embarrassment of everyone on the *Norfolk* – except Trim, who was singularly unconcerned – it became clear that the Indigenous men were actually standing upon a large, flat shoal that surrounded Green Island. They weren't paddling canoes, but driving fish into their nets, standing in a line and splashing in the water with their long sticks from one side to the other.

'Thus this hostile array turned out to be a few peaceable fishermen,' Matthew admitted, 'peaceable indeed; for on the approach of [our] vessel they sunk their canoes upon the flat, and retreated to the island, where they made their fires.'[12]

A little before midnight, Matthew was forced to anchor about five kilometres south of what is now called Peel Island, finding that the tide had gone out and the bay's deep water had become a narrow channel. The following day, 19 July, he and some of his men landed on what is now known as Coochiemudlo Island, an Anglicised version of the Yuggera words *kutchi* (meaning red) and *mudlo* (meaning stone). There were dingo footprints all over the beach but few traces of humans, except for some fireplaces. The central part of the little island, Matthew said, was 'covered with a coat of fine vegetable mould of a reddish colour'. On the south-east side of the island, an elevated part descended suddenly in a steep bank, 'where the earth was as red as blood; and, being clayey, some portions of it were nearly hardened into rock'.[13] The trees upon it were 'large and luxuriant'. In the low and sandy parts of the island there was an abundance of palm-nut trees, and evidence that the fruit was readily consumed by the locals. Cockatoos, both black and white, and the beautiful lilac-headed parrakeet, were seen, 'but there were not any marks of resident quadrupeds, apart from rats'.

At its southern end, Matthew wrote, the east and west shores of the bay came together 'in the form of a river; but the entrance was too full of shoals' to navigate.'[14]

Matthew was preoccupied with shoals for much of his time in the bay. This was understandable, since he was pressed for time and was in largely unmapped waters, where the deep water could become dangerously shallow within a boat length or two.

On Sunday, 21 July, Matthew and Bungaree were approaching Point Skirmish in the small boat when half a dozen Indigenous men came down to them unarmed, and 'by friendly gestures and offers of their girdles and small nets' endeavoured to persuade him to land. Matthew was suspicious of a trap and decided to sail on. Later in the afternoon, his men shot eight swans for their dinner. At the end of Matthew's exploring for the day, two more local men came down to the beach and motioned for him to land. Matthew did, but well out of spear range, and Bungaree traded a white woollen cap, some white cloth and some biscuit for Indigenous headbands and belts.[15]

An overhaul began on the *Norfolk* on 23 July, near what is called White Patch, about eight kilometres north of Point Skirmish. The little sloop was hauled up to a small beach and secured to trees. Over the next two days, the leak was sealed: the seam filled with oakum, the board planks were nailed back in place, and the whole area around the leak was covered with tarred canvas and sheet lead.

On the day the sloop was laid ashore, Matthew went out in the boat hunting more swans, and observed bulky animals in the water – like seals, he thought, though much bigger. They came to the surface of the water to breathe like porpoises but they did not spout, nor did they have dorsal fins. He did not know it, but these were dugongs, or sea cows. He decided to kill one to study it, and fired three musket balls into the nearest. Bungaree threw a spear into another, but both animals sank and were not seen again.

While the crew were drying the *Norfolk*'s sails, three local men appeared on the beach, unarmed, and a short distance below the vessel. Bungaree went up to them 'in his usual undaunted manner', but they would not allow Matthew or any of the Europeans to approach them without first laying down their muskets. Bungaree gave them gifts: yarn caps, pork and biscuit. When they were with just Bungaree and Matthew, the local people were 'lively, dancing and singing in concert', but the approach of more white men alarmed them.

Hoping to put them at ease, Matthew asked three of the sailors, all Scotsmen, to dance a reel for them, 'but, for want of music, they made a very bad performance', he said. The Kabi Kabi people watched 'without much amusement or curiosity', no doubt feeling their own dancing was far superior. Matthew tried to persuade the local people to visit the sloop, but there was still a great deal of understandable distrust among them for these strange, pale people who were such poor dancers. Matthew left them where they were, but, in his words, 'in a very friendly manner'.[16]

Matthew had been in Moreton Bay for the best part of two weeks, and the Glass House Mountains in the distance continued to make their siren call to him.

On 25 July, he took the *Norfolk* two or three miles further up the river, into a narrow channel where they anchored, saw a fire and heard several young female voices on the shore. Matthew was more interested in what was happening in the distance, and in the morning he, Bungaree and the oarsmen rowed upstream. From a small inlet on the western side of what is now called the Pumicestone Passage, surrounded by mangroves, they saw the more prominent of the eleven scattered peaks ahead.

Matthew, Bungaree and two crewmen headed on foot for the nearest one, a 280-metre peak now known as Mount Beerburrum. They walked for about fifteen kilometres, the first few of them through a landscape that was 'low, swampy, and brushy', full of boggy holes that covered them with mud and made their slow feet squelch. The higher, drier parts of the walk were either sandy or stony, and in these grass trees abounded.

Once they reached Beerburrum, they found the mountain was 'a pile of stones of all sizes, mostly loose near the surface'. There was decayed vegetable matter lodged everywhere in the cavities and tangles of twisted tree roots, and wild branches under a thick rainforest canopy over much of their approach.

It had taken nine hours to travel from the boat to the peak of

Beerburrum, but from the summit Matthew surveyed much of the sea and land that he had just mapped. The view was spectacular: the surrounding flat woodland with smoke rising here and there from Aboriginal people's fires, the azure Pumicestone Passage nearby, and beyond it the wide expanse of the blue-grey bay and Moreton Island guarding its entrance. Matthew noted a range of mountains stretching away to the south, which are now called the D'Aguilar Range. He could not see any big inland waterways from his position, not even the nearby Pine River, which flowed into Moreton Bay not far from the red cliffs he had noted a few days earlier.

The other ten of the Glass House Mountains were impossible to miss, though, shooting from the flat earth around Beerburrum like sharp arrowheads. The nearest one, now known as Mount Tibrogargan, was an almost sheer-sided 364-metre monolith, with a flat green roof of grass and a foreboding appearance, the setting sun casting it in a grim eerie light. The team resolved to conquer this perpendicular giant at daybreak.

They camped by what is now called Tibrogargan Creek, sleeping through the cool winter night beside a small fire, serenaded by the sounds of the local menagerie. At seven the next morning they found themselves under the steep cliffs of the flat-topped peak, their necks bent back as they surveyed the daunting task above. They had no ropes or climbing gear. The stone all around them was 'of a whitish cast, close-grained and hard, but not heavy. It was not stratified, but there were many fissures in it', making it difficult to climb.[17]

Matthew was sure these 'stupendous peaks, standing upon low flat ground', were volcanic, as the pumice stone in the nearby passage suggested, but he could not find traces of 'scoria, lava, *basaltes*, or other igneous remains'.[18] Scientists of later times would estimate the age of these volcanic plugs at 25–30 million years, and find that they were composed mostly of trachyte, formed by the rapid cooling of lava enriched with silica and alkali metals.

Matthew was an adventurer but not a daredevil, and after contemplating Tibrogargan for a while that morning, he concluded that 'the steepness of its sides utterly forbade all idea of reaching the summit'. He ordered the men back to the boat. Exhausted after an equally hard return journey, which included crossing the broad Elimbah Creek, Matthew and his men flopped onto the *Norfolk* that night. Trim and the rest of the crew were waiting with a welcome sight, a cooked black swan. Matthew was relieved, hungry and exhausted: 'a more laborious and tiresome walk of the same length would seldom be experienced'.[19] He concluded that the more inland part of this country was 'higher, better and more fertile' than the land in the neighbourhood of the salt water, but none of it seemed suitable for the production of wheat.

The following morning, Sunday, 28 July, they all sailed back down the Pumicestone Passage, seeing some Kabi Kabi people dancing and singing on the shore to attract their attention. Matthew assumed their shouts expressed goodwill, especially now that he wasn't shooting at them anymore.

Soon after, the *Norfolk* crew went on shore to cut down a local pine, in order to show the workmen in Port Jackson the quality of the northern wood. Bungaree was speaking with some of the Kabi Kabi people when the tree fell with an almighty crash. The locals, no doubt still sensitive to the sound and power of the British muskets, jumped with fright, but Bungaree eased their fears with a gift of one of his spears, and a throwing-stick of a type they had not seen before.

Bad weather kept the *Norfolk* in the Pumicestone Passage for two more days, and Matthew had occasional visits from the Indigenous people, whom he now regarded as friends. He and Bungaree watched them fish, surrounding their prey on all sides with nets.

When the rain ceased on 30 July, some of the crew went to the eastern shore to procure firewood, and for the first time a group of the Kabi Kabi people allowed the Europeans to approach them while still carrying their muskets.

They all sat down together, European and Indigenous men, and the Kabi Kabi sang songs. Matthew found the tunes 'musical and pleasing'. He said they were 'accompanied by slow and not ungraceful motions of the body and limbs, their hands being held up in a supplicating posture, and the tone and manner of their song'. The gestures seemed to 'bespeak the good will and forbearance of their auditors'.[20]

Bungaree said he would sing one of his songs in reply, but as with the dancing Scotsmen, his performance was a disaster. 'Barbarous and grating to the ear,' Matthew lamented, spicing his blistering critique with the pronouncement that Bungaree was regarded as 'an indifferent songster, even among his own countrymen'.[21] Matthew gave his hosts some worsted caps and a pair of old blanket trousers, 'with which they were much gratified'.

Each group laughed as they tried to pronounce the unfamiliar names. Like Bennelong, who always referred to 'coffee' as 'caw-be',[22] the Kabi Kabi had trouble pronouncing the letters 'f' and 's'. They called Matthew 'Mid-ger Plindah', and his brother Samuel 'Dam-wel'. Three of the local people were named Yel-yel-bah, Ye-woo and Bo-ma-ri-go – the last of which stuck in Matthew's memory because he thought it sounded like 'Puerto Rico'.[23] He concluded:

> These people were evidently of the same race as those at Port Jackson, though speaking a language which [Bungaree] could not understand. They fish almost wholly with cast and setting nets, live more in society than the natives to the southward, and are much better lodged. Their spears are of solid wood, and used without the throwing stick.[24]

On his last day at Moreton Bay, he saw a large turtle asleep on the water and guessed that the Kabi Kabi also used their large nets to capture these animals and the dugong as well.

Matthew sailed the *Norfolk* out of the Pumicestone Passage on Wednesday, 31 July 1799, and with a fresh south-east sea breeze behind him left Moreton Bay after a visit of fifteen days. By 5 p.m., the highest of the Glass House Mountains was small and faint in the distance.

Matthew often had a difficult relationship with his brother Samuel, whom the Kabi Kabi people called 'Dam-wel'. *La Trobe Collection, State Library of Victoria*

His visit to the area, the first by a European explorer, had promised much, but had not been as successful as he had likely hoped. The bay was so full of shoals, Matthew said, that he could not attempt to point out any passage that would allow a ship in without danger. In time, many others would. Matthew's preoccupation with the shoals and with staying afloat meant that he missed seeing the entrances to two more major river systems that empty into the bay, the Brisbane and Logan rivers. Although he examined what are now called Bramble Bay and Hays Inlet, he also missed discovering the Pine River. The rivers he might have found, though, were typically hidden by mudflats, sandbanks and mangroves, making their discovery difficult for a man constantly looking to avoid reefs and rocks.

FROM CAPE MORETON, Matthew followed the curve of the coast north, and the next day passed the area Cook had named Wide Bay. On 2 August he reached Sandy Cape, which

Cook thought was a peninsula but which is really the northern end of K'gari, or Fraser Island as it was known for more than a century and a half, the largest sand island in the world. Matthew saw a yellow-bellied sea snake, the same sort Cook had reported in these waters, and which Matthew had previously seen in the Torres Strait.

There were other dangers here too, and for another twenty kilometres Matthew stayed close to the dangerous Breaksea Spit. The shoals and snags were so treacherous that almost immediately the water depth could change from just three and a half fathoms (six metres) to seventeen fathoms (thirty-one metres).[25] Once past the danger, Matthew veered round to enter a bay, which Cook had named for Augustus Hervey,[26] a Lord of the Admiralty.

Matthew kept the *Norfolk* in 'Hervey's Bay' until 7 August, but again spent most of his time on the lookout for shoals. He found no waterway extending into the interior here either, though the mighty Burnett River washed into the bay. Frustrated, but with his six-week deadline on his mind, he cleared the point of Breaksea Spit on 8 August and started the journey home to Port Jackson, pushed along by a breeze from the north and the strong southerly current. Contrary to Cook's findings, Matthew believed the southward current was strongest when the *Norfolk* was between six leagues (thirty-three kilometres) and twenty leagues (111 kilometres) offshore, so he was rarely in sight of land.

Ten days later, in the region of Port Stephens, several large sperm whales played about the boat for more than two hours. It was dusk on 20 August when Matthew sailed into Port Jackson and brought the *Norfolk* into the dock alongside the *Reliance*. Trim beat everyone else off the little sloop and back up the gangplank onto the ship on which he'd been born.

Matthew considered the *Norfolk*'s voyage important and fascinating, and he produced detailed reports on the people and places north of Sydney. His were the first interaction between Europeans and the Indigenous peoples in many places he visited. He and Samuel wasted no time before writing home to tell their father about all they had seen.[27]

But Matthew also considered this trip to have been a failure, and declared that he was disappointed at not having penetrated the interior of New South Wales by 'either of the openings examined in this expedition'.

'However mortifying the conviction might be,' he wrote later, with considerable arrogance, 'it was then an ascertained fact that no river of importance intersected the East Coast between the 24th and 39th degrees of south latitude.'[28] On this, he was, of course, dead wrong. That statement was perhaps his greatest blunder in what became an extraordinary career.

Travelling north was a sad final voyage for the little *Norfolk*, too. Two years after it arrived in Sydney from Norfolk Island, the sloop was seized by fifteen convict escapees in Broken Bay while transporting a load of wheat from the Hawkesbury to Sydney in November 1800. The convicts hoped to sail the *Norfolk* north to the Dutch East Indies, but were unable to handle the sloop in bad weather. The trusty little vessel was wrecked at Pirates Point, near modern day Stockton, four weeks later. Six of the convicts escaped into the bush with the local Indigenous people, the Awabakal, but the remaining nine convicts seized another ship on the Hunter River. They were quickly recaptured. Two of the ringleaders were hanged, and the other seven were sentenced to life on Norfolk Island.[29]

By that time, Matthew was dealing with more tragedies and triumphs – even murder. He was also about to embark on the greatest adventure of his life.

Chapter 14

*'I doubt, my dear father, but that what I have to mention will
somewhat surprise you; Which is that I propose to marry before
I leave England, and take my wife out with me.'*
MATTHEW FLINDERS, TELLING HIS FATHER OF HIS LAST-MINUTE
MARRIAGE PLANS.[1]

MATTHEW RETURNED to the *Reliance* to find both
it and Governor Hunter surplus to Britain's needs. The
Admiralty had approved the purchase of two new ships to service
the colony, HMS *Buffalo* and HMS *Porpoise*, and Hunter wrote to
London to say that the *Reliance* was now 'too weak and infirm' to
be further employed in New South Wales, but that it had been
given 'such repairs as would enable her with safety to return
home'[2] under the command of Captain Waterhouse.

Hunter was unaware that, while his letter was on its way to
London, the Duke of Portland was drafting an order for Hunter
to pack his bags and return home too. Philip Gidley King would
take over the office after winning the support of Joseph Banks.

In his same letter to Nepean, Hunter reported the
circumnavigation of Van Diemen's Land and included Matthew's
chart of the island's coast. The chart was sent to hydrographer
Alexander Dalrymple, and then to Aaron Arrowsmith for
publication as *A Chart of Bass's Strait between New South Wales
and Van Diemen's Land Explored by Mattw. Flinders ... by order of
His Ex. Governor Hunter 1798–1799*. Matthew and Bass had

recommended the name Wilsons Promontory, in honour of Matthew's merchant friend Thomas Wilson from London, for the area Bass had originally called Furneaux's Land. The map from Matthew's northern adventure was published in London in January 1800, adding to the work of men who had gone before him such as Cook and Shortland. The title proclaimed that it was 'collected and arranged by Mattw. Flinders—2nd Lieut. of H.M.S. *Reliance*, Jan. 1800'.[3]

The *Reliance* was able to make two more trips to Norfolk Island, the second ending with a return to Sydney on Christmas Eve, 1799. Matthew split those trips with another appearance on a judicial bench in Sydney, this time hearing one of the earliest cases brought to trial in which white men were accused of murdering Indigenous people. Hunter tried to show that there was justice for all in the colony, black or white, even though the First Peoples had been treated dreadfully by many new settlers since the First Fleet arrived.

On 14 October 1799, five farmers from the Hawkesbury region went on trial in Sydney for killing a pair of Dharug youths known to the British as 'Little Jemmy' and 'Little George'[4] in a small settlement called Argyle Reach. The case was tried before a jury of six, comprising three army officers led by John Macarthur and three naval men, Matthew, Waterhouse and Shortland.[5]

Dharug people had been hired as trackers for a hunting expedition for Thomas Hodgkinson and John Winbow, a Second Fleet convict who was now living with a Dharug woman, who may or may not have been forced into the arrangement. Not long after the hunting expedition headed towards the Blue Mountains, Hodginkson and Winbow were hacked to death. Those responsible included the Dharug men known as Major White and Terribandy, who was the father of Winbow's 'wife'.[6]

'Little Jemmy' and 'Little George' came to Argyle Reach to return Hodgkinson's gun to his family. They were well known in the village and, until then, well liked. But now the settlers wanted revenge. Both the Dharug youngsters had their hands tied behind their backs, and as they struggled desperately against their captors

In 1801 Aaron Arrowsmith published 'A chart of Bass's Strait between New South Wales and Van-Diemen's Land explored by Matt.w Flinders 2nd. Lieut. of his Majesty's ship Reliance by order of His Excellency Governor Hunter, 1798-9'. *State Library of NSW FL18260878*

and cried out for mercy, they were slashed with cutlasses. George, who was thought to be about eleven or twelve, suffered huge wounds on his body and hip before he bled to death. Jemmy, who was estimated to be fifteen or sixteen, was shot through the body. One side of his face appeared to have also 'been much cut by a cutlass'.[7] His head was nearly severed.

Five local farmers – Edward Powell, Simon Freebody, James Metcalfe, William Timms and William Butler – were charged with murder. Apart from the deaths of Hodgkinson and Winbow, there had been other recent cases of settlers being killed by Indigenous warriors, but Hunter assured Portland in a letter that 'much of that hostile disposition which has occasionally appear'd in those people has been but too often provoked by the treatment which many of them have received from the white inhabitants'.[8]

Matthew quizzed some of the witnesses himself. He had fired on Indigenous people when threatened with spears in Moreton Bay, but he had no doubt that this action was altogether different. The five accused were found guilty of 'wantonly killing two natives'. Matthew, Waterhouse and Shortland all recommended flogging as punishment, but Macarthur and the officers of the New South Wales Corps demurred, so the case was referred to London. The five guilty men were freed on bonds and eventually pardoned.[9]

Hunter was livid at this injustice. He told Portland: 'Two native boys have lately been most barbarously murdered ... Those men found guilty of murder are now at large and living upon their farms, as much at their ease as ever ...'[10]

WITH THE TRIAL OVER, Matthew's attention turned to how he could improve his own station, as he and Samuel were to sail on the *Reliance* back to Britain. Matthew heard that Joseph Banks was pushing for him to be given command of one of the new ships coming to Sydney,[11] and Hunter rewarded his hard work and reliability by topping up his original land grant at Banks Town of 100 acres (40 hectares) to 300 acres (121 hectares).[12] But what was this real estate truly worth when

William Balmain was about to sell 550 acres in the area that now bears his name for a mere five shillings?[13]

Matthew and Samuel kept busy, making a series of astronomical observations at Cattle Point, near Bennelong's hut. On 12 February 1800, Matthew received an unexpected luxury gift hamper from Paddy Smith in Calcutta: Madeira wine, spirits, shirts, towels and silk neckerchiefs. Two days later he wrote to thank his old friend, outlining his frustrations at still being a junior officer despite what he'd achieved for the colony, especially when surrounded by wealthy traders such as Macarthur. On his arrival back in England, he said, an unnamed Irishman would be making available to him 'the command of a sum of money … which will enable me to enter upon certain speculations and give me the command of a ship'.[14]

> The thing is, my dear friend, I am tired of serving for a pittance, and as it were living from hand to mouth, whilst others with no better claim are making hundreds and thousands. The examples which have occurred in this place have opened my eyes a little to my own interest; and besides, I want to be my own master, and not to be subject to the caprices of whomsoever the Lords above may please to set over me …[15]

He told Smith: 'Between ourselves, I have some hopes that my relations in England will advance me two or three thousand pounds to forward my mercantile plans, which if they do, and moderate success should attend me, a few years will probably see me independent of the world …'[16]

A day after writing to Smith, Matthew wrote to Bass to say he was fully determined never to serve 'as a common lieutenant in a common ship' again and that if his friend wanted a business partner 'you know when to pitch upon a willing one …'[17]

He told Bass that his relations were well, 'but my father complains heavily of the fatigues of business and the failure of his constitution to go through it'.

He adds, that he must begin to contract his business, having done his duty to his family, by having provided a sum for each, by the assistance of which they may pass through life with credit and reputation. He is a good father! and I wish he would spare himself a little ...'[18]

Matthew planned to ask his father to lend him 'a thousand or two' for his business plans. But he must have been dreaming. Flinders Sr had almost choked when given the bill for Samuel's dancing lessons, and he had insisted that all his children, including his daughters, support themselves with jobs from a young age.

AFTER FIVE YEARS IN SYDNEY, Matthew was finally going home. On 3 March 1800, Waterhouse steered the *Reliance*, patched up like a badly broken hospital patient, out through the Sydney Heads, aiming for Tahiti and then Cape Horn. Matthew was now almost twenty-six, and although he wanted the wealth that the traders in Sydney were building for themselves, he could not quell his spirit of adventure and the lust for knowledge about the great landmass around Sydney. He was already thinking of making another voyage of discovery.

Years later he would reflect on his aspirations:

I have too much ambition to rest in the unnoticed middle order of mankind. Since neither birth nor fortune have favoured me, my actions shall speak to the world. In the regular service of the Navy there are too many competitors for fame. I have therefore chosen a branch which, though less rewarded by rank and fortune, is yet little less in celebrity ... although I cannot rival the immortalized name of Cook, yet if persevering industry, joined with what ability I may possess, can accomplish it, then will I secure the second place ...'[19]

Governor Hunter, still unaware he'd been made redundant, had given Waterhouse his dispatches for London, with instructions to throw them overboard if it seemed they might be captured by a French vessel.[20] The voyage proved tumultuous and at times terrifying, even without the French coming near. The patches on the hull of the *Reliance* gave way and it was taking on so much water that a pair of stowaway convicts, discovered two days into the voyage,[21] were made to work the pumps around the clock.

Still, Matthew had his cat to calm his nerves. He never tired of Trim's antics:

> [Trim] was taught to lie flat upon the deck on his back, with his four feet stretched out like one dead; and in this posture he would remain until a signal was given him to rise ... if, however, he was kept in this position, which it must be confessed was not very agreeable to a quadruped, a slight motion at the end of his tail denoted the commencement of impatience, and his friends never pushed their lessons further ...[22]

About 900 kilometres south-east of New Zealand, in the freezing sub-Antarctic waters of the Southern Ocean, the men sighted a small, previously uncharted island and several islets, which appeared to be inhabited only by seals. They compiled a sketch map; a version, 'copied by S. Flinders',[23] survives. The small cluster is now known as the Antipodes Islands.

Five weeks later, on 30 May 1800, the *Reliance* reached the mid-Atlantic island of Saint Helena, where Governor Robert Brooke, who had helped Bligh and the crew of the *Providence* eight years earlier, gave every assistance to Waterhouse, his battered ship and his exhausted men. The crew stayed on Saint Helena for three weeks, and Matthew told Brooke about the discovery of Bass Strait and how it would slash time and distance for British ships travelling to China. Governor Brooke gave Matthew a letter to take to London, in which he advocated for some of the young man's ideas.

The *Reliance* left Saint Helena with four East India Company ships for protection from the French, and from the coast of Ireland it sailed under the protection of the 32-gun frigate HMS *Cerberus*.

Matthew and Samuel finally touched English soil again for the first time in five and a half years when the *Reliance* wobbled into Plymouth Sound on 26 August 1800,[24] before docking at Portsmouth. Trim – perhaps the first cat to circumnavigate the globe – was first down the gangplank, his black bushy tail swishing proudly. But there was little mirth for the Flinders brothers when they came ashore.

The previous winter in Donington had been more severe than anyone could remember, and their sister Betsey Harvey, the mother of two small children, had died after a long and painful illness aged just twenty-four – her death, it was said, accelerated by the depression she felt from a bad marriage.[25] Flinders Sr was now fifty but looked and felt much older, and he worried how the news of Betsey's death would affect Matthew in particular, as he and his sister had been so close as children.[26] Mary Franklin, who was like a sister to Matthew, had died a few months before Betsey. To add to the family woes, the behaviour of Matthew's younger brother John Flinders – or 'my unfortunate Son John', as their father now called him – had become so uncontrollable that he would soon be admitted to the York Lunatic Asylum, where he would spend the rest of his sad life.[27]

FOR MATTHEW, THOUGH, there was no time to look back. He had to stay on board the *Reliance* for many more weeks, but he took the opportunity to advance his ideas on exploration. He would later write:

> The voyages which had been made, during the seventeenth and eighteenth centuries, by Dutch and by English navigators, had successively brought to light various extensive coasts in the southern hemisphere, which were thought to be united; and to comprise a land, which must be nearly equal in magnitude to the whole of Europe. To this land, though known to be separated from all other great portions of the globe, geographers were disposed to give the appellation of Continent: but doubts still existed ...[28]

Matthew was only back in England eleven days when, with the *Reliance* anchored at Spithead, he wrote to the one man he believed could raise him from anonymity to acclaim: Joseph Banks. It was an audacious move for a young lieutenant to cold-call the most powerful figure in British science, a man who not only had the ear of the king and the Lords of the Admiralty, but who remained the expert in all things antipodean and the authority to whom the governors of New South Wales answered.

Early in 1800, Banks had organised with Philip Gidley King to send the 60-ton brig HMS *Lady Nelson* for coastal surveys of New South Wales.[29] Matthew had heard that the ship, now under the command of a young Scot, James Grant,[30] had already sailed for Sydney and would be used for further exploration on the Australian coast. He suspected that he would be asked to continue those explorations, but he decried the vessel to Banks as too small to do the job adequately in the waters he had visited.

Philip Gidley King replaced John Hunter, becoming the third Governor of New South Wales. *State Library of NSW FL3251488*

It was seven years since Matthew had visited Banks at his country estate not far from Donington, delivering a letter for James Wiles. He knew the great man had been informed of his discoveries in New South Wales and Van Diemen's Land, and he planned on dedicating his booklet, *Observations on the Coasts of Van Diemen's Land, on Bass's Strait, etc.*, to the man he hoped would make his dreams of exploration come true.[31]

In his letter, Matthew apologised to Banks for being so presumptuous and direct, but told him it was as a result of long employment abroad and an education among 'the unpolished inhabitants of the Lincolnshire fens'.

He said a very great part of New South Wales remained 'either totally unknown, or has been partially examined at a time when navigation was much less advanced than at present'. The interests of geography and natural history, he ventured, required 'that this only remaining considerable part of the globe should be thoroughly explored ... [A] person or persons ... should examine into the natural productions of this wonderful country ... the mineralogical branch would probably not be the least interesting.'[32]

The *Lady Nelson* would be 'very inadequate' for the task, Matthew said, and he outlined a plan of circumnavigation to prove that New Holland, like Van Diemen's Land, was an island – only many times larger. 'If his Majesty should be so far desirous to have the discovery of New Holland completed ... I should enter upon it with that zeal which I hope has hitherto characterized my service.'[33]

Matthew poured out his heart in the letter, praying that the venerated botanist would see the same spirit of adventure in this young lieutenant that Banks himself had shown back in 1768. Then, Banks had turned his back on a life of opulence and leisure as one of the richest young men in Britain, opting instead to spend three years battling huge seas on the *Endeavour*, investigating people, places, fauna, flora and customs unknown to Europe.

Matthew promised to visit Banks in person as soon as his duties with the *Reliance* would allow. He waited excitedly for a response.

And waited. Days went by. Then weeks. He heard nothing. His heart must have sunk.

WITH HIS EMOTIONS IN A WHIRL at his family tragedies and his professional limbo, Matthew sought out Ann Chappelle for some emotional healing. He wasn't sure what to expect. Thomas Franklin had told him that Ann was in Boston visiting friends.

She had not answered his letter of March 1799, nor the two he had written before that. In fact, he hadn't heard from her in three years. This time he wrote to Ann from the Nore, the naval anchorage on the Thames. He told her about the grief he felt at the loss of Betsey and Mary Franklin. He called Ann his 'dear friend Annette' and confided:

> My imagination has flown after you often and many a time, but the Lords of the Admiralty still keep me in confinement at the Nore. You must know, and your tender feelings have often anticipated for me, the rapturous pleasure I promised myself on returning from this Antipodean voyage, and an absence of six years. As you are one of those friends whom I consider it indispensable and necessary to see, I should be glad to have some little account of your movements, where you reside and with whom ...[34]

Living among sailors for so long had taught Matthew little about romancing a young woman. He said he might make it to Lincolnshire if he could find the time. Everything now, he said, was 'subservient to business'.

> Indeed, my dearest friend, this time seems to be a very critical period of my life ... I have more and greater friends than before, and this seems to be the moment that their exertions may be the most serviceable to me. I may now perhaps make a bold dash forward, or remain a poor lieutenant all my life.[35]

On 7 October 1800, the *Reliance* arrived at the naval yard at Deptford. The ship's company were paid off and the vessel handed over to the dockyard carpenter to begin a confounding list of repairs.[36]

The following day, George Bass, who had arrived in London three weeks before Matthew, married Henry Waterhouse's sister Elizabeth in a small ceremony at St James's Church in Piccadilly. Waterhouse and another sister Mary were the witnesses. Matthew might have been one of the guests, though he never committed an account of it to paper.

Two months later, Bass said goodbye to his new wife and set sail from the Thames for Sydney in a copper-sheathed brig called the *Venus*, after he and a syndicate that included Charles Bishop had invested in a trading venture. The *Venus* was a two-masted 140-ton brig built of teak. It carried twelve cannons in case it was troubled by French vessels. Henry Waterhouse had helped Bass and Bishop raise £11,000 for the purchase.

SAMUEL FLINDERS LEFT the *Reliance* and went home to see his father and surviving siblings, who had last known him as an eleven-year-old boy toting a kitbag almost as big as himself. Flinders Sr was impressed by the way the now seventeen-year-old had 'grown and improved equal to my most Sanguine Expectation'. With a heavy touch of regret, though, he wrote that 'Mattw.'s affairs yet detain him in Town, but I hope to see him ere long'.[37]

Matthew took lodgings for himself at Deptford, placing the always mischievous Trim under the watch 'of the good woman of the house', who was kept at her wits' end by a cat with a mind of his own. Trim would sit at the sash window at the top of the house watching the pedestrian and horse traffic go by; when the curtain was once closed on him, he responded by jumping through the 'glass like a clap of thunder, to the great alarm of the good hostess below'. Woe to her good china, too, if Trim got into her closet chasing a mouse, because he dashed at rodents 'like a man of war, through thick and thin'. After he sent one set of her cups and saucers crashing, the landlady seized Trim to beat him soundly, but didn't have the heart after he rubbed his whiskers against her chin and got to purring.[38]

The black and white cat could only keep Matthew laughing for so long though. On 9 November, Matthew wrote to his father to discuss Samuel's career prospects, and related how the brothers had visited their sister Susanna at Dartford, where she was working as a milliner's apprentice. Matthew seemed critical that their sister was 'a milliners journey woman', and their father – whose health had been in decline for many months, and who had become increasingly irritable as a consequence – took offence at the inference that Susanna was being made to work in a position below the family's social standing.

Flinders Sr fired back at his eldest son and all his 'young folk', who, he complained, only seemed interested in him when they wanted money. He reminded Matthew that he had worked hard all his life delivering babies and potions, and had never asked anyone for a handout.[39]

'It cannot be any terrible task for a grown person in youth health and strength to obtain their own support,' Matthew Sr told his son.

> I well know it is a much harder one for a weak infirm father in the decline of life, to continue labouring for the whole to the last hour of his earthly race … Why my Childn should expect that I am able to make them all great people must be owing to their mistaking the amount of my property, and supposing that I have much more than I really have.[40]

Matthew angrily fired back, saying that except for affection, his father should now consider that he had only four children from his original family. 'I shall want no further pecuniary assistance,' he snapped.[41]

Father and son patched things up, but both were wounded by the exchange.

IN NOVEMBER, MATTHEW took Trim the ten kilometres from Deptford to London, on a stagecoach. Two gentlemen who were their fellow passengers were fascinated by Matthew's tales about the adventures both he and the cat had experienced on the far side of the world.

Matthew and Trim moved into lodgings at 16 King Street in Soho, not far from Banks' four-storey mansion in Soho Square. Trim's frolics took Matthew's mind off the fact that neither Banks nor Ann had answered his heartfelt letters. On 6 November, Samuel had left Donington to join the 64-gun HMS *Alcmene*, a move which Matthew had pushed. Samuel would be promoted to lieutenant on 1 March 1801 at the age of eighteen. He was now a small and sometimes cocky young man, with similar features to his elder brother – thick brown hair, dark eyes, a pointed nose and small chin.

Matthew met with the printers to discuss the publication of his charts and booklet. He also wrote to the Court of Directors at the East India Company, mentioning the endorsement from Governor Brooke and the savings the company would make if its ships sailed through the strait he and Bass had found north of Van Diemen's Land. He inspected the latest surveying equipment made by Edward Troughton, and strolled through London's streets, exhilarated by the rich colours of the English spring and the grandeur of St Paul's, Westminster and the naval buildings at Greenwich, which were such a contrast to the ragged village of Sydney and the wilds of New Holland.

Then Matthew's world changed in ways that took him by stunning surprise.

In the middle of November, he finally received Joseph Banks' reply. Nervously he opened the envelope, his whole career hanging on the few words on this small note.

Soho Square, Novr. 16 1800
Sir
Jos: Banks presents his Compts. to Mr. Flinders he is
sorry indeed to have been prevented by bad health from

answering a Letter he Received some time ago from
Mr Flinders will be happy to see him in Soho Square at any
time he will be so good as to Call upon him.[42]

Matthew could barely contain his joy. If there was anyone who could advance his ambitions for exploration, it was the president of the Royal Society, who had backed all of Britain's great voyages of the last thirty years.

Matthew raced to 32 Soho Square, where he was ushered into Banks' opulent office, surrounded by high shelves of centuries-old books and artefacts from his journeys across the globe. Banks was now a big, jowly 57-year-old, with a firm handshake and a mind like a sponge. He was heavier than Matthew remembered. The once dashing playboy and adventurer now weighed 109 kilograms and was beset by arthritis, but this monolith formed the bedrock of British science.

Banks listened with rapt attention to Matthew's descriptions of New South Wales and Van Diemen's Land, and the two compared notes, with Banks recalling what he had seen in the years before the colony was settled. Banks pored over the charts and maps Matthew showed him.

Within a week, Banks had the wheels in motion for Matthew to make the most celebrated journey of his life. Banks implored his friend the Earl Spencer,[43] First Lord of the Admiralty, to agree that the exploration of New South Wales was a top priority. Spencer was the man who had chosen Horatio Nelson to lead the recent British naval conflict with the French on the Nile. With the French also mounting a scientific expedition to the great landmass in the south – an expedition to be led by the veteran sea captain Nicolas Baudin[44] – Banks said it was more important than ever to keep the British territory from falling into Napoleon's hands.

Within weeks, Banks told Matthew that the appropriately named *Investigator* – 'a north-country-built ship of three-hundred and thirty-four tons' – would be put at his disposal, along with a crew of more than seventy men.[45]

The vessel very much resembled Cook's *Endeavour*, and like that ship had started its working life carting coal from Newcastle. It was 100 feet long (thirty metres), twenty-nine feet wide and had a draught of fourteen feet. The collier had been built on the Wear River in Sunderland as the *Fram* in 1795. The Royal Navy had purchased it three years later for £2350, refitting it at Deptford to serve as an escort vessel named HMS *Xenophon*. It was armed with twenty-two short cannons called 'carronades', and in 1799 had ferried the Irish rebel James Napper Tandy and some of his supporters as state prisoners from Hamburg for trial in England.

The *Investigator* was five years old, but already in such a state of disrepair that it was being saved from war service against the French. Timber that was not properly dried had been used in the construction: a report at Deptford described the lower deck beams as 'wainy and sappy'. Matthew had sailed on the *Reliance* and the *Norfolk*, though, so problems with construction and rotting timbers were nothing new to him.

Banks had arranged for the *Investigator* to undergo a refit at Sheerness, at the mouth of the River Medway, starting in November 1800. A plant cabin was planned for the collection of trees, shrubs, vegetables and flowers from New South Wales, which could then be planted at the Royal Gardens in Kew. Matthew wanted the bottom of the ship covered in copper to guard against the 'termites of the sea', the *Teredo* worms. Despite the ship's faults, Matthew considered it to be the 'best vessel which could ... be spared for the projected voyage to Terra Australis'.[46]

While the ship was being worked on, Matthew met Robert Fowler,[47] a fellow Lincolnshire man, who had served on the *Investigator* when it was still the *Xenophon*. As his former captain had spoken well of Fowler, Matthew selected him as his first lieutenant. The surgeon, chosen by Joseph Banks, was Hugh Bell, from HMS *Seagull*, who had an interest in natural history, particularly botany. But he was a blunt man, abrupt in his speech. He and Matthew would clash.

Robert Fowler, another
Lincolnshire man, became
Matthew's first lieutenant on
the *Investigator*. *La Trobe Collection,
State Library of Victoria*

THE EUPHORIA MATTHEW felt at being handed the
Investigator and a large crew sparked a frenzy of heartfelt messages
between him and Ann Chappelle. Knowing the dangers of falling
in love with a sailing man, Ann had been reluctant to respond to
the searching enquiries in Matthew's letter on his arrival back in
England, but her rosy-cheeked and blonde half-sister Belle Tyler –
who perhaps had a schoolgirl crush on Matthew – encouraged
her to write back. Belle described Matthew at the time as having
a slight figure, 'but well proportioned', and with a 'light and
buoyant step'.[48]

A week after Bass had farewelled his wife of only two months,
Matthew wrote to Ann, who was visiting relatives in Hull. He
explained that the *Xenophon* was being 'rebaptised the *Investigator*
and her guns reduced to 12', ready for him to sail for New South
Wales again, perhaps even as early as January.

'Let us then, my dear Annette, return to the "sweet, calm delights of friendship",' Matthew wrote, apparently firm in his resolve to leave romance aside.

I must call ambition to my assistance since it must be so; and in a life of activity and danger put out of my mind but that we are friends. The search after knowledge – the contemplation of nature in the barren world, the overlapping crags of utmost height and the open Fields decked with the spicy attire of the tropical climes may – nay must prevent me from casting one thought on England – on my home … And think not, my Annette, that I value thee less for the want of fortune. Heaven is my witness, that did I posses [sic] ten thousand pounds tomorrow and I found thy person and mind what I think they are, my hand should await thy acceptance … By personal conferences we may be able to come to a better and more final understanding than by letter. Let us meet as lovers, and part as friends my Annette![49]

Matthew missed spending Christmas with his family, choosing to be with Ann at Spilsby and Partney. He was torn about leaving her behind for years, and he encouraged her to make the most of her mind and her talents.

Learn music, learn the French language, enlarge the subjects of thy pencil. Study geography and astronomy; and even metaphysics, sooner than leave thy mind unoccupied. Soar my Annette, – aspire to the heights of science. Write a great deal, work with thy kneedle [sic] a great deal, and read every book that comes in thy way, save trifling novels.[50]

He raised the subject of marrying Ann, but she told him she could never wed a man who would not stay by her side.

MATTHEW ARRIVED AT HIS father's house in Donington on 2 January 1801. He asked him again for a loan because, despite

Ann's refusal, he was thinking about getting married, but he left for Soho empty-handed. Three days later, Matthew wrote to Ann again in a state of high anxiety:

> Tears are in my eyes – I am torn to pieces. Thou had promised to inform me when thou art married and I trust that the earliest opportunity afterwards will bring me the intelligence; it will be important to me. And whilst I am torn by winds and waves on various coasts and in various climes, may thou enjoy that serenity that a contemplative mind feels on surveying its own happiness. May thou meet with one whose mind and heart is worthy of thy love and whose circumstances unlike mine, can afford thee the enjoyment of life. Adieu, perhaps the last time. This excess of misery is too great to be often recalled. It is seldom that I have written a letter in tears.
> Mattw Flinders[51]

He dried his eyes promptly enough a week later to investigate the *Investigator* in dock at Sheerness. The refit was going well. The *Investigator*'s former life as a collier meant that it had only a small draught and could be used in shallow waters, while also having a wide capacity in which to store the supplies needed for such a voyage. Additional cabins were being installed for the scientists and artists whom Banks had hired. The armament was being reduced to two cannons and eight carronades to give the crew more space.

'It gives me pleasure,' Matthew told Banks, 'to say that the *Investigator* is a comfortable ship, and affords a great deal of accommodation.'[52]

He wrote to Ann three days later, his mind in a whirl. Baudin and his two ships had already sailed for New Holland, and Matthew was eager to get going as well – but could he really part with Ann as mere friends? Could he go on his grand adventure and leave her alone with her thoughts and heartache in a small Lincolnshire village, awaiting his return years later?

Flinders Sr was still miserable as well. He summed up his winter of discontent in his diary: his son John was now 'deranged' and 'Daughter Susan has arrived with Mattw' with a demand of a considerable Sum to set her in Business'. Matthew also had his hand out for a loan and his father complained that after the worst winter 'ever known in living Memory – with my Strength, Health and Spirits very much impaired, and consequently my capability of Business much abated, have almost proved too much for me … I have put [off] Matts. demand for another year. he asked for £200, but I have promised but one …[53]

Matthew now organised for Samuel, now a lieutenant, to sail with him on the *Investigator*. Samuel asked his father for £100 to pay for new uniforms but was also knocked back.[54]

With all the turmoil over his on-again, off-again romance, Matthew said he was anxious to 'arrive upon the coasts of *Terra Australis* in time to have the whole of the southern summer before me'. That meant he would have to arrive by December 1801, and time was already getting away. There were seventy-six crewmen from the *Investigator* chosen for the voyage, and he picked another eleven from the *Zealand*, after 250 had volunteered.

On 16 February 1801, Matthew was promoted to the rank of commander,[55] and on 14 March the cannons and ammunition were loaded. On 27 March he sailed the *Investigator* down to the Nore in readiness for the voyage, but Ann was paramount in his mind.

Various circumstances, he wrote, were 'retarding the departure'. One was that a passport from the French government – to prevent any interference with the voyage despite the warfare between the two countries – had not yet arrived. Largely because of Joseph Banks' diplomatic efforts to make science a subject of international neutrality, voyages of discovery were regarded as sacrosanct by both the English and the French governments. At least, Matthew hoped they were.

Although Matthew had a clear vision of what he wanted to achieve aboard the *Investigator*, he was still racked by turmoil when it came to matters of the heart. On 3 April 1801, he wrote to his father to say that, despite his previous requests for money with a view to getting married, nothing else but the circumnavigation of New Holland mattered now:

> You are acquainted with the attachment that subsisted between miss Chappelle and myself; and the reasons that have induced us to wean ourselves from it. I have no present or future intention of marrying either her or any other person, but leave England wedded only to my ship and to the service upon which I am going.[56]

Three days later, after much soul searching, he had another change of heart, and wrote to Ann from the *Investigator* at the Nore:

> My dearest friend
> Thou has asked me if there is a possibility of our living together. I think I see a probability of living with a moderate share of comfort. Till now, I was not certain of being able to fit myself out clear of the world. I have now done it; and have accommodation on board the *Investigator*, in which as my wife, a woman may, with love to assist her, make herself happy. This prospect has recalled all the tenderness which I have so reluctantly endeavoured to banish. I am sent for to London, where I shall be from the 9th [of April] to the 19th or perhaps longer. If thou wilt meet me there, this hand shall be thine forever.[57]

Matthew told Ann that he required nothing more to support her than a sufficient stock of clothes for her and a small sum from her stepfather to cover the increased expenses of two people rather than one in Port Jackson.

He specified the sum the Tylers might contribute as £200, 'or if great inconvenience will result from advancing it, I will say £150' but he assured Ann that he could see a fortune growing under him in the near future, to meet increasing expenses.[58]

He continued: 'I will write further tomorrow; but shall most anxiously expect thy answer at 26 Fleet St London on my arrival on Friday; and I trust thy presence immediately afterwards ...'

He asked Ann 'to keep this matter entirely secret', because while some sea captains had travelled with their wives before, 'I do not exactly know how my great friends might like it'.[59]

To raise the money he needed to support a wife, Matthew had decided to sell the 300 acres Governor Hunter had granted him at Banks Town. William Bowles, who had gone to school with Matthew and was now a lawyer in Boston, had agreed to buy the land for £300, less legal and conveyancing fees. He made his first payment of £100 in April 1801, with the balance, with interest, due in May 1802.[60] Matthew wrote once more to his father in Donington to say he had undergone a change of heart and now planned to marry quickly and take his new bride to Sydney on His Majesty's ship.[61]

On 15 April, Matthew requested from the Admiralty a week's extension of his leave, saying he could spend it more usefully in London than on the ship. That evening, however, he climbed into the overnight coach for Lincolnshire, and in the swaying carriage travelled the night and the next day through the green English countryside, reaching Spilsby on the night of 16 April. With Hannah Franklin, his stepmother's sister, he travelled on to Partney. Here, in the Tyler cottage, he no doubt explained his plans to Ann, convincing her that she could travel with him to Sydney and stay by his side there and then be looked after by friends while he explored the coastline in the *Investigator*. He outlined his financial prospects to Ann's parents, as was expected, and Mr Tyler reciprocated with the promise of a sum of money for Ann.

On Friday, 17 April 1801, a bright spring day, with the bells ringing merrily, Matthew Flinders married Ann Chappelle at the 600-year-old Church of St Nicholas in Partney. Ann's stepfather, William, read the wedding service himself. Hannah Franklin and Mary Hudson signed the marriage register as witnesses.

Belle Tyler was then sixteen and 'a girl of strong sense', according to Matthew, 'somewhat eccentric, but of an amiable disposition'.[62] She wrote of the joyful excitement of the wedding, for which she had 'assisted with delight in preparing the home made brides-cake',[63] and said there was never a man who looked happier than Matthew, very much in love, and putting everyone at ease distributing gifts to the bridesmaids. He gave Belle a little inkstand, and pretended to tell the fortunes of the guests by reading their palms.[64]

Matthew's father, so often the voice of calm reason, had his fears about the union when the new Mr and Mrs Flinders arrived on his doorstep the day after the wedding. Apparently he had not yet received Matthew's note about his plan to marry quickly.

With concern I note that my Son Matthew came upon us suddenly & unexpectedly with a Wife on Sat. Apr. 18, & left us the next day. It is a Miss Chapple of Partney. We had known of the acquaintance, but had no Idea of Marriage taking place until the Completion of his ensuing Voyage. I wish he may not repent this hasty step.[65]

Chapter 15

*'I go, my beloved, to gather riches and laurels with which to adorn
thee; rejoice at the opportunity which fortuitous circumstances give me
to do it. Wilt thou not feel a pride in thy M?'*
MATTHEW FLINDERS, FAREWELLING HIS NEW BRIDE, ANN, AS HE SET SAIL
FOR SYDNEY WITHOUT HER.[1]

MATTHEW HAD BEEN SAILING around the world for almost a decade, roaring down through mighty whitecaps around Van Diemen's land, gliding past one Pacific paradise after another, basking in the Jamaican sun and visiting the towering volcanic peaks of the Atlantic. But no journey he had ever undertaken had made him quite so giddy as he felt sitting in a post-chaise carriage with his new wife, Ann, and young Belle Tyler on the road from Donington to London. Belle was travelling with them to farewell the happy couple in London.

Ann was so besotted that she seemed in a trance. She had just penned a note to her dear friend Elizabeth Franklin, the niece of Matthew's stepmother:

> Thou wilt be much surprised to hear of this sudden affair;
> indeed I scarce believe it myself, tho' I have this very
> morning given my hand at the altar to him I have ever
> highly esteemed, and it afford me no small pleasure than I
> am now part, tho' a distant one, of thy family, my Betsy.[2]

Ann admitted that the suddenness of her marriage meant that her mind was all over the place, and she was 'scarce able to coin one sentence or to write intelligibly'. Matthew was chuffed, not just with the fact that Ann had accepted the rushed nuptials and was now heading to Sydney with him, but that they had been able to secure a post-chaise, which he said was the cheapest fare for the three travellers, 'and by far the pleasantest' mode of transport.[3] Matthew also had a wad of banknotes stuffed into his shoe, a wedding gift from Reverend Tyler to help meet Ann's living expenses in Sydney.[4]

The trio spent Sunday night, 19 April 1801, in the Cambridgeshire town of Huntingdon, and on the 20th reached London, where they stayed with Ann's maternal aunt. Because of the haste of the wedding, Ann had few good clothes to take on the *Investigator*, so her kindly relative helped her choose many pretty dresses at the London stores, including one standout that they both thought would be a big hit when the new Mr and Mrs Flinders dined with Governor King in Sydney.[5]

Matthew's success in winning Ann's hand at such short notice kept him in the grips of euphoria, but it also made him complacent and fed his growing arrogance. The morning after their arrival in London, Matthew appeared before Joseph Banks again, his face all business. He made no mention of his marriage, just in case taking his wife on one of the government ships without permission might cause problems. He thought he could slip Ann past any checks.[6]

Ann later wrote to a friend:

I don't admire want of firmness in a man. I love courage and determination in the male character. Forgive me, dear Fanny, but insipids I never did like ...[7]

Matthew was certainly not insipid, but he could be impulsive and careless.

The following Monday, Matthew and Ann boarded the *Investigator* and moved into the commander's cabin, spending the next few days shopping and sightseeing on both sides of the Thames, Ann dressed in the finest of London fashion, Matthew confident and handsome in his commander's uniform, his wide black bicorne hat cocked 'slightly over one eye – his sword by his side'.[8] The big city, with its sex workers and salty-tongued sailors milling about the docks, was a shock for Ann, who had grown up in the genteel home of a country reverend.

Matthew kept Banks informed about every aspect of the voyage, and he had made more changes to the layout of the vessel, insisting they needed to carry more fresh water.[9] Banks organised the East India Company to throw in £1200 of 'table money' for the voyage, telling Matthew it was 'to Encourage the men of Science to discover such things as will be useful to the Commerce of India & to find new passages'.[10]

Matthew had become quite full of himself, telling Banks:

As particular circumstances may make it necessary to sail round the island twice, I beg leave to mention the propriety of leaving me at liberty to do this. My greatest ambition is, to make such a minute investigation of this extensive and very interesting country, that no person shall have occasion to come after me to make further discoveries.[11]

Supplies were being loaded constantly, including an interesting assortment of items considered trifles to serve as presents to Indigenous peoples in the hope of cultivating friendly relations. The list included 500 pocket-knives, 500 looking-glasses, 100 combs, 100 red caps, 100 small blankets, 100 yards of coloured linen, 1000 needles, 100 shoemakers' knives, 300 pairs of scissors, 100 hammers, fifty axes, 300 hatchets, a number of medals with King George's head imprinted upon them, and some new copper coins.[12]

Reverend Tyler came to the city to preach a sermon for the London Missionary Society and to take Belle home. Before Belle left, Matthew took her out in a boat, and from alongside the ship had her hoisted on board in a chair. She had dinner with Matthew and Ann, and later walked the decks with Samuel Flinders, the second lieutenant for the voyage, and her fifteen-year-old relative

John Franklin, who would sail on the *Investigator* as one of six midshipmen as his celebrated career of exploration began.

IF MATTHEW THOUGHT he was going to keep Ann's presence on the voyage a secret, he was in for a rude shock. Lincolnshire newspapers published articles about his marriage, and the news was gradually repeated in other journals around Britain.[13]

Rumours abounded that John Jervis, who was about to succeed Earl Spencer as First Lord of the Admiralty, had paid an unannounced visit to the *Investigator*. Matthew did not have a watchman assigned to the vessel, nor a berthing party to welcome Jervis, after whom a spectacular bay south of Sydney had already been named.

According to the rumours, Jervis entered Matthew's cabin unannounced and found Ann perched on her new husband's knee. When Joseph Banks heard about Matthew's marriage, he was furious over what he saw as deceptive and disrespectful behaviour, especially after all the favours Banks had called in to win Matthew the assignment. He fired a warning shot across Matthew's bow:

Dear Sir,

I have just time to tell you that the news of your marriage which was publishd in the Lincoln papers has reachd the Lords of the Admiralty here. [I] heard also that Mrs. Flinders is on board the Investigator & that you have some thought of Carrying her to sea with you, this I was very sorry to hear & if that is the Case I beg to give you my advice by no means to venture a measure so Contrary to the regulations & the discipline of the navy for I am convinced by the Language I have heard that their Lordships will if they hear of her being in New S. Wales immediately order you to be superceded whatever may be the Consequences for all likely have order Mr Grant to finish the survey.[14]

Matthew was shattered by this development. So was Ann. They had rushed into their nuptials, shocking their families with their haste, because Matthew had to leave for Sydney as quickly as possible and wanted Ann beside him. Now, just weeks after their wedding, their union was being torn in two by the British government.

Banks had changed his tune since his days as one of the most notorious ladies' men in Britain. In 1772 he had withdrawn at the last moment from Cook's second voyage to the Pacific, angry that the Royal Navy had questioned changes he had made to the *Resolution* to suit his botanical and scientific requirements. Cook sailed without him, but when he reached Madeira he made a startling discovery – a strange character named Mr Burnett, who had been awaiting the arrival of the rich botanist for three months.

'At first [Mr Burnett] said he came here for the recovery of his health,' Cook wrote, 'but afterwards said his intention was to go out with Mr Banks ... at last when he heard that Mr Banks did not go, he took the very first opportunity to get off the Island, he was about 30 Years of age – Every part of Mr Burnetts behaviour and every action tended to prove that he was a Woman.'[15]

Now, in 1801, Banks had his reputation as England's expert on all matters scientific to protect, and he was not about to have this young upstart tarnish his good name by taking a woman on one of the King's ships.

Matthew was crestfallen and humiliated. The promises he had made to his new bride seemed hollow and cheap, and yet he had no intention of now staying in Britain with her. He believed their future together hung on the success of what he could achieve aboard the *Investigator*.

Matthew changed tack with Banks, coming clean and hoping that his status as a naval officer and explorer who had already added considerably to the map of the world might win him some special dispensation as a newly married man:

It is true that I had an intention of taking Mrs Flinders to Port Jackson, to remain there until I should have completed the purpose of the voyage, and to have then brought her home again in the ship; and I trust, that the service would not have suffered in the least by such a step. The Admiralty have most probably conceived, that I intended to keep her on board during the voyage, but this was far from my intention …

If their Lordships sentiments should continue the same, whatever may be my disappointment, I shall give up the wife for the voyage of discovery; and I would beg of you, Sir Joseph, to be assured, that even this circumstance will not damp the ardor I feel to accomplish the important purpose of the present voyage; and in a way that shall preclude the necessity of any one following after one to explore.[16]

Banks wrote an immediate reply, but it did not reach Matthew's hand until a week later. By then Matthew's fate had been sealed by a pair of catastrophes, which Banks and the Admiralty blamed on his lack of focus.

Just before the *Investigator* sailed from the Nore, three crewmen, members of a work assignment Matthew had sent to another ship, the *Advice*, deserted. By 27 May, the ship had moved out of the Nore on its way to Spithead, and the next day the pilot was discharged at the Downs. Matthew was below deck with Ann as his ship sailed slowly between Folkestone and Dungeness, on the Kent coast. He thought other officers, including Samuel, were controlling the passage of the *Investigator* adequately. But there had been confusion over their orders and each man's role. To Matthew's horror, he felt the ship lifting onto a sandbank known as 'the Roar' at Hythe Bay.

Matthew bolted onto the quarterdeck and 'the sails were immediately thrown aback'. Boats were lowered to give depth soundings around the stranded ship, and with a smooth sea and rising tide he was able to shift the *Investigator* without 'any apparent injury'.[17] The only damage was to his pride and reputation. It was a relatively minor incident, but Matthew still reported it for the benefit of other crews. The Admiralty would not let him forget the mistake.

After his mishap, Matthew sailed back to the Downs, but trouble followed him. At the Nore, Matthew had been asked to take on board a carpenter from HMS *Trent* for transportation to Portsmouth. The carpenter had been identified as a deserter but did not arrive under guard. When the *Investigator* returned to the anchorage at the Downs, the prisoner absconded in a boat.

Matthew sailed south again, dodged the sandbank and finally reached Spithead on 2 June. HMS *Buffalo* had arrived there from Port Jackson six days earlier, bringing home Governor Hunter and Matthew's letter for George Bass, who had left England for Sydney months earlier. Also on board the *Buffalo* was master's mate John Thistle, Matthew's old shipmate from the *Reliance*, who had sailed with Bass on the whaleboat voyage to Western Port and with Matthew on both expeditions aboard the *Norfolk*.

Even though he had just arrived back in England after an absence of six years, Thistle signed on to join the crew of the *Investigator* eleven days later. When Thistle was on shore, he had his fortune told. As Thistle swallowed hard, the old man reading his palm told him that he was about to go on a long voyage but wouldn't come back.[18]

THE TENSION BETWEEN Matthew and Ann must have been severe. Having been married for just a few weeks, they now feared they would be parted for years. Matthew knew that his recent calamities had ended all hopes of Ann sailing with him. Banks had promised to lay his case before the Admiralty, but Matthew wrote to him from Spithead on 3 June to say he was now 'afraid to risk their Lordships ill opinion, and Mrs. F will return to her friends immediately that our sailing orders arrive'.

> The advanced state of the season makes me excessively anxious to be off. I fear that a little longer delay will lose us a summer, and lengthen our voyage at least six months; besides that the French are gaining time upon us ...[19]

Banks replied almost immediately to warn Matthew one last time about his command. He had been entrusted with one of the king's ships, crews and a responsibility for exploration that few commanders had ever received. On 5 June, Matthew fired a salute of nineteen guns from the ship in honour of the king's birthday,[20] but Banks told him that he was the one in the firing line:

I yesterday went to the Admiralty to enquire about the *Investigator* & was indeed much mortified to learn there that you had been on shore in Hythe Bay & I was still more mortified to hear that several of your Men had deserted & that you had lost a Prisoner entrusted to your charge who got away at a time when the Quarter Deck was in charge of a Midshipman.[21]

Embarrassed but defiant, Matthew replied immediately:

The chart of the Channel supplied me by Arrowsmith, published by I. H. Moore, was the principal cause of the ship having touched the ground ... Finding so material a thing as a sand 3 or 4 miles from the shore, unlaid down in the chart, I thought it a duty encumbent upon me to endeavour to prevent the like accidents from happening to others by stating the circumstance to the admiralty and giving the most exact bearings from the shoal ... It would have been very easy for me to have suppressed every part of the circumstance, and thus to have escaped the blame which seems to attach to me, instead of some share of praise for my good intentions. I hope it will not be thought presumption in me to say, that no blame ought to be attributed to me.[22]

Matthew also objected to the Admiralty holding him responsible for the desertion of the three men who took off while assigned to another vessel.

On 10 June, the *Investigator* was towed into Portsmouth Harbour and had its copper bottom inspected for damage from the Hythe Bay incident. Fortunately there was none, otherwise Matthew might have been relieved of his command. Three days later, some of the scientific men Banks had appointed to the voyage came on board. The voyage would involve six scientists and four servants.

Ann met the naturalist Robert Brown,[23] the Austrian botanical artist Ferdinand Bauer[24] and the nineteen-year-old landscape painter William Westall.[25] The forty-year-old Bauer, in particular, was kind and attentive to Ann, acutely aware of the stress she was under. As a talented painter of flowers herself, she viewed his work with a great deal of admiration. Westall's brother Richard came on board too, to meet Matthew and Ann. Matthew explained to him that being an explorer was often a thankless task, because in his naval profession 'distinction was scarce ever to be obtaind [sic] but by the destruction of our fellow creatures' in war.[26]

The 27-year-old Brown was a short, thin and dapper man who had served as ensign and assistant surgeon of a Scottish regiment, the Fife Fencibles. He had been pressing for a botanical assignment in New Holland for three years, and had applied to Banks in 1798 when he heard that Mungo Park had withdrawn from the proposed expedition into the interior of New Holland. That expedition was scrapped, but Banks approved Brown's place on the *Investigator*, and Brown would eventually publish *Flora of New Holland*.[27]

Brown would also make a detailed study of Aboriginal words for Banks, whose interest in the anthropological study of languages would soon be stoked even further by the British Museum's acquisition of the Rosetta Stone from the vanquished French Army along the Nile.

The three other scientists on Matthew's ship were John Allen, an experienced Derbyshire miner who was appointed mineralogist for the voyage; the astronomer John Crosley,[28] an assistant at the Royal Observatory, Greenwich; and Robert Brown's assistant, the Kew-trained gardener Peter Good, who was then working at Wemyss Castle in Scotland.

Banks had employed Good to help Matthew's botanist friend Christopher Smith transport a shipment of English plants to the East India Company's garden at Calcutta, and Good had returned from there with Indian plants for Kew. William Aiton, the head

Botanist Robert Brown, as painted by Henry William Pickersgill in later life. *nla.obj-136208561*

gardener at Kew, had presented Good with Banks' offer of a posting aboard the *Investigator*, and Good wrote to tell Aiton: 'I accept with cheerfulness ... it shall be the business of my life to merit so particular a distinction.'[29]

Crosley had previously sailed as the astronomer on the *Providence*, as part of George Vancouver's expedition before the ship, which Matthew knew so well, was wrecked on a coral reef south of Okinawa, Japan, in 1797.

Banks had his scientists sign an official document at his home in Soho Square setting out the conditions of their employment, which included obedience to the ship's commander and cooperation with one another. All journals, sketches and collections were to be surrendered at the end of the voyage, but any monies derived from publication by the Admiralty would be divided between them and Matthew Flinders. Brown and Crosley were to receive £420 a year for the voyage, Bauer and Westall £315, and Peter Good and John Allen £105.[30]

Despite the bonhomie on the ship, Matthew knew that, with the *Investigator* making the last leg of its journey out of England, he was still on thin ice.

Banks wrote again, sternly, to say: 'I have at last had a long conversation with Mr Nepean on the subject of the charges brought against you & have pleaded your cause I hope effectively.'[31] On 19 June, though, Matthew was ordered to appear the next day before the solicitor of the Admiralty in London, 120 kilometres away.[32]

Uncertain of who or what he would face when he arrived at the grand Admiralty House opposite Whitehall, Matthew climbed into a post-chaise with Ann and her luggage. She had said her goodbyes to the crew and her new friends on the *Investigator*, knowing there was no turning back. In the last few weeks, Ann had suffered a great deal with nervous tension that exacerbated her seasickness, and years later Belle would refer to the men of the Admiralty who had banished her stepsister from the ship as 'those savage old Lords'.[33]

Matthew entered Admiralty House, only to discover that he was not facing official censure. His presence was required as

Astronomer John Crosley was an assistant at the Royal Observatory, Greenwich. *Wellcome Images V0001360*

Moody artist William Westall created this self portrait. *nla.obj-135716302*

a witness in a court case involving a deserter. He spent a week hoping to see Secretary Nepean and plead his case for Ann to sail with him, but Nepean begged off with ill health. Matthew returned alone and forlorn to his ship at Spithead. Friends took Ann to stay with them at Battersea until her stepfather could come down to take her back to Partney.

The newlyweds wrote to each other daily, declaring their love and promising to see each other soon, but it was a traumatic time. Matthew told Ann that he felt like 'one half of a pair of scissors without its fellow'. There was no one with whom he could share such intimate thoughts, not even his brother Samuel.[34] Trim only purred and rubbed his whiskers against Matthew's feet, in the hope of getting some meat.

Matthew assured Ann that all the turmoil of their separation would be worth it in the end: 'Rest confident, my dear, of the ardent and unalterable affection of thy own MF; he does love thee beyond everything.'[35]

On the *Investigator*, cattle, sheep, pigs, goats and poultry were taken onboard and penned, awaiting slaughter on the forward deck. There were also greyhounds and other hunting dogs, which Banks' scientists would use to bring down animals for study in New Holland. As Matthew brooded over the banning of his wife, the crew became restless and unruly while awaiting their final orders. Matthew wrote that while he constantly had the 'people exercised ... whenever the weather would permit; and endeavoured to bring the ships company under good order and government ... some of the heedless occasionally fell under the lash'.[36]

Matthew had a dozen men flogged while they waited to leave England. Tom White and Tom Smith were flogged twice, Smith receiving twenty-four lashes on one occasion for riotous behaviour and fighting.[37] Matthew also sent a boatload of men to watch the execution of one of the mutineers from the ship *Hermione*, just in case any of them entertained thoughts of an overthrow.

Knowing that life could be dangerously short on the sea, Matthew also tried to mend bridges with his father, unsure when – or if – he would see him again. Flinders Sr wrote in his diary that he had always 'foreboded' that his son would not be allowed to take his new wife to Sydney, and Matthew should have consulted 'Sir JB' before the marriage, given the importance of the naval regulations.

'If he had thought proper to have advised with me, I should have recommended him to have done so,' Flinders Sr wrote, 'but I am seldom consulted by any of my young folks, except on the head of raising Money for them.' He knew Ann was unwell and prayed that the newlyweds would survive the stress they were facing.[38]

On 10 July, as final preparations were being made to sail, Matthew wrote a heartfelt letter to his father:

I wish, my dear father, to say a little upon a former letter of yours, dated May 11, in which you seem to think that my conduct has not been altogether that of a dutiful or at least of an affectionate son. That you should think so, occasioned much uneasiness both to me and to my dear wife, for I find her so much superior in penetration and judgment to the generality of women, that there are but few occurrences upon which I do not consult her. Your succeeding letter was an affectionate, and acceptable one; and for which, my dear father, I give you my warmest thanks. It took away much uneasiness, but it served also to confirm me in a melancholy truth, that my fathers increase of years, and decline of constitution withdrew from him part of that equality of temper and mildness of disposition for which he was so justly noted. Why, my dear father, continue to fatigue yourself to the injury of your health and disposition? Your family have no right to expect it; and I think I am sure that they are very far from wishing it. If your income is not sufficient to live upon, why not pursue the plan mentioned in your letter, of purchasing an annuity? ... Now I would not only be your son, but I would be your dearest friend, would you admit me to that station ... be assured that whatever or whenever it is in my power to contribute to your happiness in any way

I shall do it … If you think me selfish, as I fear is somewhat the case, or devoid of gratitude, you do indeed wrong me.[39]

Matthew said he was 'particularly sorry' that his father was 'dissatisfied' with the marriage to Ann. ''Tis true I did not formally ask your consent to it,' he wrote, 'but when the subject [of] Miss Chappelle had been previously talked of … you made not one objection to her.'

From seeing your letters, she feels as under your displeasure, which adds to her present uneasiness. The time of my marriage, you say my dear father, is the worst part of it. It is certain I should not have married but with the idea of taking her with me. Others had been allowed this privilege, and I could not foresee that I should have been denied it. Yet I am by no means sorry for having married. If you knew her worth, you could not regret it. I am happy to add here, that her health is so far improved at this time, that she will be able to accompany Mr Tyler into the country in the beginning of next week. Her letter today from Battersea says that she is able to run upstairs.[40]

Two days later, Matthew wrote to Ann asking her to come down to Portsmouth or Gosport to see him one last time before he sailed. He could organise a feather bed for her on the *Investigator* or a snug hotel room on shore.[41] By the time his letter arrived, though, Ann was heading north with Reverend Tyler.

Banks sent his 'sincere good wishes' for Matthew's future prosperity '& with a firm belief that you will in your future conduct do credit to yourself as an able Navigator & to me as having recommended you …'[42]

Despite the cooling of their relationship, Banks always put the advancement of science at the forefront of everything he did, and he provided Matthew with a vast library of books and charts for the voyage, including the historic Dutch instructions to Abel Tasman, a twenty-volume set of *Encyclopaedia Britannica*, and

Captain Cook's book recalling the *Endeavour* voyage to Australia, in which Matthew would pen notes and correct some of Cook's coordinates.[43]

For his part, Matthew told Banks that he wanted to add his voice to the many who found the president of the Royal Society a 'friend and patron'.[44]

On 17 July, Matthew finally received his French passport, along with his orders to sail and thirteen pages of instructions from the Lords Commissioners for the voyage. These were that he make a 'complete examination & survey' of the coast of New Holland, 'on the Eastern side of which his Majesty's Colony of New South Wales is situated'. He was to stop at the Cape of Good Hope on the way to take on board 'such supplies of Water, Live Stock, Seeds etc as you may judge necessary & may be able to procure ... Having so done, you are to make the best of your way to the Coast of New Holland running down the said Coast from 130 [degrees] of East Longitude to Basses Straits ...'[45]

Matthew was to use his 'best endeavours to discover such Harbours as may be in those parts; and in case you should discover any creek or opening likely to lead to an inland Sea or Strait, you are at liberty either to examine it or not, as you shall judge it most expedient'.[46] And, 'when the season shall render it necessary', he was to sail on to Sydney Cove to refresh his crew and refit the ship and consult with Governor King.

After Matthew had examined 'the whole of the Coast from Basses Straits to King Geo the 3ds Harbour' – now King George's Sound in Western Australia – he was to explore the north-west coast of New Holland, where it was 'probable that valuable harbours' would be discovered. Then he was to carefully examine 'the Gulph of Carpentaria, & the parts to the Westward thereof'.[47]

Matthew was also to make a 'carefull investigation & accurate survey of Torres Straits' and the area around 'Timer [Timor] to mark shoal and make it safer for East India Company ships in case that passage should in future times be frequented'.

He was then 'to examine very carefully the East Coast of New Holland seen by Capt Cook', from 'Cape Flattery' (near modern-day Cooktown) to 'the Bay of Inlets' (around modern-day Mackay). He had permission to refresh his crew and give the artists on board more variety by visiting the 'Fegees [Fiji], or some other of the Islands in the South Seas'. At all times Matthew was to be most diligent in 'discovering any thing useful to the Commerce or Manufactures of Great Britain'.[48]

The next morning, 18 July 1801, with rain falling, the ship was loaded with four tons of water and three tons of beer, as well as 360 pounds (164 kilograms) of fresh beef, two barrels of peas, and some oatmeal and cheese to round out their provisions. Four men, including the oft-flogged Thomas White, were left behind after being deemed too much trouble, and the ship sailed with eighty-eight men on board,[49] including fifteen marines in their white trousers and vivid red coats. At noon the ship passed the St Helens Road anchorage, heading for the Atlantic.

Matthew checked his French passport. He did not speak French, but Robert Brown translated it for him:

> The Minister for the Navy and the Colonies directs ... that all Commanders of Warships of the Republic, all his agents in all the French Colonies, all the Commanders of Ships bearing Letters of marque, and to all others to whom it concerns, to let pass freely, and without obstruction the said Corvette *Investigator* her Officers, crew and effects ...[50]

Once again, Matthew was sailing off the map with only a vague idea of the enormous distances involved, and of whether New Holland was a single landmass. Nor had anyone any real idea of the number and size of the island groups, bays and inlets that the mysterious coastline contained.

Banks had told Matthew that he expected the voyage to take four years, but Matthew wanted to complete the circumnavigation of what he called *Terra Australis* more quickly. He wanted to sail home to his new wife as soon as possible.

Chapter 16

'The red coats and white crossed belts were greatly admired ... and the drum, but particularly the fife, excited their astonishment ... when they saw these beautiful red-and-white men, with their bright muskets drawn up in a line, they absolutely screamed with delight.'

MATTHEW FLINDERS, ON THE REACTIONS OF INDIGENOUS MEN
WATCHING HIS MARINES EXERCISING.[1]

MATTHEW STEERED THE *Investigator* towards his first stop: Madeira, some 2200 kilometres away in the North Atlantic. On 21 July 1801, he fell in with a convoy of four warships under Vice Admiral Sir Andrew Mitchell, who had recently led the naval part of the Anglo-Russian invasion of Holland.[2]

Off St Alban's Head on the coast of Dorset, Matthew found there were irregularities with his compass readings, as there had also been at different times in New South Wales and Van Diemen's Land. The readings could be out by as much as four degrees, depending on where on the ship the compass was placed. Matthew believed the variances were caused by iron on the ship and the Earth's magnetic field, with readings affected depending on the hemisphere in which they were sailing and their proximity to the equator.

His compass wasn't the only thing that was proving a problem. The *Investigator* had essentially been invalided out of war service, and on 31 July, just two weeks into the voyage, Matthew 'had the mortification to find the ship beginning to leak so soon as the channel was cleared'. It was letting in three inches of water per hour.

The *Investigator* reached Madeira on 3 August 1801 and anchored on the southern side of the island at the town of Funchal, which Banks and Cook had visited thirty-three years earlier. Funchal was a town of pristine whitewashed cottages surrounded by lush, vine-wreathed hills overlooking a deep blue sea. But Matthew was more interested in the state of his leaky boat than the magnificent vistas. His four carpenters worked around the clock over four days caulking the leaks with oakum.

Matthew first sent his brother Samuel to present his respects to the Portuguese governor, then visited the vice-regal himself and gained permission for his scientists to go about the island studying and collecting plants. The British consul had an ox killed for fresh meat, and the crew also loaded more water and wine, as well as fruit and onions. Matthew wrote to his father to say he and Samuel had enjoyed a 'pleasant and quick passage to that Island'.[3] The letter was a comfort to Flinders Sr when it finally arrived in Donington almost four months later. Matthew's father was unwell: he had lost a considerable amount of weight and could hardly stomach food.

'I am very much weaker than I ever knew myself,' he noted, 'and my flesh amazingly gone.' For some months past his health had 'been considerably on the decline', and he was now 'under considerable apprehensions' about how long he would have to 'compleat my earthly course'.[4]

Very early in the *Investigator*'s voyage, Matthew began practising a routine 'made known', he said, 'by the great captain Cook'.[5] It was in the standing orders of the ship that on every fine day, the deck below and the cockpits should be cleared, washed, aired and sprinkled with vinegar. Care was taken to prevent the crew from sleeping upon deck or lying down in wet clothes, and once every fortnight or three weeks, their beds and the contents of their chests and bags were opened out and exposed to the sun and air. On the Sunday and Thursday mornings, the ship's company was mustered, and every man appeared cleanshaven and

H.M.Sloop INVESTIGATOR 1802

Geoffrey Ingleton's 1937 etching of the *Investigator*.
State Library of NSW FL1540074

dressed. When the evenings were fine, Matthew's drummer and fife player provided tunes for dancing.

Matthew decreed that, within the tropics, the crew were to be given lime juice and sugar to ward off scurvy, and, in the colder climes, sauerkraut and vinegar. The essence of malt was reserved for the final passage to New Holland. Oatmeal was boiled for breakfast four days a week, and after the cheese ran out, boiled rice was issued on the other three days. Pea soup was prepared for dinner four days a week, and at other times there was broth full of onions and pepper as a 'comfortable addition' to the salt meat.

While the men all had full bellies, the *Investigator* was still in poor shape. The recent repairs were tested to the limit by a raging storm just out of Madeira. The foul weather continued as they neared the equator, pushing the *Investigator* much further east towards the coast of Africa than Matthew wanted, a problem he knew had plagued the voyages of Cook and the Comte de Lapérouse.

On 21 August 1801, Matthew recorded in his log that there were 'heavy claps of thunder with vivid lightening. Two electric balls of fire fell or struck near us.' The men spread the ship's awning and gathered almost a ton of water for washing.[6]

Near the equator, the ship was now taking on five inches of water an hour, as tacking to the west under high winds had worked the oakum out of the seams. This, Matthew fretted, 'indicated a degree of weakness which, in a ship destined to encounter every hazard, could not be contemplated without uneasiness'. Given the naval commitments to the war with France, no better ship could be spared from the service, and Matthew now deeply regretted that his anxiety to 'complete the investigation of the coasts of Terra Australis' did not admit of refusing the one offered.[7]

The leaky ship tottered over the equator just after dawn on 8 September. Although he held fears for his vessel, Matthew was determined to preserve the physical and mental health of his crew, and 'to promote active amusements amongst them, so long as it did not interfere with the duties of the ship'. Therefore, he wrote, 'the ancient ceremonies used on this occasion, were allowed to be performed'. There was an officer dressed as Neptune, the obligatory dunking of the newcomers to the equator and plenty of music, dancing and drinking. The sailors, Matthew said, concluded their special day 'with merriment'.[8] Peter Good noted that the men 'got drunk and turbulent at night. Some were insolent to the officers.'[9]

Matthew later admonished his crew, but took the blame on himself for having let them have so much liquor. He said they had abused that indulgence, and must not expect any further leniency. Good said the crew seemed well satisfied with his 'humane' treatment of them.[10]

Trim also provided plenty of belly laughs. He had taken a fancy to nautical astronomy, and keenly watched John Crosley observing the planets and stars. When Crosley or an officer took lunar or other observations, the black and white cat would place himself by the timekeeper and consider the motion of the hands, and the workings of the instrument, with 'much earnest attention'. Trim would try to touch the second hand, listen to the ticking, and walk all around the piece to assure himself whether or not it might not be a living animal.[11]

But the laughter at Trim's antics stopped as huge swells from the south battered the *Investigator* and made all on board fear that the timbers of the ship would break. The water started to gush in, and Matthew decided to shift the weight on the vessel to counter 'the tremulous motion caused by every blow of the sea'.[12]

Two eighteen-pound carronades – 'stern chasers' – were taken off the upper deck and put into the hold, and the spare rudder and every exterior weight possible were concentrated together to give the ship more ballast. After this was done the leaks were reduced, but still Matthew worried about how his ship would handle the Roaring Forties and the great seas of the Southern Ocean.

JUST AFTER DAWN ON FRIDAY, 16 October 1801, three months after leaving England, Matthew saw the high land of the Cape of Good Hope through the haze. He decided to bypass Table Bay, and Cape Town sheltering below it, because heavy gales invaded those waters at summertime. Instead, at 1 p.m., the ship hauled round the rocks which lie off the Cape Point and Matthew steered into False Bay.

The men were spellbound by two whales engaged in 'a furious combat' with a pair of Atlantic thresher sharks. Matthew saw one of the sinewy sharks raising its tail high out of the water to beat its adversary, and throwing 'the whole of his vast body several feet above the surface, apparently to fall upon him with greater force. Their struggles covered the sea with foam for many fathoms round.'[13]

Only a few days earlier, Matthew had cracked the whip on board when two of his men were caught fighting; both were flogged. Despite such occasional incidents of ill-discipline on board, the mood of the men was buoyant and they were in robust health. Matthew steered his ship into False Bay's inner cove, Simon's Bay, where seven British warships were anchored.

Matthew and Samuel both wrote to their father to say they had arrived safely at the Cape. There had been 'great improvement' in Samuel, Matthew said, and instead of being a 'burthen' to him, 'I now have great hope of his proving himself a very useful assistant to me in the voyage; and I believe, that opportunities will not be wanting to enable me to bring him into some repute, which would open a door for him to fame, if not to fortune also'. Matthew did not say that the brothers frequently argued, but he did express his hopes for their father:

> Long before our return I hope to hear of your having retired from business to a snug situation suited to your taste ... I have great hope of being able to assist you in accomplishing this desirable purpose soon after our return, if your circumstances should not by that time have enabled you to do it.[14]

The news lifted the spirits of Flinders Sr. He had just finished updating his will, and had recently received what he called 'a distressing letter' advising him that, in the town of St Ives, Susanna had married a 'journeyman draper', George Pearson, a 'Calvinist Methodist' whose beliefs did not sit well with Flinders Sr's Anglican ways. He doubted Pearson had the ability to adequately support his little girl, who was now twenty-three and headstrong, much like her eldest brother. He had written Susanna 'a severe letter, tho' not more than deserved'.[15]

Matthew had his ship's tattered sails repaired on board the British warship HMS *Lancaster*, and he had the *Investigator* patched up again, reprovisioned and repainted. He and Crosley set up a makeshift tent observatory on the south shore of the bay, with Samuel as their assistant astronomer. Matthew wrote three letters to Ann, praying that her love for him would not wane:

> Thou must feast me with love when I return, to recompense me for all my anxieties; and Oh, write to me constantly, write me pages and volumes. Tell me the dress thou wearest, and at what time in the morning thou puttest on thy stockings; tell me thy dreams, – any thing: so do but talk to me, and of thyself.[16]

He wrote about the voyage too, how Crosley had been feeling ill since Madeira. Matthew had swapped four crewmen for more suitable people from the *Lancaster*, while one of his men, Nathaniel Bell, 'left the ship by his own application, finding as he said that he was unfit for the service; and I have gotten a fine young Irishman [Denis de Lacy] of about 17 in his lieu, a red hot volunteer for the service'.[17]

The scientists toured the Cape, studying the wide array of plants. Brown and Good climbed Table Mountain.

After eighteen days of rest and recreation, Matthew set sail out of False Bay at daybreak on 4 November 1801. He left Crosley behind due to his ill health, which meant Matthew and Samuel would have to perform Crosley's tasks for the rest of the voyage. Crosley kept his Earnshaw watch and Troughton reflecting circle, which Matthew considered a serious loss to the mission. He wrote to Banks saying that he hoped the Board of Longitude would recompense the brothers for the extra work.

He now set a course east for 8500 kilometres, heading to the south-western point of New Holland, which the Dutch called 'Leeuwin's Land' after the first known ship to have visited the area: the *Leeuwin*, or 'Lioness', in 1622.

Having made the passage between the Cape and New Holland three times before, Matthew had no desire to pit his leaky ship against the heavy gales he'd experienced on the *Providence* and the *Reliance*. Instead he chose to stay at a latitude of 37 degrees south, believing it 'sufficiently distant from the verge of the south-east trade to insure a continuance of western winds; and … far enough to the north to avoid the gales incident to high latitudes'.[18]

From the Cape of Good Hope to Amsterdam Island, halfway across the Indian Ocean, the winds were never so strong as to reduce the *Investigator* to close-reefed topsails, while the calms amounted to no more than seven hours in nineteen days. The ship averaged 260 kilometres a day, which Matthew thought quite remarkable, given the *Investigator* 'was a collier-built ship, and deeply laden'.[19]

Matthew didn't see Amsterdam Island because of fog, but

somewhere near it in the middle of the Indian Ocean he wrote to John Franklin's older brother Willingham, who was then at Oriel College, Oxford. Matthew told Willingham that success in the voyage would make him not 'unknown in the world'. He needed help writing his journal, though, and because 'authorship sits awkward upon me', he wanted Willingham as his 'literary man' to help him when he returned. He suggested that Willingham brush up on his writing by studying different authors in readiness for the task. 'A little mathematical knowledge will strengthen your style,' Matthew explained, 'and give it perspicuity.'[20]

Matthew added that Banks, his 'acquaintance in Soho Square', would introduce Willingham to 'many of the first philosophers and literati in the kingdom', and that Matthew himself, suitably elevated to celebrity, would then be able to give Willingham 'a lift into notice' too.[21]

It was a great relief to Matthew that frequent pumping of the ship was not now required, and the health of the men remained strong due to the 'antiseptics' issued – sauerkraut and vinegar. At half an hour before noon every day, each man also received a pint of strong wort, made by pouring boiling water upon the essence of malt. This was drunk upon deck, with half a biscuit. The men's allowance of grog would be issued half an hour after dinner.[22]

The men practised with their cannons and muskets, and Matthew conducted various maritime experiments across the vast, great, grey rolling ocean. He measured the temperature of both the air and seawater carted up from the deep. He studied sea breezes and their effect on the rise and fall of the barometer. He and John Thistle conducted experiments with every compass on the ship.

ON THE MORNING OF 6 December 1801, Matthew estimated that he was twenty-two leagues (about 120 kilometres) from the westernmost islands that lay off the south-west cape of New Holland. He was cautious as he approached, having only a vague copy of a chart originally drawn by the Frenchman Antoine Bruni d'Entrecasteaux nine years earlier and supplied by the hydrographical office at the Admiralty. There were no names applied on this chart, but D'Entrecasteaux had named the islands that were supposed to be in this region the Saint Alouarn Islands, after his countryman Louis de Saint Aloüarn, who had sailed past them in the storeship *Gros Ventre* ('Fat Belly') in 1772. Only two other navigators had recorded this part of New Holland: Dutchman Pieter Nuyts,[23] aboard the *Gulden Zeepaert* ('Golden Seahorse'), way back in 1627, and George Vancouver in 1791.

At 7 p.m., in the summer evening twilight, Matthew finally saw 'Landt van der Leeuwin' about fifty kilometres to the east.

Matthew's orders were to examine closely the coast of Britain's claimed territory of New South Wales from 130 degrees east longitude – roughly from the modern-day town of Ceduna, South Australia – to Bass Strait, and then, after arriving in Port Jackson for rest and a refit, he was to double back and explore east from Cape Leeuwin. To save time, Matthew elected to begin his survey from Cape Leeuwin and to sail slowly and cautiously eastward along the coast.

> [Although] this was not prescribed in my instructions to be made at this time … the difference of sailing along the coast at a distance, or in keeping near it and making a running survey, was likely to be so little, that I judged advisable to do all that circumstances would allow whilst the opportunity offered …[24]

Matthew also feared that the French expedition – the two ships under the command of Baudin – might beat him to important discoveries. The French ships *Géographe* and *Naturaliste* had left the Isle de France – modern-day Mauritius – in March 1801, bound for New Holland, and Matthew suspected Baudin was already busy claiming territory for his ruler Napoleon.

At 7.30 a.m. on 7 December, Matthew saw hills extending behind the Saint Alouarn Islands and named the area Cape Leeuwin – the first of 240 names he would assign to geographical features while in command of the *Investigator*.

'It is a sloping piece of land of about six hundred feet in elevation, and appeared to be rocky, with a slight covering of trees and shrubs,' Matthew wrote, 'but this cape will be best known from Mr. Westall's sketch.'[25]

Matthew wanted a safe harbour where the *Investigator*'s masts could be stripped, and the rigging and sails repaired. He knew the favourable reports Vancouver and the botanist Archibald Menzies had made to Banks about King George's Sound, so he sailed east for another 350 kilometres. Guided by Vancouver's chart, he was not worried about it being dark when he anchored there, near the 'mass of granite' that is Seal Island, at 11 p.m. on 8 December. Few Europeans had ever been here, and it was with great excitement that the men and their dogs hit the beach the next morning, killing a few seals on the island for fresh meat.

Matthew and his men climbed a hill on Seal Island to search for a cairn, a bottle and parchment that Vancouver had left to commemorate his visit a decade earlier, but they could find no trace. They noticed that several trees had been felled recently with an axe and saw, and later they found 'a spot of ground six or eight feet square, dug up and trimmed like a garden'. On it was a piece of sheet copper bearing the inscription 'August 27, 1800. Chr. Dixson – ship Elligood'. That solved the mystery of the felled trees, and the disappearance of Captain Vancouver's bottle. Matthew later learnt that the *Elligood* was a whaling ship that had since sailed to the Cape; its skipper and nine men had died from scurvy.

Matthew saw that King George's Sound had two coves: Oyster Harbour in the north and Princess Royal Harbour in the west. He and some of the scientific men took the ship's cutter to Point Possession, on the south side of the entrance to Princess Royal Harbour, where the city of Albany now stands. They 'had a good view of that extensive piece of water', Matthew recalled.

> Not far from thence stood a number of bark sheds, like the huts of the natives who live in the forests behind Port Jackson, and forming what might be called a small village; but it had been long deserted.[26]

William Westall's 'View from the south side of King George's sound'. *nla.obj-135751314*

Three days later, on 12 December, Matthew moved the *Investigator* into the more sheltered waters of Princess Royal Harbour, underneath what is now Mount Clarence. The scientists fanned out into the surrounding country, examining rocks and soil and collecting new varieties of plants and seeds to take home. Snapper and mullet were hauled up, and birds shot for the table. Oyster Harbour lived up to its name, with the men eating their fill of oysters and mussels.

Matthew had tents set up on shore. One was an observatory, in which his stargazing equipment was kept under the guard of marines under Samuel's supervision. The positions of various stars were noted, and a record kept of temperature and barometric readings and tidal variations. A gale blew away the covering over the observatory; the instruments were drenched but not damaged.

EVIDENCE OF THE COUNTRY being inhabited was everywhere, but it wasn't until 14 December that Robert Brown and other scientists at last met with some Menang warriors – of the Indigenous Noongar people – who seemed 'shy but not afraid'. One of the warriors they met 'was admired for his manly behaviour' and the men from the *Investigator* gave him a bird which had been shot, and a pocket-handkerchief, but 'these men did not seem to be desirous of communication with strangers and they very early made signs to our gentlemen to return from whence they came'.[27] The next morning, however, the crew were agreeably surprised by the appearance of two more warriors, and afterwards by others, on the side of the hill behind the tents.

> They approached with much caution, one coming first with poised spear, and making many gestures, accompanied with much vociferous parleying, in which he sometimes seemed to threaten us if we did not be gone, and at others to admit of our stay … On the 17th, one of our former visitors brought two strangers with him and after this time, they and others came almost every day, and frequently stopped a whole morning at the tents. We always made them presents of such things as seemed to be most agreeable, but they very rarely brought us anything in return; nor was it uncommon to find small mirrors, and other things left about the shore; so that at length our presents were discontinued.[28]

Matthew and his men spent Christmas in their new surrounds, sweltering in the summer heat. The officers and scientific staff ate Christmas dinner with Matthew, while the crew had a day off work. Peter Good said 'some got completely drunk'.[29]

By 30 December, the crew had taken on all the firewood and fresh water they could fit, and Matthew was satisfied with the repairs to the rigging and sails. Young William Westall had written home to his family from the Cape, telling them of the kindness Matthew had shown to him. Richard Westall later wrote a letter of thanks to Matthew, hoping it would find him sometime in Sydney. He said he hoped Matthew would 'acquire honour and celebrity by adding to the stock of Human knowledge', and that he would return from the voyage 'safe and prosperous'.[30]

There were still occasional problems with discipline. Matthew ordered thirty-six lashes for a sailor named William Donovan for repeated drunkenness and fighting following the raucous Christmas party,[31] the second time in a month Donovan was flogged. First Lieutenant Robert Fowler, who had sailed on the *Investigator* when it was still the *Xenophon* and probably thought he knew more about the ship than the young captain, had been difficult in the early stages of the voyage but gradually became more agreeable.[32]

Matthew wished he could say the same about Samuel, who had shown himself to be clever in mathematics and in nautical calculations, but already had a reputation for being 'idle, selfish and quarrelsome'.[33]

The distance between us has widened considerably. [Samuel] is satisfied with being as much inferior to other officers as I would have him superior to them ... John Franklin ... is capable of learning everything that we can shew him, and but for a little carelessness, I would not wish to have a son otherwise than he is.[34]

Menang tribesmen continued to come to the tents and Matthew ordered the party of marines on shore, to be exercised in their presence as a treat. The vivid red coats and white crossed belts of the marines appeared to delight the local people, resembling the red and white markings with which they adorned their own bodies. When the marine drummer began to beat out a tune, the Indigenous men were fascinated, and they were particularly excited by the fife. When the marines formed up in a line, their bright musket barrels gleaming in the sun, the Menang men cried out with apparent delight.[35]

Several of the local people mimicked the movements of the troops. One elderly man placed himself at the end of the rank, with a short staff in his hand, which he shouldered, presented and grounded in imitation of the redcoats. Matthew explained to them that the guns would make a huge noise so that the volleys 'did not excite much terror',[36] although Samuel Smith, a crewman from Manchester, recalled that for a moment or two after the gunfire the warriors moved about and hollered 'unmercifully'.[37] Smith also recalled that some of these Menang people 'rubbd their skin against ours, expecting some mark of white upon theirs, but finding their mistake they appeared surprised'.[38]

There were great differences in the languages of the people of Port Jackson and those around King George's Sound. Matthew found it difficult to pronounce their Indigenous words, and while the Menang could pronounce several English words perfectly, they, like the people on the east coast of New South Wales, had trouble with the letters 'f' and 's'. 'Finger' was pronounced 'bing-gah', 'ship' became 'yip', and of 'King George' they made 'Ken Jagger'.[39]

Matthew compiled a chart of some of their words, and compared these with terms used by the other groups, including the people he had met in Van Diemen's Land.

After four weeks at King George's Sound, and with a new year having begun, Matthew felt he had won the trust of the Menang, although they were not really friends and the British held all the power. He later noted that 'one of our best proportioned visitors' had stood patiently and calmly as Hugh Bell, the *Investigator*'s surgeon, made more than fifty anatomical measurements of his frame, from his full height (five feet and seven and a half inches) to the circumference of his head, to the length of his tibia, the distance from his 'inner ancle' to the tip of his heel, and the 'end of the great toe'.[40]

ON 3 JANUARY 1802, with the tents having been packed and the men preparing to sail, Matthew steered the *Investigator* back and forth across the sound, trawling with a net overboard. A variety of small fish were hauled in, and while most were of little use as food, they – together with shells, seaweed, coral and a pretty kind of seahorse – were sources of fascination for the scientists.

Plant boxes were brought on board, and Brown wrote to Joseph Banks to say that, in twenty-four days at the sound, he had collected nearly 500 plant species. Ferdinand Bauer had been 'indefatigable', he said, and had exerted infinite pains in making his drawings of them as intricate as possible. Peter Good had proved 'a most valuable assistant; a more active man in his department could hardly, I believe, have been met with'.[41] And 'Captain Flinders' had 'upon all occasions' given him every opportunity of collecting.[42]

Matthew left a bottle on Seal Island containing a parchment stating the arrival and departure times of the *Investigator*, and on 5 January a strong, wet westerly wind pushed the ship into the Southern Ocean.

Matthew now endeavoured to keep the *Investigator* so close to land that from the ship's deck he could see breaking waves on the low sandy shore for the next few days. With this method, there would be little error in the distances on his charts, and no river or inlet could escape being seen – or so he thought. Whenever possible, he landed the scientists and artists to inspect the coastline for its treasures. When safety forced him to stay further out to sea, Matthew was up a mast, hanging on for dear life in sunshine and rain, using his telescope and calling out measurements to Samuel or another of his assistants below.

At the same time, another crewman would be swinging the lead, using chains and a lead weight to measure the depth of the water, while other crewmen acted as lookouts for navigational hazards, swirling eddies or changes in the colour of the water. All the bearings and notes were laid down as soon as they were taken, and at night, by the flickering yellow glow of a swaying lantern, Matthew completed his rough chart for each day, along with his log and his astronomical observations. This meticulous record-keeping, he said, required 'constant attention and much labour but was absolutely necessary to obtaining that accuracy of which I was desirous'.[43]

Just before sunset on 8 January, near what is today the town of Esperance, Matthew sighted the westernmost island of an archipelago D'Entrecasteaux had named after his ship *Recherche* while searching for the missing Lapérouse. The maze-like collection of 105 islands, some almost 200 metres in height, stretch for 230 kilometres and are up to fifty kilometres off the shore. 'The French admiral had mostly skirted round the archipelago,' Matthew wrote, 'a sufficient reason for me to attempt passing through the middle.'[44]

Throughout the following day, Matthew guided the *Investigator* through the labyrinth of rocks and islands, but by early evening he saw breakers and rocks all around. There was no open water in which the ship could stay safely for the night. Then he breathed a heavy sigh as between the islands he saw a beach. He steered the *Investigator* through a narrow gap in the rocks to enter a small, sheltered inlet. Because of the 'critical circumstance under which this place was discovered', Matthew gave it the name 'Lucky Bay'.[45]

At the request of the scientists, the *Investigator* stayed in Lucky Bay for four days, allowing them to explore the countryside. There was an abundance of shrubs and plants that yielded a delightful harvest to the botanists, but John Thistle and some of the other officers were taken ill after eating the poisonous nuts of the zamia palm. On a rock on the side of a hill Matthew found a large nest, similar to those seen at King George's Sound. He presumed it belonged to a giant eagle, because in it were masses of hair from what looked like seals and 'the scaly feathers of pinguins', along with the bones of birds and small animals.[46]

Seals, ducks and geese were shot, and Matthew saw what he called 'three monstrous sharks' beside the ship. One was harpooned and dragged on board: it measured twelve feet and three inches (3.7 metres) long, though its bloated body was fully eight feet (2.4 metres) in circumference.

Amongst the vast quantity of substances contained in the stomach, was a tolerably large seal, bitten in two, and swallowed with half of the spear sticking in it with which it had probably been killed by the natives. The stench of this ravenous monster was great, even before it was dead; and when the stomach was opened, it became intolerable.[47]

The explorers met no Indigenous people in this part of New Holland, but Matthew counted nine fires burning on the land, and there was so much smoke that it was difficult to make out much of the countryside.[48]

The *Investigator* left Lucky Bay on 14 January – and their luck almost ran out, as thirty fathoms (55 metres) under the keel suddenly shoaled to three (5.5 metres), and Matthew could see the floor of the sea just under the ship.[49] Just as it had done at Hythe Bay, the *Investigator* touched bottom, though the only damage was a slight loosening of the copper sheath.[50]

Matthew found another safe anchorage at Middle Island, the largest in the archipelago, and the scientists explored it and the surrounding islands for another two days. Some of the 'little, blue pinguins, like those of Bass Strait, harboured under the bushes' and the men killed or caught more than fifty geese.[51] They had no luck, though, trying to catch the small and agile tammar wallabies, which darted through the thick brush and across the loose stones.

The *Investigator* sailed away on 17 January, and for the next ten days passed hundreds of kilometres of huge, perpendicular cliffs, which Matthew overestimated to be 500 feet high (160 metres) in parts – the highest was just over 300 feet – but they were still so steep that he could not see beyond them, to what would later prove to be the flat, barren country of the Nullarbor Plain. The lower two-thirds of the cliffs were limestone, and the upper layer darker sandstone and 'almost as level as the horizon of the sea'.[52]

As Westall sketched the seemingly endless coastline wall, complaining about the uninspiring, barren landscape, Matthew speculated over what had formed it. He regretted not attempting a landing to further explore the great cliffs and what lay beyond, 'notwithstanding the great difficulty and risk'.[53]

On 26 January, the *Investigator* had crossed the longitude of 130 degrees east, the point marking the start of the British territory of New South Wales, and where Matthew's survey was originally designated to begin. The ship was in a large open bay that Flinders would later call the Great Australian Bight.[54]

The next day, the wall of cliffs gave way to a long strip of sand, with a ridge of low scrub behind it. He had to proceed cautiously along the coastline: the French charts from the Admiralty had often been inaccurate, and they had already encountered countless unmarked rocks, shoals and reefs.

On the evening of 28 January, Matthew anchored at what he said was 'the extremity of the before known south coast of Terra Australis'.[55] He later named the place Fowler's Bay, after a first lieutenant who had now learnt to do as he was told. Over the next few days, he named the Nuyts Archipelago, the Franklin Isles and the Investigator's Group of Islands, calling the largest Flinders Island, in honour of his brother.

The heat was intense, with Peter Good's thermometer at one stage registering 130 degrees Fahrenheit (54.4 degrees Celsius), but as a heat haze enveloped the ship, Matthew could still see smoke on the mainland. On an island that Flinders recorded as St Francis, assuming it to be the island so named by Nuyts nearly 200 years earlier, the sandy ground was full of muttonbird burrows. The heat made walking across the island exhausting, but the exploration was made even more difficult as the men frequently stumbled or sank into the bird holes. As they lurched and fumbled about, they still killed more than 300 birds, so that each man on the ship received four to eat.

Fresh water was scarce, though. Matthew thought he'd found the source of some in Denial Bay, given the number of insects about the ship that tended to inhabit lakes, but the water proved as salty as the ocean.

On 20 February, the *Investigator* rounded a headland and Matthew saw what appeared to be a strait running north-east. There was a strong outgoing tide, with no land visible in the direction of the strait. Every man on the ship was suddenly infused with 'new life and vigour', and all the talk on board was of 'large rivers, deep inlets, inland seas, and passages into the Gulph of Carpentaria'.[56]

The joy would turn to grief, though, and Matthew soon named the headland Cape Catastrophe.

Chapter 17

'These birds sit watching in the trees, and should a kanguroo come out
to feed in the day time, it is seized and torn to pieces
by these voracious creatures.'

MATTHEW FLINDERS, ON THE SIGHT OF GIANT EAGLES HUNTING
KANGAROOS IN SPENCER GULF.[1]

D ANGER WAS EVERYWHERE in this territory that
was virtually unknown to European visitors. Early on
Sunday, 21 February 1802, Matthew and John Thistle carried
their surveying instruments onto an island off what Matthew had
marked on his chart as Inlet 12.

The island was about seventeen kilometres long and between
one and four kilometres wide. There were seals upon the beach
and, further on, traces everywhere 'of the kanguroo', as well as
evidence of 'a conflagration of the woods' – or bushfire.[2] Matthew
later called the land 'Thistle's Island'.

As the two men huffed and puffed up a steep hill, they found
speckled yellow carpet snakes everywhere. Two were killed, one
measuring seven feet nine inches (2.4 metres), but when Matthew
found one asleep he pressed the butt of his musket on the snake's
neck as Thistle moved in to sew its mouth shut with a sail needle
and twine. The snake quickly woke up. Matthew wanted to take
it on board alive for Robert Brown to examine and young Westall
to sketch.

Matthew and Thistle were heading back to the *Investigator* with their angry serpent when a white eagle 'with fierce aspect and outspread wing' bounded towards them. Stopping short twenty metres away, it flew into a tree. Another eagle made a motion to pounce but backed off. Matthew guessed that the giant birds took the men 'for kanguroos', having probably never before seen an upright animal on the island.[3]

Despite the abundance of snakes, eagles and kangaroo tracks, no water could be found. Matthew returned to the *Investigator* intending to land on the mainland to continue the search, but he was delayed by having to retake bearings, after a difference in longitude between his calculations and Samuel's. John Thistle was sent in a cutter to look for water on the mainland about ten kilometres away, taking midshipman William Taylor and six able seamen with him.

At dusk, the cutter was spotted halfway between the shore and ship, returning to the *Investigator* under sail.[4] Peter Good watched the boat for fifteen minutes from the deck, studying the effect of the current on its passage. He looked away for five minutes, busy with other tasks, and when he looked back in fading light, there was no sign of the little boat or the eight men on board.[5]

Lieutenant Fowler, who had been out in another boat, went searching for the cutter at the place it was last seen, using a lantern in the dark. Hours passed as Fowler rowed and sailed up and down the inlet, with no sign of the missing men. Growing ever more anxious for the safety of his friend and the men with him, Matthew paced the deck, finally ordering the firing of a gun to signal an end to the search. Fowler returned to the ship at 11 p.m. as a pall descended over every man on board. He told Matthew that when he was near the place where Thistle's cutter was last seen, he met with 'so strong a rippling of tide that he himself narrowly escaped being upset'.[6] Only two out of the eight people in Thistle's boat were strong swimmers, and Matthew 'feared that most of them would be lost'.[7]

The search continued at daybreak, as Matthew steered the *Investigator* across the treacherous water he now called Thorny

Passage, keeping an officer at the masthead with a telescope. There were many strong 'ripplings' caused by the narrowing of the waterways between the islands. Matthew anchored in a small cove and sent Fowler in a boat, hoping that the men had made it to safety on one of the smaller islands, although crewman Samuel Smith noted that 'ravinous shirks' were 'very Numerous'.[8]

Before long, Fowler was towing the remains of Thistle's upturned cutter back to the shore. It was smashed in on all sides, evidently having been dashed against the rocks. Some of the oars and belongings of the men were later found floating in the current, with Thistle's compass and binnacle remarkably undamaged.[9] On a high point on the mainland, a man with a telescope was positioned to scour the area, and Robert Brown led a party walking north looking for survivors along the coast, while Matthew led another group south. They found footprints in the sand where Thistle and his men had looked for water the previous afternoon, but no other traces.[10] The crew kept searching for three more days, until Matthew decided on 24 February that even recovering bodies was now doubtful.

Matthew had a sheet of copper engraved with the names of the eight men lost and nailed to a stout post at the head of what he now called Memory Cove. He named all the islands nearest to Cape Catastrophe after the lost crewmen: Thistle, Taylor, Grindal, Hopkins, Williams, Smith, Little and Lewis. He wrote that he had known Thistle since 1794, and had undergone many perilous adventures with this brave man, who had a 'zeal for discovery'. Taylor was a young officer who, he said, promised to become 'an ornament to the service'. The six young crewmen were all 'active and useful' and their loss was 'heavily felt'.[11]

Matthew later wrote to Ann: 'Thou knowest how I valued him; he is however gone, as well as Mr Taylor and six seamen, who were all drowned in a boat.'[12]

Before Matthew quit Memory Cove a boat was sent to haul a seine net upon the beach, bringing in so great a catch that each crewman was allotted two fish dinners.

The loss of Thistle and his men turned Matthew's thoughts to his ailing father and his loved ones at home, and on 25 February he rounded a headland which he named Cape Donington.

Only a week earlier, Matthew's father had celebrated his fifty-second birthday in that little town, dining with his two brothers, John and William. 'I wish I could note the restoration of my Health,' Flinders Sr wrote, 'but alas that is not the Case.'[13]

Matthew would name other places on his charts from ones that had stirred his heart in Lincolnshire – Boston Island, the Bicker Isles, Surfleet Point, Stamford Hill, Spalding Cove, Point Bolingbroke – as well as the Sir Joseph Banks group of islands. The 'most interesting part of these discoveries', he said, was the large natural harbour he called Port Lincoln,[14] where the weather was pleasant and they were 'not incommoded by noxious insects'.[15]

Still thirsting for water on 26 February, Matthew and some of his men carried spades three kilometres to a lake they saw from Stamford Hill. The water was brackish, but in a moist place 100 metres from the head of the port the men dug through a metre of whitish clay and found water that was perfectly sweet, though thickish and discoloured. They began the long, laborious task of rafting their casks to and from the boat, eventually loading sixty tons of water. Samuel Flinders set up another makeshift observatory on shore, and the naturalists and their dogs went hunting for specimens.

Matthew had difficulties with his surveying work and could find no convenient beach or open place where a baseline could be measured. Knowing the exact coordinates of his ship, he went to the south-east of Boston Island with a pendulum made from a musket ball slung with twine, which he knew swung every half-second. He gave directions for three cannons to be fired at given times, and from the instant that the first gun flashed to the time of hearing the blast, he counted eighty-five swings of the pendulum. The result was the same for the two succeeding blasts. Calculating the distance by the speed of sound – '1142 feet in a second of time' – he made his baseline 8.01 miles (12.89 kilometres).

After a few days at Port Lincoln, a boat's sail and yard, which no doubt came from Thistle's cutter, were seen floating in the current, and on 3 March Matthew sent Fowler back to Memory Cove for one final search along the shores and around the islands in Thorny Passage. Westall, meanwhile, sketched the countryside and a hut made of bent branches, but although there were many such huts across the shore, they saw just two Indigenous people during their stay over the next week, and then only from a distance.

'I had always found the natives of this country to avoid those who seemed anxious for communication,' Matthew wrote, 'whereas when left entirely alone, they would usually come down after having watched us for a few days.'[16]

> What ... would be the conduct of any people, ourselves for instance, were we living in a state of nature, frequently at war with our neighbours, and ignorant of the existence of any other nation? On the arrival of strangers, so different in complexion and appearance to ourselves ... the first sensation would probably be terror, and the first movement flight ... but if, on the contrary, we saw them quietly employed in occupations which had no reference to us, curiosity would get the better of fear; and after observing them more closely, we should ourselves seek a communication. Such seemed to have been the conduct of these Australians ...[17]

'These *Australians*,' Matthew called the Indigenous people. It was one of the first times the word had been used, but with his urging, it would gradually catch on.

MATTHEW STAYED AT Port Lincoln to watch a long-awaited solar eclipse. The morning of 4 March 1802 was cloudy and brought a little rain, but towards noon the weather cleared and Matthew had the satisfaction of observing the celestial phenomenon 'with a refracting telescope of forty-six inches focus, and a power of about two hundred'.[18]

The following morning, Matthew anchored under Cape Donington at the entrance to Spalding Cove, and in the evening

Fowler returned from two days of searching the nearby waterways without finding any sign of his lost companions.

At 10 a.m. on 6 March, the *Investigator* finally set sail out of Port Lincoln and prepared to follow the unmapped coast towards what they speculated might be the Gulf of Carpentaria or some vast inland sea. That afternoon, as naval custom dictated, the small personal items from most of the dead men were collected for their families, and their clothing was sold at the mast. Thistle's belongings were sold the next morning.[19]

After two days, the ship reached a quickly narrowing gulf, which had 'a want of boldness in the shores'. That, combined with the shallowness of the water, indicated it was not a channel 'capable of leading us into the Gulph of Carpentaria, nor yet to any very great distance inland'.[20] In fact, the prospect of 'a channel or strait, cutting off some considerable portion of Terra Australis, was lost'.[21]

On 9 March, still in the gulf, the crew saw a ridge of mountains to the east – 'the most considerable' they had yet seen in New Holland[22] – and Westall got busy with his sketchbook. The next morning, Matthew took a cutter and, accompanied by Surgeon Bell, went to explore the head of this gulf, while Brown, Bower, Westall, Good and some servants headed off to climb the highest mountain they could see, wading through swampy water and mud for more than a kilometre and a half before they reached the *terra firma* of *Terra Australis*.

The servants were overcome with heat and fatigue but the scientific men kept climbing until, after twenty-five kilometres of hard slog, they reached the summit, with the golden glow of sunset over the vast empty plains before them. Good wrote:

We had a most extensive view, probably the most extensive ever had in New Holland, being elevated full 3000 feet above the level of the Sea, and it may be said 100 miles in the heart of the country ... having gratified ourselves with viewing this extensive and boundless desert, we begun to descend with all expedition, but were soon overtaken by darkness & some of party, being overcome with fatigue on arrival at the bottom of a deep Gullie, we thought fit to spend the night.[23]

But it was spent very uncomfortably. They had little water, and had trouble finding any firewood in the cold, dark night. The ground was full of stones and so uneven that they could not lie flat.

Matthew, the surgeon and their crew spent the night sleeping rough in a narrow inlet amid mangroves surrounded by ducks and other waterfowl. At 10 a.m. their oars touched the mud on each side and it was not possible to proceed further. 'It seemed remarkable, and was very mortifying,' Matthew wrote, 'to find the water at the head of the gulph as [salty] nearly as at the ship.'[24]

His return to the *Investigator* was slowed by chasing black swans, ducks and other birds through the mangroves, but when he returned, he and Brown discussed the countryside, with Brown telling him that, from the top of the mountain:

… neither rivers nor lakes could be perceived, nor any thing of the sea to the south-eastward. In almost every direction the eye traversed over an uninterruptedly flat, woody country; the sole exceptions being the ridge of mountains extending north and south, and the water of the gulph to the south-westward.[25]

Matthew rewarded the botanist's tireless efforts by calling the peak Mount Brown. The chain of mountains would later be called the Flinders Ranges.[26] Matthew called the vast body of water 'Spencer's Gulph'.[27] The city of Port Augusta would eventually grow near its head.

Once again, Matthew and his men met no Indigenous people, but traces of their fires were all about, even on the high points of Mount Brown, so that Matthew noted: 'it should therefore seem that the country here is as well inhabited as most parts of Terra Australis'.[28]

Trim, as always, kept the ship's crew amused. 'His desire to gain a competent knowledge in practical seamanship,' Matthew wrote, saw him in the middle of whatever bustle was on deck, because 'he was endowed with an unusual degree of confidence and courage'.[29]

> He knew what good discipline required, and on taking in a reef [sail], never presumed to go aloft until the order was issued; but go as soon as the officer had issued the word – 'Away up aloft!' up he jumped along with the seamen, and so active and zealous was he, that none could reach the top before, or so soon as he did …[30]

On 13 March, Matthew sailed the *Investigator* down the eastern side of Spencer's Gulph along what he named the Yorke Peninsula, after Charles Philip Yorke,[31] Britain's Secretary of War. Just before 8 a.m., the water shoaled suddenly from four fathoms (7.3 metres) to two fathoms (3.6 metres), and the ship hit a mudbank covered with grass, three or four kilometres from the shore. A kedge anchor was needed to free the ship, and within half an hour it was gliding south again.

The men travelled south-east for another week, during which Matthew celebrated his twenty-eighth birthday while charting as many features of this unmapped terrain as he could. On 20 March, he saw land distant in the south and steered for there. By midnight the *Investigator* was being lashed by a gale. The storm lasted throughout the night as the ship heaved around and up and down in the huge seas, but by daylight Matthew realised he had reached an island where the land stretched east and west as far as he could see. He hoped to anchor at what he called Nepean Bay, but as the winds were strong he instead anchored about a kilometre and a half from a headland he called Kangaroo Head.

Although the *Investigator* had passed along more than a hundred kilometres of this large island's coast, the men saw no fires or other signs of any inhabitants.

It was too late in the evening to go ashore, but every telescope in the ship was pointed to land. Several of the crew, perhaps aided by their ration of spirits that night, said they saw black lumps, like rocks, in motion. This 'caused the force of their imaginations to be much admired', Matthew noted, but the next morning, 22 March 1802, the men went on shore and found a large mob of big, brown kangaroos feeding on grass by the side of the bush.

In Matthew's experience, kangaroos were usually as timid as wild deer, but the landing of the men from the *Investigator* gave these docile animals no cause for alarm, since they had apparently had little or no interaction with humans. That would not last, as Matthew wrote:

> I had with me a double-barrelled gun, fitted with a bayonet, and the gentlemen my companions had muskets. It would be difficult to guess how many kanguroos were seen; but I killed ten, and the rest of the party made up the number to thirty-one, taken on board in the course of the day; the least of them weighing sixty-nine [thirty-one kilograms], and the largest one hundred and twenty-five pounds [fifty-six kilograms].[32]

Matthew felt a pang of guilt about the savage 'butchery' of this unsuspecting wildlife – 'for the poor animals suffered themselves to be shot in the eyes with small shot, and in some cases to be knocked on the head with sticks' – but there was no telling when the next meal would present itself so conveniently for his hungry men.[33]

The whole of the ship's company was kept busy all that afternoon skinning and cleaning the kangaroos, and they did it gleefully, because after four months without much fresh meat since leaving the Cape, they could now have as many kangaroo steaks as they could eat.

'In gratitude for so seasonable a supply,' Matthew wrote in his log, 'I named this southern land Kanguroo Island.'[34]

Peter Good went exploring the next day and found the kangaroos had already become much warier of man. He saw 'four Emoos' too, but they all outran the dogs and guns.[35] Matthew began surveying from 'Kanguroo Head', but nothing could be seen to the north; the sole feature of importance was a high hill on the mainland. Matthew named the 710-metre peak Mount Lofty.

Thick bush covered almost all of the island visible from the ship, but everyone was puzzled by the absence of Indigenous people and the large number of fallen trees. There had been traces of a large fire, which, judging by the size of the trees, Matthew estimated to have occurred ten to twenty years previously. He wondered if the missing man Lapérouse had visited here on his expedition.[36] Brown guessed that the trees had simply fallen down because of age and weather.

After three days Matthew set sail again to the north-east, following the low coastline into another large body of water, which he initially named No 14 Great Inlet, but would later call 'the Gulph of St Vincent' after Admiral John Jervis, also known as the 1st Earl of St Vincent. Again Matthew sailed to the head of the gulf, hoping that there might be a great river leading into the interior, but he sailed out again knowing there was not. The country had a pleasant appearance around the inlet, he said, but was poor in fertility. The trees grew but in patches and the grass was thin. With the 'examination of the gulph of St. Vincent now finished',[37] he steered back to Kangaroo Island on 2 April so his men could harvest more meat, although the kangaroos had become exceedingly shy.[38]

Adjustments were made to the chronometers after Samuel took some readings, and the scientists continued to explore all that the island had to offer. Soon it was found that the timekeepers had stopped because someone had forgotten to wind them.[39] That was the responsibility of Samuel and his assistant, John Franklin.

At a place Matthew called Pelican Lagoon, he visited four small islands. On two of these he found many young pelicans unable to fly:

Flocks of the old birds were sitting upon the beaches of the lagoon, and it appeared that the islands were their breeding places; not only so, but from the number of skeletons and bones there scattered, it should seem that they had for ages been selected for the closing scene of their existence. Certainly none more likely to be free from disturbance of every kind could have been chosen, than these islets in a hidden lagoon of an uninhabited island, situate upon an unknown coast near the antipodes of Europe; nor can anything be more consonant to the feelings, if pelicans have any, than quietly to resign their breath, whilst surrounded by their progeny, and in the same spot where they first drew it.[40]

Matthew needed seal skins to repair the rigging, and a shooting party was directed to kill some. They returned with a boatload of skins, but one of the sailors, Richard Stanley,[41] 'having attacked a large seal incautiously' – he had tried to kill it with a stick – had a large piece of his leg ripped away by the animal and was left lame. At least thirty emus were seen at different times, but none could be brought down.[42]

The approach of winter added to Matthew's fear that he might not complete his charting of this unknown part of the coastline before his men ran out of food. He reluctantly abandoned further explorations of the island, and with the rising of a breeze he left 'Kanguroo Head', at two in the afternoon on 6 April, heading north.[43]

Contrary winds and difficult tides slowed the ship's progress, but on the morning of 8 April it was in a wide bay about eight kilometres off the low, sandy shore of what is now called the Coorong National Park. In the afternoon, the crew harpooned a porpoise to add to their kangaroo steaks,[44] and at 4 p.m. the lookout cried out to beware of a large white rock in the distance. It looked ominous.

MATTHEW SAILED CLOSER to this 'rock' cautiously, as clouds obscured the afternoon sun. The man aloft shouted out that the monolith looked like a white pyramid, but on coming nearer, Matthew was startled to see that it was a ship – the first he had seen since leaving the Cape. At first he thought it might be a vessel from Sydney involved in a whaling or sealing expedition. Then his hackles rose as he saw a French flag. France and England had signed a peace treaty in the city of Amiens two weeks earlier, but Matthew and the French crew were unaware of that. Matthew shouted commands quickly. 'Clear the decks for action!' Men rushed about, loading the heavy guns and their muskets. Swords were by their sides. Matthew recalled that the French ship now raised a white flag:

> At half past five, the land being then five miles distant to the north-eastward, I hove to; and learned, as the stranger passed to leeward with a free wind, that it was the French national ship *Le Géographe*, under the command of captain Nicolas Baudin. We veered round as *Le Géographe* was passing, so as to keep our broadside to her, lest the flag of truce should be a deception.[45]

Matthew knew that Baudin's two-ship expedition had left Le Havre back on 19 October 1800, nine months before the *Investigator* had left English waters. Baudin had fought against the English during the American Revolution and had bitter memories. He had been wounded and taken prisoner in Canada in 1778 before escaping, and he was later recaptured in another battle off Haiti. He had since made a name for himself with important scientific expeditions, first to the Indian Ocean to collect plants and specimens in 1792, and then in 1796 to the West Indies.

Now in command of the ships *Géographe* and *Naturaliste*, Baudin had with him a number of scientists, including his chief zoologist François Péron[46] and the cartographer-surveyor Louis de Freycinet.[47] Problems had beset their voyage. Many of the twenty-two scientists who set sail with him were the sons of French aristocrats who had managed to keep their heads in the revolution, and they had scant regard for their captain, who was born a commoner.

The ships had taken six months to reach Isle de France (Mauritius), where, owing to shipboard arguments, sickness, austere conditions and a lack of provisions from the local government, forty-six sailors and ten of the scientists abandoned the voyage. Péron stayed, but clashed repeatedly with his commander, while Freycinet could not hide his contempt for his skipper.

Baudin had sighted Cape Leeuwin on 27 May 1801, seven months before Matthew's arrival, and he had anchored in what he called Geographe Bay three days later. The two ships had then lost touch with each other, as Baudin defied his orders and sailed the *Géographe* north along the western coast of New Holland for Timor, in order to replenish his supplies, repair his ship and refresh his crew. He named Cape Leveque (near present-day Broome) after his hydrographer,[48] and reached Timor on 21 August, with the *Naturaliste* joining him there a month later.

On 13 November 1801 the two ships had left Timor, reaching Van Diemen's Land two months later, while Matthew was negotiating the shoals and currents of Lucky Bay. Baudin and his team had surveyed around D'Entrecasteaux Channel for more than a month. They had long, extensive and peaceful interactions with the Indigenous Tasmanians.

Then, on the afternoon of 8 April, sailing west, Baudin was even more surprised than Matthew to see a ship sailing towards him. At first he thought it was the *Naturaliste*. Baudin had no inkling that a British voyage was underway, nor was he aware of its approval by Napoleon's government.

As the sun was going down, Matthew had a boat hoisted down from the *Investigator* to row out to the French ship, but because he did not speak French, he also took Robert Brown. As soon as the two climbed on board the vessel, they could see evidence of scurvy in the Frenchmen surrounding them. The men looked wan and listless. They had not hoisted any topgallant masts because they did not have the energy.

Matthew was conducted to Baudin's cabin, where he met a 48-year-old man regarded as a favourite of the French ruler.

A watercolour miniature portrait of Flinders made shortly before he sailed for Australia on the *Investigator* in 1801. Artist unknown. *State Library of NSW a069001h*

Flinder's sextant, presented to the State
Library of New South Wales by his grandson
Sir William Matthew Flinders Petrie in 1922.
State Library of NSW FL1133847

The bicorne, or cocked hat, worn
by Flinders in his circumnavigation
of Australia. *State Library of NSW FL1044307*

A box chronometer, Serial
No 520, made by Thomas Earnshaw
of London and used by Flinders on the
Investigator. *Museum of Applied Arts & Sciences, Australia H3940*

Captain Thomas Pasley, Flinders' great mentor early in his career. Painted by Lemuel Francis Abbott. *National Maritime Museum, Greenwich*

Third Lieutenant George Tobin described the echidna as an 'Animal of Adventure Bay … about the size of a roasting pig'. *State Library of NSW FL1606693*

George Tobin's sketches of Flinders and the *Providence* crew at Adventure Bay in lutruwita/ Tasmania, camping at an Aboriginal hut and collecting wood. *State Library of NSW FL1606698, FL1606697*

Matavai Bay in Tahiti provided a base for repairing the *Providence* and *Assistant*, bartering with the locals and, most importantly, collecting breadfruit to feed Caribbean slaves. *State Library of NSW FL1606719*

Flinders was appalled by the firepower used in a clash with Torres Strait Islanders in their war canoes. *State Library of NSW FL1606800*

Above: The warships of Lord Howe blast away at the French in the battle known as The Glorious First of June. Painted by Philip James de Loutherbourg.

National Maritime Museum, Greenwich

John Hunter was the second governor of NSW and a man who saw great potential in both Flinders and George Bass. Portrait by William Mineard Bennett.

State Library of NSW

Flinders' friend, interpreter and guide Bungaree, whom he described as 'worthy and brave'. Portrait painted by Augustus Earle in 1826, 'Portrait of Bungaree, a native of New South Wales, with Fort Macquarie, Sydney Harbour, in background'. *National Library of Australia*

nla.pic-an2256865

The Glass House Mountains fascinated Flinders. *Martin Valigursky / Alamy*

A model of the HMS *Investigator* made by Roland Michel Laroche in Sydney in 1980. Although it was ready to be scrapped after the voyage around Australia the ship lasted for almost 60 years after Flinders' death. *Museum of Applied Arts & Sciences, Australia B2371*

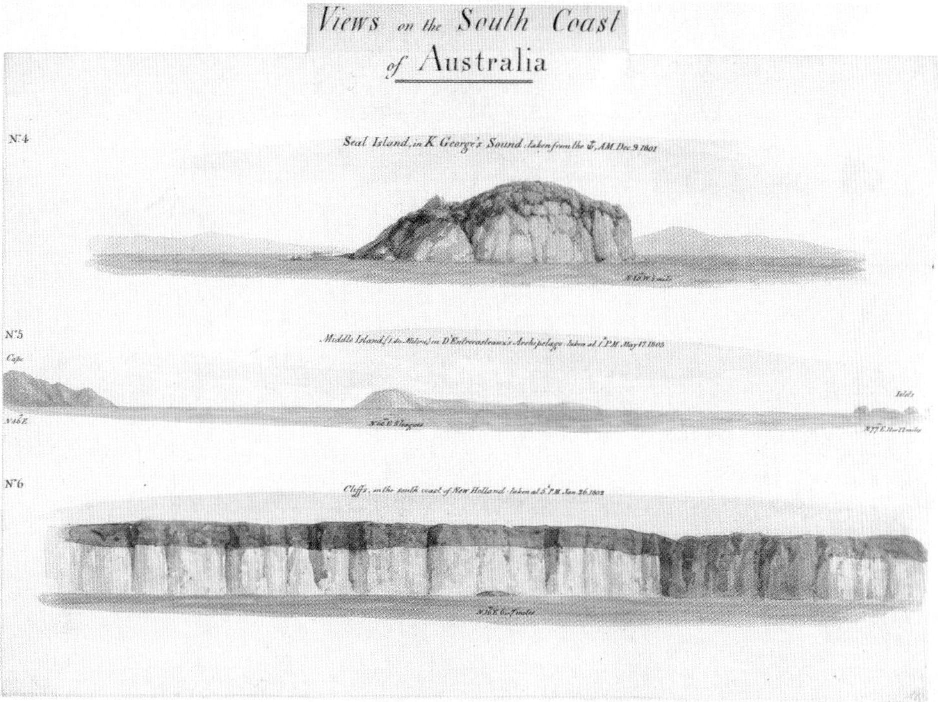

Landscape artist William Westall's 'Views on the South Coast of Australia'. *nla.obj-138893194*

A carpet snake on what Flinders would call Thistle Island, at the southern reaches of the Spencer Gulf. *nla.obj-138876274*

Flinders named Encounter Bay, in what is now South Australia, after his meeting with the French crew under the command of Nicolas Baudin. *William Robinson/Alamy*

William Westall's 'View of Port Bowen', in what is now Queensland, August 1802. *National Maritime Museum, Greenwich, ZBA7941*

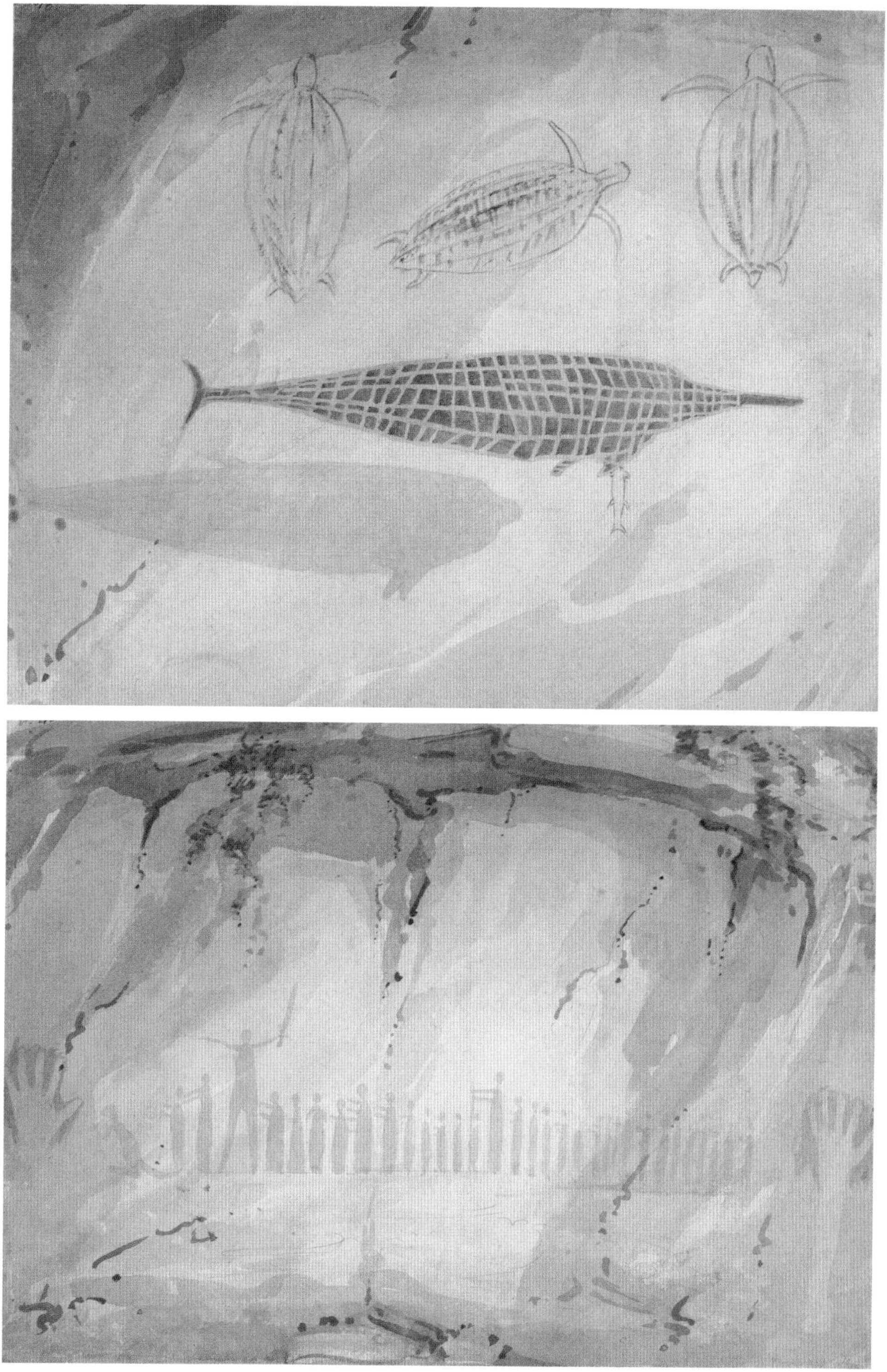

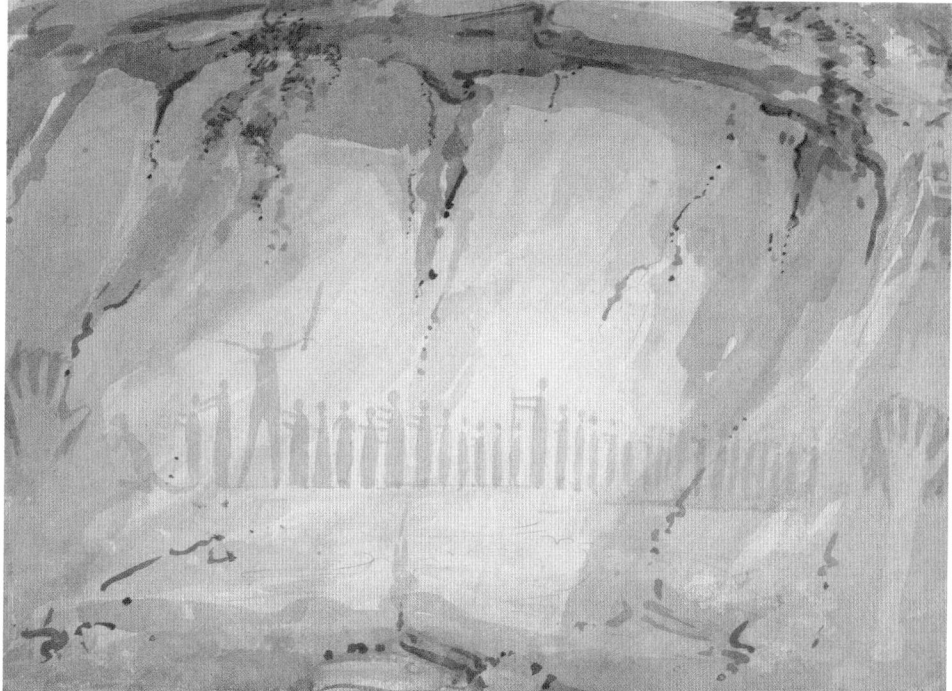

The extraordinary Aboriginal wall paintings on Chasm Island, north of Groote Eylandt, as copied by William Westall. *nla.obj-138890799, 138890494*

'View of Malay Road from Pobassoo's Island', off the coast of Arnhem Land, by William Westall, February 1803. *National Maritime Museum, Greenwich, ZBA7936*

Ferdinand Bauer's natural history paintings on the *Investigator* voyage included this study of a koala and joey. *nla.obj-136418125*

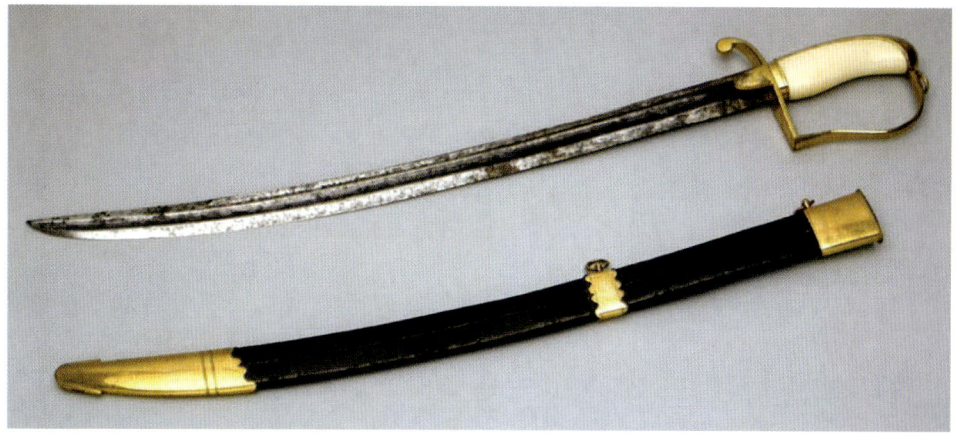

Flinders' jailer,
General Charles
Mathieu
Isidore Decaen.
Wikipemedia Commons

Flinders was forced to surrender his sword on Mauritius. *State Library of NSW FL1123356*

Flinders' widow Ann (right) with daughter Anne standing and Ann's half-sister Belle Tyler (left) about 20 years after his death.

Rachel Flinders Lewis

Flinders' daughter Anne in about 1830. *Rachel Flinders Lewis / Clive Aylard*

Flinders' daughter, Mrs Anne Flinders Petrie in 1861, with her son who became the leading archaeologist Sir William Matthew Flinders Petrie.

A memorial to Flinders and his cat Trim in the marketplace of Donington, Lincolnshire, where he was born. *Matt Limb/Alamy*

The statue of Flinders (with Trim looking on from a window ledge) outside the State Library of NSW. *Grantlee Kieza*

An image issued by Britain's HS2 high-speed rail project shows the lead plate placed on top of the coffin of Matthew Flinders in 1814. *MOLA Headland/HS2*

Matthew Flinders will be reburied in Donington's Church of St Mary and the Holy Rood where he was baptised. His remains, lost for more than two centuries, were found during work on England's high speed rail network in 2019. *Alamy*

The Duke of Cambridge admires the statue of Flinders at Australia House in 2014. *PA Images/ Alamy*

Baudin was usually an affable man, if uninspiring for his crew.[49] He was about the same height as Matthew, with thick brown hair and a prominent Gallic nose. The French captain was sick and under siege from all the internal pressures of the voyage and the weight of expectation on him back in Napoleon's court.

Matthew was detached and aloof. He was intensely competitive, and no doubt saw this esteemed rival explorer as far more dangerous to his ambitions than any wartime enemy. He requested that Baudin show him his passport from the Admiralty, even though he knew he'd been issued with one, and he then checked it like an officious customs official. Matthew offered his own passport from the French, but Baudin gave it back without looking at it.

Speaking in a mix of broken English and fractured French, Matthew and Baudin talked about their voyages. Matthew later said the conversations were mostly carried on in English, 'which the captain spoke so as to be understood', through Brown spoke about the 'badness' of the Frenchmen's knowledge of the language.

Baudin told Matthew he had spent some time examining the southern and eastern parts of Van Diemen's Land, where his geographical engineer, with the largest boat and a crew of eight, had been left and was 'probably lost'. In Bass Strait, Baudin had encountered a heavy gale and become separated from the *Naturaliste*, but having since had fair winds and fine weather, he had explored the coast from Western Port to this place of their meeting. But he had found no river, inlet or other shelter that might afford a safe anchorage. Baudin had a copy of Matthew's 1800 chart of Bass Strait and said he 'found great fault with the north side of the strait' – until Matthew pointed out that the north side had only been mapped in an open boat by George Bass, 'who had no good means of fixing either latitude or longitude'. Matthew told Baudin that next morning he would give him a copy of the charts which had been published since then, in his booklet *Observations on the Coasts of Van Diemen's Land*.

While Baudin and Matthew were talking, the men of both ships also swapped information. Peter Good learnt that as well as the men lost in the long boat in Bass Strait, many of the Frenchmen had succumbed to disease in Timor.[50] One of the French sailors told Samuel Smith that the ship was poorly manned, and that many of the crew had simply run away.[51]

There were many misunderstandings due to language. Baudin later related that after the accident that had befallen his dinghy, he had asked Matthew to give the men all the help he could if he came upon them, and Matthew had told him that 'he had met with a similar misfortune on his Kangaroo Island, where he had lost eight men and a boat'.[52] The men had in fact been lost off Cape Catastrophe, at the foot of Eyre Peninsula. In his log, Baudin also mentioned Matthew being separated from a 'companion ship … during the equinoctial gale'.[53] He was probably referring to the *Lady Nelson*, which Matthew might have mentioned as he outlined his future plans, and which Baudin might have misunderstood to be engaged in surveying work with the *Investigator*.[54]

The next morning, having learnt from his men that the *Investigator*'s mission was also discovery, Baudin was more energetic and inquisitive. In a conversation lasting more than an hour, Matthew told him about the two gulfs and Port Lincoln, pointing out where fresh water could be found. Pointing to the kangaroo-skin caps worn by his crewmen, Matthew let the Frenchman know about the large island where there was plenty of fresh meat to be had.[55]

Matthew and Baudin did not know that much of the territory along the south coast of what is now the state of Victoria had already been mapped by James Grant and John Murray in the *Lady Nelson*,[56] but Baudin had mapped almost 300 kilometres of previously uncharted territory, from what Grant had named Cape Banks, near the present-day Victoria–South Australia border, to the point where he had encountered the *Investigator*.

Baudin told Matthew about a dangerous hazard, now known as Margaret Brock Reef, 200 kilometres further along the coast, off Cape Jaffa. When they finally parted, Baudin asked Matthew to keep an eye out for the men who had been lost from the French

French skipper Nicolas Baudin's expedition beat Matthew's to *Terra Australis*. *State Library of NSW FL3202909*

François Péron was the chief zoologist on Baudin's mission and one of his enemies on board. *nla.obj-136095433*

longboat, and to tell the skipper of the *Naturaliste*, if he saw him in Sydney, that Baudin would sail for there also as soon as bad weather set in. Matthew later related that when he asked Baudin the name of the captain of the *Naturaliste* – Jacques Hamelin[57] – Baudin asked Matthew's name, 'and finding it to be the same as the author of the chart which he had been criticising, expressed not a little surprise; but had the politeness to congratulate himself on meeting me'.[58]

That was Matthew's version, anyway. He may have misrepresented the facts to the advantage of his reputation. Baudin may only have been double-checking the English pronunciation of 'Monsieur Flandaire'. It seems implausible that Baudin, by all reports a courteous and gracious man, would have hosted two meetings on his ship with an English sea captain without first asking his name, or that Matthew would not have introduced himself earlier.

Baudin remembered their meeting this way:

The English captain, Mr Flinders – the self-same Flinders who discovered the Strait which ought to bear his name …

came aboard, expressing his delight at making such a pleasant encounter, though he was extremely reserved about everything else. As soon as I learnt his name, I paid him my compliments, informing him of the great pleasure it gave me to make his acquaintance and of all that we had done systematically up till then in terms of geographical work.[59]

Matthew named the place of their meeting as Encounter Bay.

Later, Baudin's zoologist, François Péron, laid a claim for France to the discovery of all the parts between Western Port and the Nuyts Archipelago, calling the coastline Terre Napoléon. Péron also hatched a plan for Napoleon's forces to invade Sydney and capture New South Wales for France.[60] The French ships took home 18,414 zoological specimens, 2542 of which were deemed new to science, their haul including live kangaroos and emus, black swans, wombats, parakeets and a lyrebird.[61]

AT HALF PAST EIGHT ON the morning of 9 April, Matthew steered the *Investigator* south while the French ship sailed north-west. Hummocks of sandy land were seen from the masthead, but while Matthew had been searching for an inland river since sighting Cape Leeuwin four months earlier, he remained unaware that the longest river on the continent emptied into the Great Australian Bight through a narrow opening in coastal sand dunes only about thirty kilometres away. The place is now called Murray Mouth.

Onward they sailed slowly, struggling against adverse winds and currents as they passed a low, sandy shore. On 13 April, Matthew named rocks off Cape Jaffa the Géographe's Rocks in his logbook, later amending them to Baudin's Rocks – though it was not the reef Baudin had warned him about, as the *Investigator* had passed that hazard during the night. Four days later, they passed two large rocks that the French had named 'The Carpenters', and from Cape Banks they could see two inland mountains to the north-east; the nearer, Mount Schanck, was 'of a flat, table-like form', and the further one, Mount Gambier, was peaked.[62]

As the winds picked up, the *Investigator* ploughed on through rolling grey ocean, past what are now the towns of Portland, Warrnambool and Port Campbell. The night of 19 April brought squalls of wind with hail and rain, and the next morning the shore was scarcely perceptible through the haze. When the night came on and they were off what James Grant had called Cape Otway, Matthew was 'uncertain of the trending of the coast'. At 8 p.m. a short lull in the rain and a bright moon shining across the water revealed a high and dangerous headland, and Matthew called for all hands on deck to set the fore and mizen topsails and reefed mainsail so the ship would be pushed away from the dangers of what might be a rocky shore.

The *Investigator* was now in Bass Strait, and Matthew went looking for the large island that Captain William Reid of the boat *Martha* had reported visiting after he arrived at Port Jackson on 14 December 1799 with more than 1000 seal skins and thirty large barrels of seal oil. Two years later, young John Black,[63] skipper of the *Harbinger*, had named the island after the colony's new governor, Philip Gidley King.

'Of this I was ignorant at the time,' Matthew wrote, 'but since it was so very dangerous to explore the main coast with the present south-west wind, I was desirous of ascertaining the position of this island before going to Port Jackson, more especially as it had escaped the observation of captain Baudin.'[64]

Matthew first saw King Island on 22 April, and the following afternoon the *Investigator* anchored by a sandy shore on its north-east. Behind the beach there were great ridges held together by thick grasses. Matthew and the 'botanical gentlemen' were the first ashore, shooting a couple of wombats, a seal and a kangaroo for the table, and bagging a few more animals the next day.

Matthew sailed back across Bass Strait on 24 April, and on the 26th came to a prominent headland. On the west side of the rocky point there was a wide bay with a small opening that had breakers surging across it. Matthew decided to investigate and steered into the bay at half past one, telling every man to be ready for tacking at a moment's warning.[65] The inrushing tide pushed the *Investigator* across the surf into what is now called Port Phillip, and which now has the city of Melbourne at its head.

At 5 p.m. the speed of the water and the shallowness of the sea floor meant that 'the flood tide set [the ship] upon a mud bank, and she stuck fast'.[66] Again a kedge anchor was used to get the boat moving again. Samuel Smith wrote that 'we got off without damadge. Here we caught a Shirk which measured 10 feet 9 inch in length; in girt very large.'[67] Matthew at first suspected he was in Western Port, which Bass had explored, but he soon realised this was a place he had not seen on any maps, and he was astounded that Baudin had missed it. He devoted the next few days to exploring it in a cutter, unaware that James Grant in the *Lady Nelson* had beaten him to it and already named many of the prominent features.

Matthew, Brown and Westall took a boat to the eastern end of the bay and climbed a 314-metre hill, which Matthew called Bluff Mount, unaware that James Grant had already called it Arthur's Seat after the towering rise above Edinburgh in his native Scotland. Matthew could see Western Port in the distance, but even more exciting was the fact that he could not see the boundaries of a harbour that would prove to have a coastline of 260 kilometres.[68] The hikers celebrated their climb with oysters from the beach.

Good, Bauer and John Allen, the mining expert, took their boat to the area of what is now Portsea and found a 'neck of land' that resembled 'a Gentlemans Park in England, being covered with fine Green grass and Numerous Trees and Bushes'.[69] The surrounds teemed with black swans and pelicans; there were oysters and shellfish for the taking, and dingo footprints in the sand. But although Good saw the fires of Indigenous people, they remained out of sight.

Matthew hoped to explore the vast bay in the *Investigator*, but the troubling winds and shoals forced him to send it back to the entrance under Fowler's command. Matthew provisioned a cutter for three days and set off on 29 April with Brown, midshipman Lacy and a rowing crew to survey as much of the bay as time would allow. They rowed and sailed for fifteen kilometres north of Arthur's Seat, and then headed west across the bay for thirty kilometres or so until they came to what Matthew called Indented Head, which is now known as the Bellarine Peninsula, the site of modern-day Geelong.

The following day, while Matthew was taking angles for his surveying work from a low point at the north-easternmost part of Indented Head, he saw a party of the Wadawurrung people about a kilometre away. The crew rowed Matthew over to the spot, and while they found a hut with a fire in it, the people were gone. Matthew left some strips of cloth 'of their favourite red colour' hanging about the hut, and proceeded westward along the shore to examine that part of the bay.

A little later, three of the Wadawurrung approached the boat 'without hesitation'. Matthew swapped a bird he had shot and a few items for some of their spears and boomerangs.

They afterwards followed us along the shore, and when I shot another bird, which hovered over the boat, and held it up to them, they ran down to the water side and received it without expressing either surprise or distrust. Their knowledge of the effect of fire arms I then attributed to their having seen me shoot birds when unconscious of being observed; but it had probably been learned from Mr. Murray [on the Lady Nelson].[70]

At dawn on 1 May, Matthew set off with three of the boat's crew towards a range of hills later called the You Yangs, aiming to climb the highest point, which he called Station Peak but which now bears his name.

The men reached the 343-metre summit at 10 a.m. and Matthew left the ship's name on a scroll of paper inside a small pile of stones there for posterity. He finally made it back to his tent at 3 p.m., 'much fatigued, having walked more than 20 miles [32 kilometres] without finding a drop of water'.[71]

They camped at Indented Head for the night and reboarded the ship the next day, 2 May, after it had once again hit a mud shoal without damage. They weighed anchor at daybreak on 3 May to leave what Matthew called an 'extensive, but obscure port'.[72]

> On the one hand it is capable of receiving and sheltering a larger fleet of ships than ever yet went to sea; whilst on the other, the entrance, in its whole width, is scarcely two miles, and nearly half of it is occupied by the rocks lying off Point Nepean, and by shoals on the opposite side. The depth in the remaining part varies from 6 to 12 fathoms; and this irregularity causes the strong tides, especially when running against the wind, to make breakers, in which small vessels should be careful of engaging themselves ...[73]

Matthew sailed along the southern coastline of modern-day Victoria and then turned north-east in squally winds. After two days the *Investigator* passed Cape Howe and Matthew changed course in a northerly direction for Sydney. He sighted Mount Dromedary, or Gulaga, as the First Peoples called it, through the rain on 7 May. At dusk the next day, Matthew saw the flagstaff upon Port Jackson's South Head about twelve kilometres away. He hoped to sail into the port that night, 'being sufficiently well acquainted to have run up in the dark had the wind permitted'. But it didn't.

In the morning, Westall made a sketch of the entrance to Sydney's harbour, and at one o'clock Matthew sailed through the heads, a pilot came on board, and soon after 3 p.m. the *Investigator* was anchored in Sydney Cove, ten months after having left England.

Matthew was not shy about praising the way he held scurvy and other ship-borne diseases off his vessel:

> There was not a single individual on board who was not upon deck working the ship into harbour ... the officers and crew were, generally speaking, in better health than on the

day we sailed from Spithead ... Several of the inhabitants of Port Jackson expressed themselves never to have been so strongly reminded of England, as by the fresh colour of many amongst the *Investigator*'s ship's company.[74]

As he anchored, Matthew's thoughts turned to home. While Matthew had been sailing from Encounter Bay to Bass Strait, his father had been compiling his taxation accounts for the month of April, noting his payment to the milkman, the purchase of some rum and some cash to his wife to pay the household bills.

Then, on 1 May 1802, while Matthew had been surveying Port Phillip, his stepmother noted in her husband's diary in a shaky hand:

I have to Remark & to lament the Death of my Dear Departed Husband on May the 1st 1802 after a long & Trying affliction, I may say for more than 12 Months. It is easyer to conceave than it is in my Power to Discribe my feelings upon this most maloncholy occasion. Elizabeth Flinders.[75]

She would later write that Matthew Flinders Sr, the apothecary of Donington, was her 'Beloved Partner' of nineteen years 'and one of the Best of Men'.

Matthew would not learn the news of his father's passing for several months. For now, he had a new governor to meet – and a wife back home in Lincolnshire, whom he was aching to see as soon as he could.

Chapter 18

'Oh my love, my love, how much do I sympathize in thy sufferings.
That I could but transport myself to thee, and soothe thee,
whilst thou shouldst rest upon my fond bosom.'

MATTHEW FLINDERS, WRITING TO HIS AILING WIFE,
ANN, IN LINCOLNSHIRE.[1]

THE *INVESTIGATOR* WAS surrounded by ships on its arrival in Sydney Cove. His Majesty's armed vessel *Porpoise*, the whaler *Speedy*, a privateer called *Margaret* and Baudin's second ship the *Naturaliste*, commanded by Captain Hamelin, were all anchored there.

The little brig *Lady Nelson* bobbed about nearby, but her original commander Lieutenant Grant had sailed home to Britain six months earlier, carrying a report from the besieged new governor, Philip Gidley King, about the continuing machinations of John Macarthur. Matthew had formed a friendship with Macarthur and his wife, Elizabeth, when in Sydney with the *Reliance*, but he thought it best not to mention that to King or William Paterson, who was back in Sydney as lieutenant governor – especially since Macarthur had shot Paterson in the shoulder in a duel and had since been shipped off to London[2] for a court martial.

As soon as the *Investigator* reached Sydney, Matthew went to meet Governor King, whom he already knew well from his voyages to Norfolk Island.

Matthew walked up the hill from the Governors Wharf to the two-storey whitewashed Government House, which had just undergone an extension to its lower level. Sydney had long been Matthew's second home, but even in the two years since he was last here, the infant town had expanded. The skyline was now dominated by two windmills and the colony's clock tower on what became known as Church Hill.

John Franklin thought Sydney was about the same size as Spalding in Lincolnshire, while Peter Good noted that although the northern and western shores of the harbour were 'entirely uncultivated', in the south and east there were 'pretty Snug houses', each of which 'had a considerable space of Garden ground so that the town spreads over a great space'. Some of the principal houses were built with 'Brik and white washed, others with wood painted'. They all had roofs made of wood, cut in the form of tiles that resembled English slate.[3]

Governor King, now forty-four, was heavier and balder than Matthew remembered, but the First Fleet officer was eager to hear about the discoveries the *Investigator* had made. Matthew explained

William Westall's sketch of Sydney's Government House in 1802. *nla.obj-138873721-1*

the circumstances that had seen eight of his men lost at sea, and King deemed that there was no need for an official enquiry. Matthew had not found any major rivers, but he had proved that Terra Australis was one huge landmass stretching for thousands of miles from the Indian Ocean in the west right across to the Pacific in the east.

Matthew told King he had a lot more to uncover, and he handed over his orders from the Admiralty, which directed the governor to place the *Lady Nelson* under his command to accompany the *Investigator*. King told Matthew that he would give him every assistance, and he provided Cattle Point, the site of Bennelong's hut, on the eastern side of Sydney Cove, as a base where the explorer could refit the *Investigator* and prepare for the rest of the circumnavigation.[4]

Matthew then went to see Captain Hamelin on the *Naturaliste*, telling him of Baudin's intention to join him at Port Jackson if bad weather set in on the south coast. The *Géographe*'s missing boat had been picked up in Bass Strait by William Campbell's brig *Harrington*. If only John Thistle and his men had been so lucky.

At Cattle Point, Samuel Flinders and John Franklin set up their timekeepers and telescopes beside the ship, with marines to guard the equipment. Matthew got busy making improvements to the *Investigator*. Four convict carpenters were employed to lower the barricade on the quarterdeck, to make it easier for Matthew, with his shortish stature, to take bearings. Previously he had to balance on top of the ship's binnacle for the survey work, which was difficult and dangerous in rough weather.

To replace John Thistle's cutter, Matthew paid £30 to have a new one built, based on the whaleboat in which Bass had sailed to Western Port. It was twenty-eight feet and seven inches (about nine metres) in length, flat-floored and fitted to row up to eight oars, though it was intended for six in most cases. Matthew insisted on using banksia in the construction, saying it was more durable than mangrove. The planking was of cedar. The boat was constructed by Thomas Moore, who had replaced Daniel Paine as master builder to the colony, and who would later run a large farm on the Georges River called Moorebank. Matthew said the vessel proved 'to be excellent in a sea, as well as for rowing and sailing in smooth water'.[5]

The day after the *Investigator* arrived in Sydney, the British whaling ship *Venus* – not be confused with George Bass's brig of the same name – had arrived back at Port Jackson with the news that peace had been declared between Britain and France.

At the time, Bass and his ship were sailing across the Pacific, under contract to Governor King to bring salt pork from Tahiti into Sydney at sixpence a pound during a time when meat remained scarce.[6] Bass and Charles Bishop had arrived in Sydney the previous year to find that despite their hopes of huge markups on goods from Europe, the market had already been flooded with cheap imports. They put their merchandise into safe-keeping in the government stores, awaiting the return of eager buyers, and entered the meat importation business.

Matthew would not be able to see Bass, because the *Investigator* was due to leave on the next stage of its voyage before Bass and Bishop returned to Sydney with fifty-seven tons of salted pig carcasses that netted the partners £3000.[7]

The *Investigator* had never taken on more than three inches of water in an hour after leaving the Cape of Good Hope, so not much caulking was required, but Matthew's crew were still kept constantly at work, stripping and re-rigging the masts, and preparing the hold to receive a fresh stock of provisions and water. The crew also worked on fitting the greenhouse on the quarterdeck, and cutting planks to make boxes for all the plants Brown wanted for Kew.

A prefabricated greenhouse had been taken on board at Sheerness and stowed away in pieces. Matthew was worried that the upper parts of the ship might become overloaded with boxes of earth, and that in bad weather he might have to throw the plants and the greenhouse overboard for the safety of his men. He proposed reducing the greenhouse to two-thirds of its size, and Brown agreed that it could still serve its purpose. The plants already collected on the voyage from Cape Leeuwin had been landed in good order in Sydney, and could be kept in the governor's garden until Matthew was ready to sail home.[8]

Robert Brown, Peter Good and the other scientific men made excursions into the interior from Port Jackson, collecting samples, while the artists recorded their finds. Matthew busied himself constructing the charts of his discoveries and examinations and reporting to the Admiralty on the success of the mission so far.[9]

He also wrote letters home. He penned a love letter to Ann in 'a moment snatched from the confusion of performing half a dozen occupations':[10]

> … I am safe and well, and have done every thing thus far that I could have expected to do, is to tell thee something. How highly should I value such short information reciprocated from thee! but alas, my dearest love, I am all in the dark concerning thee, I know not what to fear or what to hope. Pray write and releive [sic] my anxiety.[11]

He wished that Ann could have joined him in Sydney and made friends with two women he admired, Anna King, the governor's wife, and Elizabeth Paterson. 'These would have been thy choicest friends, and for visiting acquaintances there are five or six other ladies, very agreeable for short periods, and perhaps longer'.[12] Trim was 'very well', he told Ann, and Matthew's young servant John Elder was proving 'good and faithful'.[13]

He was gaining some measure of fame: 'my situation makes me of some consequence in the eye of the world', he told his wife, 'and this should extend to thee, and have its influence in regulating thy appearance and mode of conduct'.[14] He asked Ann to save any newspaper cuttings about the voyage. His agent was sending her £40 a year to help with expenses in his absence. This was not a lot of money, but his father had taught him that every penny counted.

Matthew wrote to Joseph Banks to announce the success of the voyage thus far, and said that, with Governor King's assistance, the ship would be ready to sail again for the 'Gulph of Carpentaria' in a few weeks. Matthew and King had both agreed it was safer in winter to travel north than risk the freezing winds of the Southern Ocean.[15]

'It is fortunate for science that two men of such assiduity and abilities as Mr. Brown and Mr. Bauer have been selected,' Matthew told Banks, 'their application is beyond what I have been accustomed to see.' With the report of a truce between Britain and France, Matthew told Banks that he hoped 'the difficulty in obtaining promotion which usually follows a peace will not extend to the *Investigator*'.[16]

The winter in Sydney throughout Matthew's stay in 1802 was mostly 'dull and rainy', but on 4 June the town celebrated the king's birthday with what he said was 'due magnificence'. The *Investigator* was covered with coloured ribbons, and Matthew ordered every man to wear 'his best apparel, and to make himself merry'. He attended Governor King's ball, with fifty-two gentlemen and ladies present; Matthew told Ann that the only thing missing was his wife by his side.[17]

The governor was in a particularly chipper mood. Two days earlier, the Bidjigal warrior Pemulwuy, who had been a constant

Governor Philip Gidley King saw the Bidjigal warrior Pemulwuy as a constant menace to the European settlers. *nla.cat-vn2312357*

menace to the European settlers since he had speared Governor Phillip's gamekeeper John McEntire twelve years earlier,[18] had been shot dead. Pemulwuy had mounted a campaign of guerrilla warfare against the Europeans, and King had put a bounty on him,[19] claiming he was guilty of two murders. Matthew's crewman Samuel Smith said the killer was the hard-drinking, extremely violent Henry Hacking, now the first mate on the *Lady Nelson*.[20]

Like all the early governors, King sent regular shipments of the colony's flora and fauna to Joseph Banks. On one shipment aboard the *Speedy*, the governor included a preserved platypus and, almost as an afterthought, Pemulwuy's severed head preserved in spirits. To King, the Indigenous warrior was little more than a scientific specimen. The governor told Banks that although he was 'a terrible pest to the colony', Pemulwuy had been a 'brave and independent character'.[21]

Matthew did not have time to finish his charts before the *Speedy* sailed for England, but the ship carried his request for provisions for his men to Evan Nepean, as Sydney was in short supply. Matthew also sent a long, detailed report to the Admiralty of all that had transpired on the voyage: the places visited and marked, the loss of Thistle and his men, and the bright news that the *Investigator* had arrived in Sydney with all its officers and crew in good health, 'four men excepted'.[22]

When the *Speedy* arrived in London that November, the officers of the Admiralty were disappointed that Matthew had disobeyed orders by not surveying the coast around Western Port first, because they had hoped to start a settlement in Port Phillip as quickly as possible and did not have his detailed maps. Once again, Joseph Banks had to make apologies on behalf of the young skipper. When the charts were finally finished, the lack of British ships sailing for England delayed their arrival further, so that Banks finally wrote to Governor King to say it was a 'great misfortune to the voyage.' The Admiralty was unimpressed by the non-appearance of the charts; Banks said Matthew's 'enemies suspect idleness on his part'.[23]

SOON AFTER MATTHEW WROTE his first letter to Ann from Sydney, he received two from her, one of them written in January 1802 that related how she had just undergone an eye operation to save her sight. She felt abandoned and humiliated after returning home to her little village just a few weeks after the wedding, with her husband expected to be absent for several years.

'How melancholy, how severe are these pieces of information to a fondly affectionate husband,' Matthew wrote back. 'Oh my poor wife! Often do I visit thee in my dreams, sometimes with delight, but oftener with that confused and uncertain feeling which partakes too much of my troubled thoughts concerning thee.'[24]

Ann had also told Matthew that, by putting his profession first so soon after the wedding, he had shown 'poor proof' of his affection. He explained in reply that without the assignment on the *Investigator*, they would 'have barely existed in England', 'debarred of even necessaries', unless they were reduced to begging from friends. He figured he would now make at least £1500 from the voyage.

Heaven knows with what sincerity and warmth of affection I have loved thee, let us ... look forward with our best hopes for the good which is in store for us. See me engaged, successfully thus far, in the cause of science and followed by the good wishes and approbation of the world: – see rising out of this employment a moderate competence for thee and myself.[25]

EVEN THOUGH FRESH MEAT remained scarce in Sydney, the food was certainly better than that being served by Baudin on the *Géographe*. On 20 June 1802 the teetering vessel lumbered through Sydney Heads; Matthew sent a boat from the *Investigator* to help tow it up to the cove.

It was grievous, Matthew said, to see the miserable condition to which both officers and crew were reduced by scurvy, with just twelve men out of 170 on board capable of doing their duty.[26] Matthew may have been exaggerating the distress of the Frenchmen to amplify British benevolence and his own success in preserving the health of his crew. But Governor King told Baudin soon after: 'Although last night I had the pleasure of announcing that a peace had taken place between our respective countries, yet a continuance of the war would have made no difference in my reception of your ship, and affording every relief and assistance in my power.'[27] The Governor had the sick taken to the colonial hospital, and sent the French ships everything the colony could supply, including fresh beef even though it was a rare commodity.

'The distress of the French navigators had indeed been great,' Matthew noted, 'but every means were used by the governor and the principal inhabitants of the colony to make them forget both their sufferings and the war which existed between the two nations.'[28]

Baudin acknowledged this when he wrote that 'among all the French officers serving in the division which I command there is not one who is not, like myself, convinced of the indebtedness in which we stand to Governor King, Colonel Paterson, and the principal inhabitants of the colony'.[29] François Péron was received warmly at Government House, later writing that King and his secretary spoke 'our language well', and that 'the commandant of the troops of New South Wales, Mr. Paterson, a member of the Royal Society of London, a very distinguished *savant*, always treated me with particular regard. I was received in his house, as one might say, as a son.'[30]

Before he set sail, Matthew hosted a number of dinners on board the *Investigator*, celebrating peace in Europe. His guests included King, Paterson, Baudin, Hamelin and Péron, the Frenchmen arriving under eleven-gun salutes.

Trim was always first to the table, 'commonly seated a quarter of an hour before any other person'.[31] After everybody else was served, he put in his request, with a 'gentle caressing meow'. It was no use trying to refuse him either, Matthew said, for Trim was enterprising in time of need, and if the diners were not careful he would whip food off their forks with his paw, with 'such dexterity and an air so graceful, that it rather excited admiration than anger'.[32]

Matthew showed Baudin one of his charts of the territory along the south coast of *Terra Australis*, containing the part first explored by the French commander, and distinctly marked as his discovery. After examining Matthew's chart, Baudin said he did not produce his own maps: they were not constructed on board the ship. Rather, Baudin sent all his bearings and observations to Paris, with a regular series of sketches of the land, and the charts were made at a future time.

'This mode appeared to me extraordinary, and not to be worthy of imitation,' Matthew wrote, 'conceiving that a rough chart, at least, should be made whilst the land is in sight, when any error in bearing or observation can be corrected.'[33]

MATTHEW HAD THE *LADY NELSON* at his service under the command of young Scotsman John Murray.[34] The small ship was fitted with three sliding keels so that it was perfectly adapted for going up shallow rivers or beaches with shoals. Murray's crew was mostly composed of convicts, and with no trustworthy officer to rely on, Matthew lent him his red-haired midshipman Denis Lacy as first mate.

The price of fresh beef in Sydney was so exorbitant in 1802 that it was impossible for Matthew to think of purchasing it on the public account. Instead he paid £3 a head for sheep; pigs were bought at ninepence per pound when weighed alive, geese at ten shillings each, chickens at three shillings. He also ordered fifteen tons of hard biscuit, three tons of flour and four tons of kiln-dried wheat. The convict ship *Coromandel* arrived in Sydney on 13 June with most of the supplies he had ordered from Spithead. From two American vessels in Sydney he purchased 1483 gallons of rum. He was confident the food and drink would last for the twelve months and 20,000 kilometres of exploration ahead.

Matthew also managed to recruit crewmen to make up for John Thistle and the others lost in Spencer Gulf, as well as the sailor bitten by the seal on Kangaroo Island, and another marine who was invalided. John Aken, the chief mate of the *Hercules*, came onto the *Investigator* as the new master, along with five other sailors. King also gave Matthew nine well-regarded convicts, some experienced on the sea, with seven of them hoping for a reduction in their life sentences.

He sought out his 'worthy and brave' friend Bungaree, too, while Governor King authorised Matthew to take the Cadigal youth named Nanbarry,[35] 'a good-natured lad'.[36] Nanbarry was in his early twenties and spoke 'pretty good English'.[37] He was a nephew of the Cadigal leader Colebee,[38] and had already travelled with Matthew and Bungaree to Norfolk Island on the *Reliance*.

The European settlement in Sydney was only three months old when Nanbarry, then a child of about nine, was brought into the town, seriously ill from the smallpox which had killed his parents. Eruptions covered him from head to foot,[39] but he recovered after treatment by surgeon John White, who employed him to shoot

Nanbarry, pictured far right, was a nephew of the Cadigal leader Colebee, and had his upper right front tooth knocked out at an initiation ceremony at Farm Cove, now part of Sydney's Royal Botanic Gardens. Illustration from 'An account of the English colony in New South Wales' by David Collins, 1798 *State Library of NSW FL3729340*

small game.[40] Nanbarry was also one of fifteen Aboriginal youths who had their upper-right front tooth knocked out at an initiation ceremony in February 1795 at Farm Cove, now part of Sydney's Royal Botanic Gardens.[41] In an engraving by James Neagle, Colebee is shown pressing a cooked fish against Nanbarry's mouth to comfort him.[42]

Before leaving Sydney, Matthew gave Governor King two copies of his chart of the south coast of *Terra Australis*. He asked King to send one copy with his letters to the Secretary of the Admiralty, Evan Nepean; the other was to remain in his hands until Matthew's return, or until King should hear of the loss of the *Investigator*, when it too should be sent to the Admiralty.[43]

WITH THE REFITTING, refreshment and preparation of the *Investigator* complete, the harbour pilot came on board before sunrise on 22 July 1802 and fired a gun to alert all crew that it was time to sail. Under a cloudy sky and with fresh breezes from the south-east, Matthew sailed the *Investigator* past Bradleys Head and out through the sandstone cliffs guarding Port Jackson. He turned left.[44] John Murray and his crew in the *Lady Nelson* were close behind.

Matthew had already charted this coastline as far north as Breaksea Spit, off Hervey Bay, filling in much of the detail Captain Cook had missed due to his haste and lack of a chronometer, but he and the crew still kept a close eye on the earlier maps. By the end of their first day's sailing, they had travelled 170 kilometres and were off Port Stephens. By the second day, *Lady Nelson* was having trouble matching the *Investigator*'s pace.

The ships passed Cape Byron and then Cape Moreton on 26 July, and the following day Matthew saw a few Kabi Kabi people in the distance at Double Island Point, as well as one of the two whaling ships operating in the area.[45]

Further on, past the shoals of Wide Bay, as many as fifty of the Kabi Kabi followed the distant ship along the shore. Matthew figured the area was heavily populated with Indigenous people, and when he arrived at the northern end of K'gari (or Fraser Island),

at Indian Head – named by Cook for the many 'Indians', as he
called the Butchulla people – Matthew saw what looked like a large
population. 'There were smokes upon every part near the shore, the
country seeming to be on fire; and at night the fires were useful in
shewing us our situation,' he noted.[46]

Again the *Lady Nelson* lagged behind, out of sight, thwarting
Matthew's hopes of closer exploration of the bays he was visiting.
He anchored the *Investigator* just inside the great sand island's
Sandy Cape on 31 July and waited for John Murray to catch up.
As Peter Good surveyed the lush forests around him, he remarked
that 'it was astonishing to see what Vegetation is produced from
Sand'.[47]

Men from both ships went ashore in three parties: the
naturalists in one, the woodcutters in another, while a group
of six, including Matthew and Bungaree, walked to the north-
eastern part of the cape. Several Butchulla in the distance stopped
and angrily waved branches in their hands, signalling for the
interlopers to go away. Bungaree, now more than ever crucial to
the negotiations, went to them – 'singly, unarmed, and naked' –
and although he could not understand a word they were saying,
he managed to convince them that the visitors came in peace.
After receiving some presents, twenty Butchulla men followed
Matthew and Bungaree to the boats that the crew had rowed
ashore. Matthew gave each of them a large slice of porpoise
blubber – which had been prepared for such gifts – and they ate
with 'apparent satisfaction'.[48]

Peter Good didn't see any of the Butchulla women up close,
but he heard that they wore their long hair tied back.[49] Bungaree
amused the men, Good said, by 'throughing his Spear, which
seemed to surprise them and it appears they are unacquainted
with the use of the Wumora'.[50]

After washing the decks on 1 August – and again without
having seen the mouth of the Burnett River among the
mangroves – the two ships set sail into territory Matthew had not
seen before. He closely followed and corrected Cook's charts as
they travelled north along the lower reaches of the Great Barrier

Reef – 2300 kilometres of coral, rocks and small islands. On 3 August 1802, they entered Bustard Bay, where Cook and Joseph Banks had gone ashore thirty-two years earlier, and two days later they reached what Matthew later named Port Curtis, after Vice Admiral Sir Roger Curtis,[51] who had been in charge of the British station at Cape Town when Matthew was last there. Matthew observed 'many natives and some canoes upon the west shore'.[52]

Robert Brown and the scientists went on shore but were met by a group of the Goeng people, who rushed at them from bushes, shouting what Matthew took to be a war song and hurling stones and sticks at them, apparently for want of spears. After the intruders fired their muskets into the air, the local people ran away.[53] The naturalists found part of a turtle hanging on a tree and saw that the Goeng used the same kind of scoop nets favoured by the people of Hervey Bay. They counted seven canoes made of bark.[54]

When leaving the port, the *Lady Nelson* hit the Facing Island Reef[55] and damaged its keel. On the *Investigator*, when the anchor was hauled up, it became clear that an arm had been broken off because of a flaw in the iron. The negligence in the anchor's construction, Matthew recognised, could have caused the loss of the ship.[56]

Both vessels pressed on with their handicaps, and later that day they crossed the Tropic of Capricorn and entered Keppel Bay. As the *Lady Nelson* was damaged, Matthew used the *Investigator*'s whaleboat to explore the many surrounding inlets over the next eight days. He and Brown traversed the shore and were ravaged by mosquitoes and sandflies. The mangroves were so thick that Matthew mused it might have been easier to climb trees and swing across vines than to try to penetrate 'through the intricate net work in the darkness underneath'.[57]

On Sunday, 15 August, Brown, Peter Good and Ferdinand Bauer led a party exploring on a small island in Keppel Bay. Others went ashore with two dogs and guns, intending to chase the emus they had seen. They had hardly started their hunt when some of the local Wapaburra people appeared in what Good recalled as 'a hostile manner, which caused them to neglect the Dogs & Emoos & attend to their own safety – however after a considerable parley they behaved friendly [and] accepted some trifling presents & went away satisfied'.[58]

Soon after, fifteen of the Wapaburra came down to the beach, put their spears down, and spent time among Matthew's crew 'in a friendly manner'. Some of the officers went ashore and gave them trinkets. It was then discovered, with great alarm, that master's mate Thomas Evans and one of the crewmen were missing. They had been last seen near Cape Keppel, and their absence, as darkness fell, caused considerable unease for Matthew and his men. A boat and crew remained onshore overnight, and Matthew ordered the frequent firing of a musket to help the missing men find their way back to the beach if they still could. But there was no sign of them.

At daybreak, Matthew ordered the firing of a cannon. A musket shot was heard from the mangroves. Another cannon blast roared out but there was no further response. Matthew took a boat to Cape Keppel, while Fowler went in the direction of the musket shot. Still there was no sign of them.

Fowler was getting together a search party at noon when in the distance he saw the two missing men coming along the beach, surrounded by a large number of the Wapaburra, who had saved their lives. Peter Good said the Indigenous men were 'painted with a great variety of figures & colours'. They wore necklaces of 'reed & shells'. Some were more than six feet (183 centimetres) tall.[59]

Good said the bedraggled crewmen made 'a ludicruous' sight. They had been wading for hours through muddy swamps, their 'cloaths all rags without Shoes or Stockings having all Stuck in the mud'.[60]

Thomas Evans told Matthew that he and his companion had 'incautiously strayed away by themselves' and become entangled in a mangrove swamp. The Wapaburra had found them, and – after getting over the shock of seeing these muddied, miserable white men on their land – offered them roasted duck, fish and fresh water before kindly guiding them back to the beach. Fowler gave the rescuers some gifts – mirrors, hatchets and red nightcaps – but the Wapaburra were startled by the mirrors and could not bear to look at their reflections. Fowler also sent the ship's musicians ashore to play an impromptu concert, but the fife, fiddle and drum were caustic to the Wapaburra ears and they ran off until the screeching stopped.[61]

All along the coast Matthew kept a lookout for any 'refuse' that might have been thrown up by the sea, believing that if there was any wreckage from the final voyage of the 'unfortunate Lapérouse', it was likely to wash up somewhere along this great reef. He found none.

On 21 August,[62] with the *Lady Nelson* struggling to stay within sight of the *Investigator*, Matthew reached a place which offered excellent anchorage – a place Cook had missed. He later named it for Captain James Bowen, a naval war hero who had been master of Lord Howe's flagship at the Glorious First of June.[63] The crewmen immediately began cutting down pine trees on the shore to repair the *Lady Nelson*'s keel. Matthew offered to name the highest point in the vicinity after the man who could climb the peak first. It became Mount Westall.

Matthew defied his instructions – which were to reach Torres Strait as quickly as possible – though he felt the Admiralty had given him leeway to rely on his own judgement, and he was enjoying the chance to correct the maps that Cook, his idol and exemplar, had made.

'It has before been said that captain Cook had no time keeper in his first voyage; nor did he possess many of our advantages in fixing the positions of places; it cannot therefore be thought presumptuous, that I should consider the *Investigator*'s longitude to be preferable,' he wrote later.[64]

Matthew steered his ship into Shoalwater Bay through a narrow passage off what Cook had named Townshend Island – for Tommy Townshend, Lord Sydney. Matthew called the ten-kilometre-long strait 'Strong-Tide Passage', and it proved dangerous. After a day of explorations, the new cutter was being hoisted on board the *Investigator* when it flipped in the rapid waters and the boatman was flung overboard. He was rescued by the men in the *Lady Nelson*'s boat, but the cutter was carried away at great speed by the fast-flowing tide. Two men pursued it in another boat, but after spending the night in treacherous waters, they returned with no sign of the cutter.[65]

There was trouble among the crew, too. Matthew suspended Charles Douglas, the boatswain, for drunkenness and for fighting with William Job, a crewman. He issued Job with twelve lashes, after 'explaining the heinousness of striking a superior officer'.[66]

Trim was unconcerned about rank, regarding himself as master of all. Whenever the hunting dogs were playing on the deck, Trim would go in among them with a stately air and brush them out of his way, warning them to keep their distance by extending his claws. He would also chase the dogs away if he wanted their spot in the sun.[67]

The relationship between the visitors and the local Darumbal people of Shoalwater Bay was far more cordial. The ships spent a week there, and the scientists who landed on the south shore met a party of nineteen of the Darumbal. These people fished from bark canoes and had harpoons for hunting turtles, with the points made from neatly fitted sharp quartz.[68]

MATTHEW LEFT Shoalwater Bay at noon on 4 September 1802, and the next morning sailed into Thirsty Sound. At Pier Head, Matthew speculated that the stone of the area had played havoc with Cook's compass, as he had reported great variations in his readings. The ships spent almost three weeks in Thirsty Sound and the adjoining Broad Sound as carpenters shaped the pine logs from Port Bowen to repair *Lady Nelson*'s keel. Matthew planned to spend a few days 'adding to the accuracy and minuteness' of Cook's chart, while 'the botanical gentlemen would have a fresh field opened to them'.[69]

Broad Sound, fifty kilometres long and twenty kilometres wide, has the greatest tidal range on Australia's east coast – around nine metres – and is ringed by a low-lying coastline of mangroves and mudflats. These tides played havoc with the two ships. On 13 September, the *Investigator* was caught on a bank of quicksand and spun around, 'heeling considerably', Matthew said, 'from the rapidity of the tide'.[70]

As the water rose, sails were struck and the best bower anchor was let go, helping the ship to right itself. The *Lady Nelson* struck another shoal, and this time lost two sheets of copper from its sheathing and part of its sliding keel, which floated away. At low tide it was surrounded by a stinking web of dry mud.[71] Samuel Flinders vexed his brother by again forgetting to wind the marine chronometers, the devices crucial to establishing the ship's longitude, as Matthew noted:

> This fresh difficulty was very embarrassing. To go away for Torres' Strait and the Gulph of Carpentaria without good rates, was to cripple the accuracy of all our longitudes; and on the other hand, the expected approach of the contrary monsoon on the North Coast admitted of no longer delay in Broad Sound.[72]

On 28 September, the two ships reached the Northumberland Group of islands. They spent the next five days in a heavily forested tropical paradise – Matthew called it 'one of the prettiest little places imaginable' – a small group of islands that provided them with fresh water, good hauls from the fishing net and pine logs. Matthew called the cluster the Percy Isles after the family name of the dukes of Northumberland.[73] While heating water for washing, one of the servants accidentally set off a huge grassfire on one of the islands and it spread rapidly, lighting up the night sky for miles around and presenting an eerie spectacle.[74]

For the next two weeks, Matthew sailed along the great reef, searching for a way to cut to the open sea through this seemingly endless line of treacherous coral and shoals that had almost killed Cook and his crew on the *Endeavour*.

The narrow channels through the coral created raging tides, while heavy breakers smashed into the reef in other areas. In many places the coral was barely submerged. Lookouts had to deal with cloud cover and glare, trying to spot clear water and a myriad of hazards as the dazzling reflection on the sea half-blinded them. The whaleboat went ahead of the two ships, and its crew told Matthew that what often looked like coral were sometimes mere shadows caused by the clouds and eddies.

Still, by 8 October 1802, Matthew felt that he was surrounded by reefs. They were of dazzling shapes and hues: 'wheat sheaves, mushrooms, stags horns, cabbage leaves ... glowing under water with vivid tints of every shade betwixt green, purple, brown, and white ... but whilst contemplating the richness of the scene, we could not long forget with what destruction it was pregnant.'[75] Matthew aimed to carry his men to the 'windward of the high breakers'.[76]

THREE DAYS LATER, Matthew tried to get through a narrow gap in the coral. It was a disaster. Both ships lost anchors, as the anchor ropes were unable to cope with the tides running 'with extraordinary violence' over the jagged rocks and coral.

He was not too proud to admit he'd misjudged the situation badly: 'My anxious desire to get out to sea, and reach the North Coast before the unfavourable monsoon should set in, had led me to persevere amongst these intricate passages beyond what prudence could approve.' Now he knew that 'certain destruction' awaited if he tried again – and that even if it delayed him reaching the gulf, it was safer to stay between the shore and the larger reefs, 'until a good and safe opening should present itself'.[77]

On the night of 12 October, Matthew anchored the *Investigator* in about fifty-five metres of water, but during a change in watch before dawn, the anchor was dragged by the tide more than three kilometres north-east among the reefs, where it again caught on the coral. This change of place had not been perceived, Matthew admitted, 'but it might have been attended with fatal consequences'.[78] Matthew blamed his careless brother for the mishap.

THE *LADY NELSON* HAD BECOME more of a hindrance than a help. Of its three anchors, one had been lost and another badly damaged, and the brig now 'sailed so ill, and had become so leewardly since the loss of the main, and part of the after keel, that she not only caused us delay, but ran great risk of being lost'. Instead of saving the crew of the *Investigator* in case of accident, which was one of the principal objects of her voyage, it was probable, Matthew felt, that the *Investigator* would soon have to save the *Lady Nelson*.[79]

Reluctantly, he sent Murray and his crew back to Sydney. Nanbarry decided to return home with Murray, while midshipman Denis Lacy left the *Lady Nelson* to sail with Matthew again. Matthew kept the *Lady Nelson*'s boat to make up for the *Investigator*'s lost cutter.

The ships parted on 18 October 1802, Murray taking letters back to Sydney for Matthew in a slow ordeal that saw him using a makeshift anchor made of two swivel gun cannons lashed together. The *Lady Nelson* finally arrived in Sydney on 22 November with Matthew's mail, including a report for Governor King and the Admiralty, whose members would not be pleased, he knew, with his slow going over territory Cook had already mapped.

In one of the letters Matthew told Ann – who was living the life of a widow in a small English village – that she was 'not one day forgotten'. 'Be happy, my beloved, rest assured of my faith,' he wrote, 'and trust that I will return safely to sooth thy distress, and repay thee for all thy anxieties concerning me.'[80]

Matthew might have been promising more than he could deliver.

Chapter 19

'The two old men appeared, to my surprise, to have undergone circumcision; but the posture of the youngest, who remained sitting down, did not allow of observation being made upon him.'

MATTHEW FLINDERS, ON MEETING KAIADILT PEOPLE IN
THE GULF OF CARPENTARIA.[1]

THE HIGH SURF POUNDING against the great reef and the fast, shifting tides were like deadly traps among the jagged rocks just below the surface of the sea. As Matthew and his lookouts continued to search for an escape through this immense coral barrier to the safety of open water, things went from bad to worse.

On 19 October 1802, the wind blew fresh from the east all night, raising a short swell that tested the ship's battered timbers more than anything Matthew had encountered since leaving Sydney. The *Investigator* was now leaking badly again, taking in five inches of water an hour.[2]

Matthew had been inside the reef for two weeks and had sailed for 800 kilometres along it, looking for a way through its fatal traps. Finally, not far from Holbourne Island, on the morning of 20 October, as his anxiety rose over the state of his leaking ship, he turned left and steered northward, 'round the west end of the great reef'. There was a heavy rain shower, the first since leaving Sydney, and in the rough weather some of the scientific men were seasick.[3] There were still many small reefs in sight, and huge breakers pounding in – but those breakers created a hope for Matthew that the ship might be near an opening in the coral.

'Although caution inclined to steering back towards the land,

312

this prospect of an outlet determined me to proceed,' Matthew wrote. '... We were successful. At four, the depth was 43 fathoms [seventy-nine metres], and no reefs in sight; and at six, a heavy swell from the eastward and a depth of 66 fathoms [120 metres] were strong assurances that we had at length gained the open sea.'⁴ His route to safety, north-east of the modern city of Townsville, is now known as Flinders Passage.

Matthew figured that the great coral fence stretched all the way to Torres Strait and New Guinea, 'and was not to be equalled in any other part of the world'. He named this immense natural wonder the Great Barrier Reefs,⁵ and warned other captains coming this way: 'If he do not feel his nerves strong enough to thread the needle, as it is called, amongst the reefs, whilst he directs the steerage from the mast head, I would strongly recommend him not to approach this part of New South Wales.'⁶

FOR THE NEXT WEEK, Matthew sailed slowly and cautiously north towards Torres Strait, not travelling at night in case of more barely submerged coral. Matthew named a chain of coral banks the Eastern Fields. He also thought of the other dangers awaiting, remembering all too well the skirmishes with the Indigenous people when he had sailed with Bligh on the *Providence* a decade before. Matthew had his men practise daily with the heavy cannons, the swivel guns and their muskets.⁷

In 1792, Matthew had spent nineteen days watching Bligh and his two ships crawl through the labyrinth of shoals in Torres Strait, and this time he elected to take a different route – that favoured by Captain Edward Edwards of the *Pandora* in 1791. On 29 October, he reached the largest of the Murray Isles,⁸ the most easterly group in the strait, about 185 kilometres north-east of Cape York.

A number of poles were standing up in various places between the islands; from a distance they appeared like the masts of canoes. Matthew swallowed hard and his brow furrowed: he feared the local people had gathered a fleet. On sailing closer, though, he saw that the poles were upon the reefs, and were probably to help with fishing.

The *Investigator* had scarcely anchored when 'between forty and fifty' Torres Strait Islanders came rushing towards the ship in three canoes. From a short distance, they held up coconuts, joints of bamboo filled with water, plantains, bows and arrows, all the time shouting out, 'Tooree! Tooree!' – their word for iron.

> A barter soon commenced, and was carried on in this manner: a hatchet, or other piece of iron (tooree) being held up, they offered a bunch of green plantains, a bow and quiver of arrows, or what they judged would be received in exchange; signs of acceptance being made ...[9]

The barter was friendly enough but Matthew remained wary, remembering the attacks on the *Providence* and the *Assistant*, and recalling that a year later, in 1793, the islanders killed six of the crew from the ships of William Bampton and Matthew Alt after they had been caught polluting Darnley Island's only source of fresh water. In retaliation, Bampton and Alt ordered the destruction of huts, canoes and gardens of the local people, and several islanders were killed. Matthew kept his men under arms, but thankfully 'no intention of hostility was manifested' by the local people.

Matthew and his crew engaged in vigorous barter with the Torres Strait islanders at the Murray Isles as captured by William Westall. *nla.obj-135753549-1*

He described them as being dark in colour; they were 'active, muscular men, about the middle size, and their countenances expressive of a quick apprehension … they also go quite naked; but some of them had ornaments of shell work, and of plaited hair or fibres of bark, about their waists, necks, and ancles'.[10] Some wore earrings of pearl shell, and others had large shells covering their 'privy parts'.[11] Matthew noted that Bungaree 'could not understand any thing of their language, nor did they pay much attention to him; he seemed, indeed, to feel his own inferiority'.[12]

Soon after daylight the next morning, 30 October, seven canoes came up to the *Investigator* as Matthew was preparing to sail on. About twenty islanders came on board to barter for more iron, bringing decorative pearl-oyster shells, necklaces and baskets. One traded a bow, which had attached to it a piece of blue striped cotton cloth, apparently of 'European manufacture'.[13]

THE *INVESTIGATOR* WAS now leaking by as much as fourteen inches (thirty-five centimetres) an hour in a fresh side wind,[14] as the men worked the pumps feverishly and Matthew set a course west by south-west for the 'Gulph of Carpentaria'. There was nowhere suitable among the islands to turn the vessel on its side so it could be inspected by the carpenter, Russell Mart, and caulked with oakum, but wherever possible Matthew anchored so his scientists could study the fauna and flora on some of the 274 islands in Torres Strait, even though the boat was almost swamped in high surf on 31 October.

The next day they reached the lee of Wednesday Island, named by Captain Bligh when he sailed in the open boat to Timor,[15] and Matthew and Samuel took bearings which, they believed, showed that Cook's chart for the area was out in some places by more than 110 kilometres.[16] The crew were puzzled by tall, pale brown conical structures that they saw on some of the islands. They looked like sentry boxes or the spires of churches, and for a while Matthew suspected they might be huts. The following day, 2 November, just east of Bligh's Thursday Island, Matthew named 'Good's Island, after Mr. Good, the botanical gardener'.[17] Good and the other scientists went ashore and discovered that the conical structures were 'ant hills, of eight or more feet high'.

'The insects which inhabit, and I suppose erect these structures,' Matthew wrote, 'are small, reddish, with black heads, and seemed to be a sluggish and feeble race.' The crew found the black flies 'excessively numerous here' too, and almost as troublesome as Dampier had described them on the north-west coast of New Holland more than a century earlier.[18]

A day later, Matthew left the Prince of Wales Channel and sailed into the Gulf of Carpentaria, having crossed Torres Strait in six days – although now, with a proper understanding of the danger spots, he believed he could do it in three.[19] He would have spent longer surveying the strait but was worried about the threat of the approaching monsoon, though he knew that, once again, he had defied the orders of the Admiralty, who had insisted on a thorough investigation of that vital waterway. The route could take a month and a half off the journey from Sydney to the British East India Company's base in Calcutta. The company ships generally sailed north of New Guinea, though Matthew conceded that, even with accurate charts, this new route was 'not without difficulties and dangers'.[20]

MATTHEW WAS NOW IN THE Gulf country, a place of mangroves, mud, crocodiles and big rivers. The *Investigator*'s visit was the first major exploration of the area since Jan Etienne Gonsal[21] led two Dutch East India Company ships to the gulf in 1756. Gonsal was investigating reports from a Chinese trader who had drifted far south from Timor and reported meeting friendly people in the gulf, 'both male and female, wholly naked and unarmed ... people of an above-average size and sturdiness, very black and hair frizzy, but rather long'.[22]

Matthew began sailing down the east coast of a gulf more than 500 kilometres wide and 600 kilometres deep, staying about seven kilometres off its low, sandy shore. He was cross-checking features with a 1663 Dutch map, and on 5 November 1802 marked off a shallow estuarine bay as the Batavia River on his chart. The opening is now called Port Musgrave, and it is the place where two major rivers, the Wenlock and the Ducie, discharge.

The next day, Bungaree approached a group of spear-carrying Indigenous men of the Wik people near the sandy beach of a small inlet. They retreated, though, and all his endeavours to communicate with them were unsuccessful. The *Investigator* had reached a river mouth that marked the first recorded landfall in Australia by a Dutch explorer, Willem Janszoon,[23] in 1606. Janszoon named it 'River with the Forest'; it is now called the Pennefather. Nearby, Matthew noted the Coen River on his Dutch chart.

He sailed on, and on 9 November entered a wide shallow bay, naming Duyfhen Point in honour of Janzsoon's vessel[24] and Pera Head after the ship of another Dutchman, Jan Carstensz, who sailed into the gulf in 1623. The land was remarkable, Matthew said, 'for having some reddish cliffs in it'.[25] The bay is now called Albatross Bay and is home to the major bauxite-mining town of Weipa.

Matthew continued south, and on 10 November was off the place Janszoon had name Cape Keerweer, before the Dutchman turned his ship around and ended his exploration of the gulf. Sometimes the water along the coast was so shallow that Matthew had to stay a long distance out to sea. Often the land was barely visible, let alone the rivers marked by the Dutch. He and Robert Brown thought that many of the rivers marked by earlier expeditions 'were nothing more than creeks or lagoons'.[26]

By 14 November, the land was trending west. What was marked on Matthew's Dutch chart as Van Diemen's River was an extensive piece of flat land that was nearly dry. 'If this place had any title to be called a river in 1644,' Matthew noted, 'the coast must have undergone a great alteration since that time.'[27] There was little hope that the gulf would lead the *Investigator* into the interior of *Terra Australis*, and before long the land was trending north-west.

On 16 November, after having surveyed more than 700 kilometres of the gulf's eastern coast, Matthew named Sweers

Island in honour of Cornelius Sweers, one of the signatories to Abel Tasman's sailing orders. Matthew called the high point on the island Inspection Hill, because it was here he found a beach where the *Investigator* could at last be checked for rotting timber. The extent of the damage would turn his stomach.

WHILE THE AILING *INVESTIGATOR* was being examined, Nicolas Baudin's two French ships left Sydney on 18 November 1802. In case Napoleon thought to send his navy south, Baudin's men made reports on the defences at the penal settlement and the growing volatility among the Irish convicts being brutalised by their British jailers.

Baudin granted permission for the Sydney surgeon James Thomson and his family to sail on the *Naturaliste* back to Europe, and in return Governor King granted Baudin permission to buy a twenty-ton locally built schooner to help with his mapping. The French commander called the little vessel the *Casuarina* after the wood used in its construction. The *Naturaliste* would sail for France carrying the expedition's thousands of natural history specimens, while the *Géographe* would continue its surveys along the southern coastline of New Holland.

The day before his departure, Baudin sent a letter of thanks to King for his hospitality, and gave King's wife a donation of £50 for her new orphanage. He also took with him an eighteen-year-old convict girl, Mary Beckwith, as his mistress.

Some of the crew were aghast at Baudin taking the girl with him on the ship, and according to one of the artists on the voyage, Charles Alexandre Lesueur, other members of the crew also took advantage of the teenager. When the *Géographe* reached Kupang, Baudin wanted to leave Mary there, but, now drinking heavily, she threatened to commit suicide. Deeply distraught, Mary eventually returned to the captain's cabin.[28]

Baudin had also left Governor King with an original and twelve copies of a letter addressed to the administrators of the French settlements at Mauritius and Reunion Island, relaying the generosity shown to the French expedition in Sydney and endorsing any commander who carried the letter for special treatment from French officials if they should stop at those places.

Baudin left blank spaces for the names of the ships and the captains that King wanted to recommend. The governor apparently put the letters in his drawer and forgot about them.

MATTHEW'S CHIEF CARPENTER, Russell Mart, did some patchwork repairs on the ship while Matthew searched for the best place to begin a major overhaul. Matthew surveyed some other islands near Sweers. He named one for Lieutenant Fowler, and one for the expedition's miner, John Allen. He named another after Lord William Bentinck, then the British governor of Madras.

On one of the smaller islands, he saw some Kaiadilt people, and walked after three men who were dragging small rafts made from straight mangrove branches lashed together. The Englishman and the Kaiadilt sat down together and exchanged gifts so that 'a friendly intercourse' was established. Matthew recalled:

> Two of the three men were advanced in years, and from the resemblance of feature were probably brothers. With the exception of two chiefs at Taheity, these were the tallest Indians I had ever seen; the two brothers being from three to four inches higher than my coxswain, who measured five feet eleven [180 centimetres]. They were not remarkable for being either stout or slender; though like most of the Australians, their legs did not bear the European proportion to the size of their heads and bodies … Their features did not much differ from those of their countrymen on the South and East Coasts; but they had each of them lost two front teeth from the upper jaw. Their hair was short, though not curly; and a fillet of net work, which the youngest man had wrapped round his head, was the sole ornament or clothing seen amongst them.[29]

After five minutes together, the old men asked to see Matthew's boat – which was a rather more solid construction than their rafts –

and they began walking towards it 'hand in hand'. But the Kaiadilt men stopped halfway, and Matthew realised they were trying to lead the Europeans away from the Kaiadilt women, who were 'very quietly picking oysters' in the distance.

'It was not my desire to annoy these poor people,' Matthew recalled, 'and therefore, leaving them to their own way, we took an opposite direction to examine the island.'

Russell Mart needed to examine the *Investigator* much more closely, and Matthew found a place close to the beach, in a channel between Sweers and Bentinck islands. It was a safe anchorage that Matthew called Investigator's Road. Lieutenant Fowler established a camp on shore with sailors and marines, and the carpenters began caulking the ship. But Mart and John Aken, the ship's master, reported that the *Investigator* was rotting all over – 'the planks, bends, timbers, tree-nails'.[30] As the crew worked on the ship, one of the team carved the legend 'Investigator 1802' into the trunk of a native hackberry tree on Sweers Island – to which other explorers would add in later years.[31]

On Sweers Island, Matthew's men found seven human skulls and many bones lying together, near extinguished fires. On the western beach, there was a piece of teak, seven feet (two metres) square, which, according to Mart, had been a quarterdeck carling of a ship. On Bentinck Island, there were the stumps of at least twenty trees, which had been felled with an axe or some sharp instrument made from iron; not far away were the broken remains of an earthen jar. Matthew deduced that a ship from the East Indies had been wrecked here, most likely two or three years earlier, and that some of the crew had been killed while others had escaped to the mainland on log rafts. He was anxious to trace the route and look for survivors. But with a dilapidated ship, he knew that would be difficult.

On 26 November, Mart and Aken presented their captain with a detailed report on the extensive damage to the ship. It was damning. Out of ten top timbers on the port side, near the fore channel, five were entirely rotten and the ship 'would immediately go to pieces' if forced on shore.[32]

Mart and Aken told Matthew: 'From the state to which the ship seems now to be advanced, it is our joint opinion ... that if she remain in fine weather and happen no accident, she may run six months longer without much risk.' But, they warned him, 'in twelve months there will scarcely be a sound timber in her'.[33] They were fearful that the ship would fall apart in 'Boisterous Weather'.[34] Even Trim the cat was suffering – from anxiety, from the heat and from want of his usual fresh food. 'This worthy creature became almost grey, lost much of [his] weight, and seemed to be threatened with a premature old age,'[35] Matthew explained.

MATTHEW WAS STUCK. To sail on west would likely pitch him into the north-westerly monsoon. Trying to go back to Sydney the way he came presented all manner of traps among the reefs and shoals of Torres Strait and the Great Barrier Reef.

For the time being, Matthew decided to stay in the relative protection of this part of the coastline, out of the savage winds. He was 'determined to go on in the examination of this gulph' – so long as the north-west monsoon did not prove too great a hindrance.[36] If the weather proved foul after the *Investigator* left the gulf, Matthew would steer for the nearest Dutch port in the East Indies.

So, on 28 November 1802, after the carpenters had made all the repairs they could, Matthew sailed on for the north-west, exploring the islands adjacent to Sweers, in what he called the Wellesley Islands. He named the largest of them Mornington, after Richard Wellesley, the Earl of Mornington. He named another pair the Bountiful Islands, after the feeding frenzy that gripped his men on 3 December when they saw turtle tracks on the beach.

The impatient sailors caught three large turtles from the boat, breaking Matthew's harpoon in the process. The sandy beach was full of turtle holes, and from one they filled a hat with turtles eggs. From another they 'took a swarm of young ones, not broader than a crown piece', which Matthew later found crawling in every part of the boat. After sunset, his men had filled the boat with turtles and Matthew hastened to the ship. He ordered Fowler and a party of men to remain on shore all night in the hope of catching more.

Their success was so great that it was necessary to hoist out the launch as well; and it took nearly the whole day to load their catch. They returned many lucky turtles to the beach, but still stowed forty-six on the *Investigator*, the least of them weighing 250 pounds (114 kilograms), and the average about 300 pounds (136 kilograms).[37] Turtle soup became the flavour of the month, although they also caught seven tiger sharks, as the savage sea creatures waited in the water to feed on the turtle hatchlings. Bauer, Brown and Good were deceived by the pleasing appearance of nuts from cycad palms, and spent a night violently ill.[38]

A week later, and with Matthew constantly worried about his ship falling apart, Samuel Flinders again forgot to wind the timekeepers, throwing Matthew's navigational calculations awry. Matthew continued to steer the ship with his heart in his mouth and his eyes peeled for shallows. The fragile vessel twice hit submerged rocks, only shifting because it was under full sail.[39]

Westall's drawing of an Indigenous hut in an area of hilly islands, which Matthew called the Sir Edward Pellew Group. *nla.obj-135754240-1*

The *Investigator* held together, though, and while the low, swampy coastline prevented Matthew from seeing the mouth of what would be called the McArthur River, there were other astonishing sights that gripped his attention.

The ship was now on the western shore of the gulf, and Matthew named an area of hilly islands the Sir Edward Pellew Group, after the British naval hero. He called the largest of the islands Vanderlin, after a name on his Dutch chart. The summer heat was stifling. Trim was often flat out on the deck, exhausted, and the men made wide-brim cabbage-tree hats from a new species of the plant they found in the week that Matthew surveyed the area.

Matthew had originally conjectured that the skeletons discovered among the Wellesley Islands belonged to shipwrecked Dutchmen, but he now believed they were most likely the remains of Chinese sailors who had visited the area to harvest wild nutmeg growing on the islands, which Matthew thought was no competition to the valuable spice from the Moluccas.[40]

Besides pieces of earthen jars and trees cut with axes, the crew found remnants of bamboo latticework, and palm leaves sewed with cotton thread into the kind of hats the Chinese wore. There were also the remains of blue cotton trousers. There was a wooden anchor and three boat rudders. What puzzled Matthew the most was 'a collection of stones piled together in a line, resembling a low wall, with short lines running perpendicularly at the back, dividing the space behind into compartments'. In each of these were the remains of a charcoal fire, and all the wood nearby had been felled by an axe. On another island Robert Brown saw a similar construction, with thirty-six partitions. More than an acre and a half (two-thirds of a hectare) of mangrove nearby had also been cut down.[41]

On Christmas Eve 1802, Peter Good and the other scientific men found a skeleton in a cave. There were wisps of brown hair on the skull. The skeleton had been tied into a roll of bark, and the colour of the hair convinced them it belonged to a European man.[42]

Death would also come calling soon enough for Matthew's crew.

Chapter 20

*'We had scarcely reached the ship, when the report of muskets was
heard; and the people were making signals and carrying some
one down to the boat, as if wounded or killed.'*

Matthew Flinders, recalling a fatal skirmish with Yolngu
people near Groote Eylandt.[1]

THE NEW YEAR OF 1803 started with driving winds,
thunder, lightning and rain – a portent for another disaster
ahead.[2] With careful sailing, the *Investigator* was holding together,
but on 3 January a heavy wind filled the whaleboat with water
and two men were thrown out. William Murray, who had been
flogged earlier in the voyage for fighting, failed to surface, but
William Job, who had been whipped for hitting a superior officer,
was hauled to safety.[3] The whaleboat was dragged back to the
ship, intact but much damaged.

The accident happened not far from the place marked on the
Dutch chart as Groote Eylandt, which translates as 'Great Island'.
At fifty kilometres long and sixty kilometres wide, it would prove
to be the largest island in the Gulf of Carpentaria, and the fourth-
largest in New Holland.

Matthew started sailing around the island to map it the next
day, while the botanists went on shore, finding freshwater lakes
and a burial ground of the Anindilyakwa people. Each corpse had
been placed in the hollow of an upright tree, though a number
of the trunks had fallen over with time. Some of the bones were

painted with streaks of red.[4] Matthew spent a week charting the island, and on 14 January decided to take bearings from the high cliffs of an islet at Groote Eylandt's northern end. But there were so many deep chasms on the rocky outcrop that it proved impossible for him to reach the top in the short time he had to spare. He called the place Chasm Island.

Inside some caves, his men found drawings on the stone walls 'made with charcoal and something like red paint upon the white ground of the rock'.

> These drawings represented porpoises, turtle, kangaroos, and a human hand; and Mr. Westall, who went afterwards to see them, found the representation of a kangaroo, with a file of thirty-two persons following after it. The third person of the band was twice the height of the others, and held in his hand something resembling the whaddie, or wooden sword of the natives of Port Jackson; and was probably intended to represent a chief ...[5]

Westall made copies of this extraordinary cave art.

GUNFIRE AND SCREAMS shattered the morning stillness on 21 January in a place Matthew called Blue Mud Bay, north-west of Groote Eylandt.

He had anchored the *Investigator* off a long, narrow island he named Woodah, because it was shaped like the 'whaddie, woodah or wooden sword used by the natives of Port Jackson'. Matthew sent a party of men to a small island to the west to cut wood, and another to haul the seine. The botanists also landed, and Matthew went to take bearings from the western end of the island. Most of his men were armed, as footprints were seen on the sand and Matthew was unsure what sort of reception the newcomers would receive from the Indigenous people.

Matthew finished his calculations and began walking through thick brush back to the boat. Westall was with his servant, making sketches, when he saw six of the Indigenous Yolngu-speaking people arrive in a canoe from Woodah. He and the servant retreated towards the wooding party, but the Yolngu followed.

When the warriors appeared on the brow of a hill, John Whitewood, the master's mate, and some of his woodcutters 'went to meet them in a friendly manner'. At this point Matthew and his men appeared from the brush; perhaps suspecting they were being surrounded, the Yolngu fled.

The Yolngu had spears, 'but from the smallness of their number, and our men being armed', Matthew did not sense any danger and headed back to the ship with his party. Then he heard gunfire.

> I immediately despatched two armed boats to their assistance, under the direction of the master [John Aken]; with orders, if he met with the natives, to be friendly and give them presents, and by no means to pursue them into the wood. I suspected, indeed, that our people must have been the aggressors …

At 5 p.m. Whitewood was brought on board, with four spear wounds to his body.[6]

Peter Good told Matthew that Whitewood, who had a loaded musket, and John Allen, who was unarmed, went up to the Yolngu and the locals 'divided themselves 3 to each, those nearest Mr Allen had laid down their spears & had exchanged a green bough with him'.[7] Whitewood, he said, was still holding his musket but put out his hand to take what he thought was a spear being offered as a gift. The Yolngu man apparently thought he was trying to take it by force, and rammed the spear into Whitewood's chest. Whitewood tried a shot at his assailant but the gun misfired.[8]

He and Allen ran back to their companions. Allen dodged the spears coming his way but Whitewood was hit repeatedly; he had a spear sticking out of his side as he got a shot away before his men helped him onto the boat. Two more shots were fired by Whitewood's companions, but they missed. The Yolngu ran off, but not without taking away one of the crewmen's straw hats, which had been dropped.

Also helped into the boat was one of the marines, Thomas Morgan, who had worked in the summer heat cutting wood all day without a hat, before collapsing from the effects of the sun.

When John Aken, the ship's master, arrived on shore and learnt what had happened to Whitewood, he hurried as fast as he could in the whaleboat to the eastern end of the island, intending to seize the Yolngu canoe. Disregarding Matthew's orders, Aken sent Denis Lacy with the woodcutters into the scrub to intercept Whitewood's assailants. At dusk, they fired on three of the Yolngu as they pushed off in their canoe. One fell, and the others leapt out into the water.

The sailor who claimed to have hit the dead man found him at the bottom of the canoe, still with a straw hat on his head 'which he recognised to be his own'.[9] But in picking up the hat and showing his friends, the sailor upset the narrow vessel and the body fell into the water and sank. The canoe was towed to the shore, and Aken returned with it at 9 p.m. It was thirteen and a half feet (four metres) long and just two and a half feet (0.8 metres) wide, and was made from two pieces of bark sewed together lengthwise.

Matthew was intrigued by the construction of the canoe, but furious that Aken had 'acted so contrary' to his orders by chasing the Yolngu into the scrub, with fatal consequences. Matthew lamented that there was nothing he could do now that 'the mischief was done'.

The next morning, 22 January, a boat was sent to search for the Yolngu man's body, with Brown and Dr Hugh Bell wanting to examine it 'for anatomical purposes', and Westall wanting to sketch it for posterity.

> The corpse was found lying at the water's edge ... with the head on shore and the feet touching the surf. The arms were crossed under the head, with the face downward, in the posture of a man who was just able to crawl out of the water and die. A musket ball had passed through the shoulder blade, from behind; and penetrating upwards, had lodged in the neck.[10]

John Whitewood survived his ordeal, but Thomas Morgan didn't. He died 'in a state of frenzy, the same night'.[11] Matthew named Morgan's Island in his honour.

Matthew wrote that the fatal skirmish did not accord 'with the usually timid character of the natives of Terra Australis', but suspected their aggression was perhaps a carryover from previous confrontations between the local people and 'Asiatic visitors',[12] who, he learnt, had also carried firearms.[13]

Matthew made sure that Bungaree was at the forefront of all future meetings with the Indigenous people of the area as the *Investigator* reached the north-western corner of the gulf. As they entered Caledon Bay on 3 February, Matthew sent Lieutenant Fowler and Bungaree on shore to communicate with some of the Yolngu, and to search for water. There was 'a friendly intercourse in which mutual presents were made', and Fowler returned to the ship with the information that fresh water was plentiful.

TWO DAYS LATER, ON 5 FEBRUARY 1803, George Bass sailed out of Sydney's harbour on the *Venus* bound for Tahiti.

Bass may also have planned to engage in contraband trade in Chile, where only Spanish merchants were permitted to import goods. Matthew's friend carried a diplomatic letter from Governor King saying that his sole purpose on the western coast of South America was to buy provisions for the settlement in New South Wales.

As Bass and his crew of twenty-five headed out into the blue Pacific, Matthew was stepping onto the shore at Caledon Bay, delighted when about a dozen of the Yolngu ran out to greet him. He said they expressed 'much joy, especially at seeing [Bungaree]'.[14] Two of them helped haul in the seine net, and received a portion of the fish as a reward. Bungaree could not understand their language, but Brown and Dr Bell made a list of about fifty words they used. Matthew had two tents erected, and the Yolngu guided the 'botanical gentlemen' into the bush 'arm in arm'. Then things deteriorated.

One of the Yolngu snatched a hatchet and they all ran off. Matthew was willing to sacrifice it for peaceful relations. The Yolngu returned, but when one of them took a musket and ran, Matthew demanded that his men retrieve it. A musket was fired as a warning but it only made the culprit run faster, and his companions disappeared into the undergrowth.

When some of the Yolngu were spotted, Matthew made it known to them that he would reward the return of the gun with another hatchet, and the musket was eventually brought back. Even though the stock was broken and the ramrod was missing, Matthew still gave the hatchet as a reward. But the next day, when the Yolngu took a wooding axe, essential for the ship, Matthew emulated Captain Cook's last, fatal decision and captured two of their people, a man and a boy of fourteen who said his name was Woga. Matthew kept the boy but let the man go, telling him that unless the axe was returned, Woga would be carried away on the ship.

In the evening, Matthew and his men rowed Woga to a place where the Yolngu gathered and he again demanded the return of the axe. Woga seemed to be imploring his people to give it back – but as best Matthew and Bungaree could interpret their reply, the Yolngu explained that the thief had been beaten before running away.

Two of the Yolngu came forward carrying a young girl in their arms. Matthew later wrote that they seemed to be offering her to Bungaree to entice him onto the shore 'for the purpose, apparently, of seizing him by way of retaliation'.[15]

Instead, Woga was carried on board the *Investigator*, with a 'great deal of crying, entreating, threatening, and struggling on his part'. But Matthew ensured that he ate heartily, and soon the boy was laughing and noticing everything, 'frequently expressing admiration at what he saw, and especially at the sheep, hogs, and cats'.[16] Perhaps his apparent happiness was the survival strategy of a boy who had been kidnapped by people who must have seemed like aliens.

The next morning, 8 February 1803, the scientific team went ashore and they were quickly surrounded by angry Yolngu men with poised spears. The pointing of muskets in return stopped their fearsome threats only momentarily, and when it appeared that the spears were about to be launched, Matthew's men fired twice. The Yolngu ran, but Robert Brown still thought it unsafe for his men to stay onshore and returned to the ship. None of the Yolngu dropped from the gunfire, but since the muskets were loaded with wide-spreading buckshot, rather than the deadlier musket balls, it was supposed that some must have been wounded.[17]

By the second evening of Woga's captivity, Matthew had given up on the axe being returned. He decided it was unjust and provocative to keep the boy any longer.

Woga was taken onto the shore. Matthew gave him some clothing and presents, and Woga walked away leisurely for 200 metres – but then, looking behind him at his captors, 'took to his heels with all his might'.

On 10 February 1803, the *Investigator* set sail again, and two days later Matthew named a 'smooth, grassy' cape at the western edge of the Gulf of Carpentaria after the Dutch ship *Arnhem*, which had visited there in 1623.

At Melville Bay, the mosquitoes and sandflies were 'numerous and fierce' – as were the small green ants that made their nests in the bushes, and which came out in 'squadrons' when disturbed. In forcing their way through the undergrowth, Matthew and his men would often be forced to perform a frenzied dance, ripping off their clothes and tearing at their hair as the ants stung them.[18] But after 105 days, the exploration of the gulf finished in heavy rain on 17 February.

MATTHEW NOW STEERED west in his crumbling ship and with his weary men along 'Arnhem's Land'. They cleared a narrow passage between Cape Wilberforce and a group of small islands, which Matthew named after Reverend John Bromby of Hull. They were then met with gusting winds and heavy rain.

Matthew steered through a narrow channel between a coastline rimmed by mangroves and a series of large, high islands to starboard. Just off the closest island Matthew was stunned to see a canoe full of men, and at the island's southern tip six more vessels covered over, 'as if laid up for the bad season'.[19]

Alarm spread through the *Investigator*. Matthew's crew had seen traces of Asian sailors throughout the gulf, and opinions flowed as to the identity of these people. Perhaps they were Chinese fisherman, or maybe murderous 'Ladrones' – pirates of the South China Coast.

Matthew eased the *Investigator* closer to the small fleet, so that they were in range of the British muskets. He had his ship's pendant and ensign hoisted, while the other vessels each hung out a small white flag. Matthew sent a nervous Samuel in an armed boat to learn who they were, as the men of the *Investigator* prepared for battle. But Samuel returned promptly to tell his brother that the boats were proas from Macassar – now Makassar, Indonesia – from the island of Sulawesi. Six Malay commanders then came to the *Investigator* on a canoe; fortunately, the ship's cook, Abraham Williams, spoke their language.

Their chief was a short, elderly man named Pobassoo, and he told Matthew that scattered along this coastline were sixty proas belonging to the Rajah of Boni, and 1000 men sailing on them. Matthew wrote that the chiefs were 'Mahometans', or Muslims; on seeing the *Investigator*'s pigs, they 'expressed great horror … nevertheless they had no objection to port wine, and even requested a bottle to carry away with them at sunset'.[20] Pobassoo was on his seventh voyage to the area, and had first come to 'Arnhem's Land' twenty years earlier. He had never seen any ship here before.

The Malays were in these waters to collect *trepang*, also known as *bêche-de-mer* or sea cucumbers, marine animals which Peter Good explained were 'of a gelatinous substance & somewhat of the Shape of a Cucumber, which when dried is a great delicacy with the Chinese'.[21]

Matthew remained wary of the Malays, as each carried a short 'kris' dagger by his side. His crew had their muskets ready, and at one point fired them in an exercise at the request of the Malay chiefs. The proas each carried twenty to twenty-five men armed with muskets; Pobassoo's vessel also had two small brass cannons, obtained from the Dutch.[22]

Pobassoo told Matthew that his men harvested the sea cucumbers in depths of from five to fifteen metres, bringing up as many as ten at a time. Each proa then carried as many as 100,000 sea cucumbers to Timor, where they were sold to Chinese traders.

When William Westall made this sketch, Malay commander Pobassoo was on his seventh voyage to 'Arnhem's Land'.
nla.obj-138888247-1

Matthew later went on board Pobassoo's proa and was told that one like it had been lost a year earlier. Matthew revealed that he had seen pieces of wreckage; he later showed Pobassoo the canoe's rudder, and the chief recognised it as belonging to the proa that had gone down.

The Malay sailors sometimes had skirmishes with the Yolngu, who fiercely defended their land; Pobassoo himself had once been speared in the knee. Another of their men had been slightly wounded since their arrival this time. The Malay chief seemed keen to help Matthew, and he delayed his squadron's departure for a day to answer a multitude of questions patiently 'and with apparent sincerity'.

He and his men had no knowledge of any European settlement in New Holland, and on learning the name 'Port Jackson' wrote it down in their language. Their only navigational instrument was a small Dutch pocket compass. They carried a month's water, in joints of bamboo; and supplies of rice, coconuts and dried fish, with a few fowls for the chiefs.

Westall sketched Pobassoo, his men and his boats, and Matthew gave the Malay visitors gifts of iron tools, which they were anxious to obtain, and – at the chief's request – a British Union Jack, which Pobassoo carried at the head of his boats. Matthew also gave Pobassoo a letter of introduction, should he meet other European ships.

Matthew named Malay Road and Pobassoo's Island in honour of the meeting, and he called the island chain the English Company Islands as a tribute to the financial help the British East India Company had given his voyage.

The expedition pressed on, Matthew and the scientific men taking boats to examine the surrounds of Arnhem Bay. Bungaree kept them fed by spearing fish, though he turned up his nose at some of the catch, Matthew relating that, like his countrymen at Port Jackson, he had 'a prejudice against all fish of the ray kind, as well as against sharks'. While Bungaree would 'devour with eager avidity the blubber of a whale or porpoise, a piece of skate would excite disgust'.[23]

On 5 March, the *Investigator* left Arnhem Bay and headed north-west, through what was marked on Matthew's Dutch chart as Wessel's Eylandt, but which he now renamed Wessel's Islands.

MATTHEW AND HIS MEN HAD now survived three months of sailing since the damning report by Aken and Mart on the *Investigator*'s future. In addition to the 'rottenness of the ship', most of the crew were now debilitated 'from the heat and moisture of the climate, from fatigue, and from the want of nourishing food'.[24] There were twenty-two men with signs of scurvy – loose teeth, lethargy and sores on their limbs – and many had diarrhoea. Matthew was disabled by fatigue brought on by the stifling heat, and he had also developed painful scorbutic ulcers on his feet because of the prolonged lack of fresh vegetables and fruit. It was difficult for him to climb the masthead or even to board a small boat to make closer inspections of the islands and coastline. With unfavourable winds and the fear that his men might no longer have the strength to work the pumps in bad weather, Matthew reluctantly postponed the continued charting of the coast along the north of New Holland and steered north-west for rest and refreshment at Kupang, Timor, 1100 kilometres away. The Dutch had sided with France in the war against the British, and Matthew hoped that the Peace of Amiens still held. He wanted to send Lieutenant Fowler back to London so he could request a replacement ship from the Admiralty to finish the survey. Matthew planned to keep the *Investigator* in the northern waters during the calm winter weather, before meeting the new ship in Sydney.

Heartsick, he started a loving, rambling letter to Ann back in the vastly different world of Partney:

Our voyage has hitherto gone on prosperously upon the whole, but the poor ship is worn out, she is decayed, and rotten both in skin and bone … I am indeed my only beloved, thine with all the fervor with which I loved thee after marriage … In evenings I oft take a book, then reclining on my little couch, and running over some pleasant tale or sentiment, perhaps of love, my mind retraces with delight, our joys, our conversation, our looks, our everything of love.[25]

'With delight', he had 'at last perused' Milton's *Paradise Lost*, and with that epic work in mind, he told his wife: 'in thee I have more faith than Adam had, when complying with Eves request … how much dearer art thou here than our first mother'.

But all the way to Timor, Matthew lamented that if the *Investigator* was beyond repair, he might never complete his mission, and that he was showing none of the spark from 'that ethereal fire with which the souls of Columbus and Cook were wont to burn'.[26]

On 31 March 1803, Matthew saw Kupang for the first time since he had sailed there with Bligh in 1792. He sent Samuel as his envoy to the governor, 'Mynheer' Johannes Geisler, and soon thirteen-gun salutes were exchanged.

The Dutch sent a boatload of refreshments, and the next day Matthew went ashore with three of his officers to meet the governor. To his dismay, Matthew learnt that another ship bound for England had sailed from Kupang just ten days earlier. There was no certain way to send Fowler back for help, but then a further inspection of the *Investigator*, which involved boring into the timbers, showed that the damage had not worsened much. Matthew decided to complete the survey in the only vessel he had.

The skipper of a Dutch ship said he could take mail to Batavia – modern-day Jakarta – where it could then be forwarded to England, and Matthew began wearing out quill after quill with quickly written letters. He added to the love letter he had started to Ann, asking her to write to his father for him to tell him how the voyage was preceding.[27] Matthew was unaware that his father had been dead for almost a year.

He wrote to his agent, Osborne Standert, to organise some more expenses,[28] and to Joseph Banks, in the hope of ensuring that his efforts would not go unrewarded.[29]

He told his cousin Henny, who had been instrumental in getting him on Captain Pasley's ship all those years earlier, that he would always love her and was sending her £20,[30] which would in a small way repay 'some part of the kindnesses done me in my childhood'. 'My brother and myself are in tolerable health,' he told Henny, 'and as far as we can see into futurity, have fair prospects of some little share of eminence. It is now our harvest, and the labour is both heavy and tedious; we hope the fruits will be adequate.'[31]

Matthew longed for fame and fortune. He told Christopher Smith, who was still working for Banks in Calcutta, that he was not looking beyond promotion to post-captain as a result of the voyage, but 'think not however my good friend, that I have lost my ambition, no, truly, for if to covet honour be a sin, my spirit is indeed a wicked one'.[32]

After a week in Kupang, and having purchased water and food to continue the voyage, Matthew readied the ship for departure. Some of the crew went on shore for one last look at the Dutch town, and two of them – the cook, Abraham Williams, and a boy from Port Jackson named Mortlock – deserted. Matthew sailed without them on 8 April. Heavy rain, thunder and lightning followed the ship, and the health of the men deteriorated again. Matthew suspected that the Timor water had brought on more diarrhoea. Matthew's immediate task was to find the Tryal Rocks, on which the British East India Company's ship *Tryall* was wrecked in 1622 while sailing from the Cape of Good Hope to Batavia. Using a chart published by Arrowsmith with uncertain coordinates for the rocks, Matthew spent twelve days looking for the shallows off the coast of New Holland, but could find only deep water.[33]

With the Dutch having already mapped much of what is now the coast of Western Australia, he now sailed for Cape Leeuwin, keeping hundreds of kilometres off the coast, travelling for eight days on a south-west course before heading south-east and reaching the south-western corner of New Holland on 14 May 1803. But his delight at reaching a landmark he knew well was overshadowed by the gloom on the ship. Boatswain Charles Douglas died of dysentery on 17 May after weeks of agony. Matthew named some islets in his honour. It was in recording this death that Peter Good, who was also seriously ill, made the final entry of his journal.[34]

The following morning, Matthew sent a party of men to kill geese and seals upon some rocky islets in Goose Island Bay, so that the men could have fresh meat for the table and oils for their lamps. The geese were few and far between, and when Matthew's men dug a hole in a sandy gully to fill their casks with fresh water, they found it was too salty to drink.[35]

Matthew set sail again at daybreak on 21 May, but a fresh north-westerly breeze drove the *Investigator* towards rocks on Middle Island, where Douglas had been buried. Two bower anchors were lost before the ship, with its fragile hull, was brought under control.

Matthew's quartermaster William Hillier, 'a quiet good man who had been ill mostly since leaving England', died that day from a lung complaint.[36] Fourteen others were unable to do their duty through severe illness, and Matthew made haste for Sydney.

On 26 May, as the crew bartered for the belongings of the deceased William Hillier, marine sergeant James Greenhalgh – a man who had shown Matthew 'zeal and fidelity' in constantly fulfilling his duties – also succumbed to dysentery.[37]

With more deaths likely, Matthew and Surgeon Bell were at each other's throats, with Bell formally telling Matthew to get a move on, putting his concerns about the ship and the health of the men in writing.

Determined to show he was the man in charge, and to preserve his honour, Matthew told Bell that the tone of the letter 'ought to enduce me to make all possible speed to port'.[38]

> Had the health of the people been the great object of my duty as it is of yours, and I had been permitted to follow my own plan for their preservation, I should certainly have kept them on shore in their native country, and not have exposed them to the danger of the seas and enemies, and to pernicious changes of climate …[39]

Matthew had wanted to explore more of Hunter's Isles and the north coast of Van Diemen's Land, but now he had eighteen sick men, 'several of whom were stretched in their hammocks almost without hope', and he knew their lives depended upon a speedy arrival in port.[40]

Another quartermaster, John Draper, 'one of the most orderly men in the ship' died from dysentery on 2 June, and six days later, as the *Investigator* neared Botany Bay, Thomas Smith, one of the convict sailors provided by Governor King, succumbed to the same illness only a few miles from his home. At daybreak on 9 June, Matthew sailed the *Investigator* through Sydney Heads, and at noon he anchored off Garden Island.

It had been an extraordinary voyage. Though his efforts for a more detailed examination of the coast had been thwarted by the poor health of his ship and his men, Matthew had completed the circumnavigation and filled in the gaps in the long, continuous line that was the map of *Terra Australis*. He had proved it was an island continent.

He could hardly wait to set sail again on another dangerous voyage.

Chapter 21

'Everything that I have ever said or done that was displeasing to you now strikes upon my mind like moral guilt. I had indeed a strong propensity to independence of mind of thought and action ...'
MATTHEW FLINDERS AFTER LEARNING OF HIS FATHER'S DEATH.[1]

MATTHEW WAS EAGER TO set sail as soon as possible but twelve of his men, including the dangerously ill Peter Good, were unable to move because of dysentery.[2] Their captain was not much more mobile, hobbling about on ulcerated feet as he presented an official report to Governor King. Some of the report was reprinted in the new government newspaper for the colony, *The Sydney Gazette and New South Wales Advertiser*,[3] a journal edited by 'Happy' George Howe,[4] a convicted English shoplifter who was more than glad to swap a death sentence for transportation to the prison colony.

The pain in Matthew's feet was nothing compared to the pain in his heart when he began to open the mail that had been waiting for him in Sydney for months. There were six letters from Ann, written between December 1801 and September 1802, and one from his stepmother Elizabeth, written with a wavering hand. The announcement of his father's death a year earlier flattened him.

He wrote back immediately to the woman he called 'My dearest mother':

The death of so kind a father and who was so excellent a man is a heavy blow and strikes deep into my heart. The duty I owed him and which I had now a prospect of paying with the warmest affection and gratitude, had made me look forward to the time of our return with increased ardor. I had laid such a plan of comfort for him as would have tended to make his latter days the most delightful of his life, for I think an increased income, retirement from business, and constant attention from an affectionate son whom he loved would have done this … Oh my dearest, kindest father, how much I loved and reverenced you, you cannot now know.[5]

Matthew had parted from his father with unreconciled differences over the sudden marriage to Ann and his requests for money. He told his stepmother: 'I thought indeed that I was certainly not acting wrong, in anything I did, but I was, in not making your ease and happiness the first rate of my conduct I have not acted right.' He later scratched out those words.

Matthew put the letter aside as eleven of his men were carried to the hospital and into the care of the Irish-born naval surgeon Thomas Jamison. Peter Good was too ill to be moved. The next day Matthew continued his letter:

As for Samuel and me, we are going on very well, laying a sure foundation, I trust, for future fame and fortune. Our voyage has gone on prosperously, my credit with the great friends who have pushed me forward seems to increase, and I am getting something richer every year. At this time I consider our business to be nearly half done; therefore, as we left England in July 1801 we ought to arrive again about the same time in 1805 …[6]

Even though the unifying force in their lives had passed, Matthew told his stepmother they would still be a family, 'and still be respectable'. 'That you will cultivate a good understanding with my dear wife I cannot doubt. Those who love or care for me, will love my wife, who is the dearer half of me; for I cannot have any friend who is not a friend to her.'[7]

Matthew Sr had left an estate of about £6000, with £600 for his eldest son. Matthew asked that the money be banked until he arrived home, and that the five per cent interest be put to the education of his two half-sisters. Again he put the letter aside to finish later; ultimately it would take almost two years to reach Donington.

Peter Good died on the ship the next day, and was buried on shore. Matthew and other officers stood with heads bowed as marines fired three volleys over the grave.[8] Matthew asked Robert Brown to sell Good's clothes and belongings to the men, and Brown added the young gardener's plant collection to his own. Before long, sailors John Simmonds and Oloff Wastreen, and Robert Chapman of the marines, joined Good in the primitive Sydney cemetery.[9]

To keep the rest of his men alive, Matthew paid for a regular supply of vegetables and fresh meat; such was the sudden prosperity in the colony that pork and mutton were obtained at just ten shillings per pound, half the price of a year before. Governor King supplied a pint of wine a day to the men on the ship, as well as those in hospital, provided Jamison thought it would help.[10] Seven of the eight surviving convict sailors King had assigned to the *Investigator* were granted their freedom. Matthew refused to recommend one of them because of his poor behaviour, so he went back into his prison bonds.

Waiting in Sydney for Matthew was the brilliant young scholar James Inman,[11] who three years earlier had been the top mathematics undergraduate at the University of Cambridge. The Board of Longitude had sent Inman as the replacement astronomer after John Crosley had left the voyage at Cape Town, but he had arrived too late for the circumnavigation.

Five days after the *Investigator* returned to Sydney, Matthew began the inspection of his ship, along with Acting Lieutenant William Scott from the ship *Porpoise*, Captain E.H. Palmer of the *Bridgewater* and Thomas Moore, Sydney's master boatbuilder. The 'state of rottenness' in the timbers startled them, and the *Investigator* was found 'incapable of further service'.[12] The opinion was that if the ship had met 'a severe gale of wind in the passage from Timor', it would have been 'crushed like an egg and gone down'.[13]

Governor King decided to have the *Investigator* permanently moored in Sydney as a storeship hulk. He offered Matthew any ship in the colony to complete the detailed survey, though the best vessel at his disposal, the *Buffalo*, was sailing through the East Indies for India to buy cattle and horses and would not be back for a year.

Robert Cumming,[14] the captain of the privately owned convict transport *Rolla*, which was in port, wanted a jaw-dropping £11,550 for the vessel – and Matthew calculated that it would still take six months to prepare it for the voyage.[15] He liked the idea of the *Porpoise*, which was sailing back from Van Diemen's Land. The ship was built in Spain and captured by the British off Portugal in 1799. It was a twelve-gun sloop and, at 308 tons, slightly smaller than the *Investigator*, but in company with the repaired *Lady Nelson*, Matthew felt, it could be adequate for his needs.

While Matthew awaited the return of the *Porpoise*, he wrote heartfelt letters to his wife. He told Ann he was grateful for her recovered health, and that she was his 'dearest love', but also that his heart ached not just for her, but for the 'dreadful havock that death is making all around'. So many of the men on his ship had perished, and the news of his father's passing was a dreadful blow.[16] The sores on his feet had made him lame for the past four months, and he was 'much debilitated in health', though 'Deaths hand' was now 'staid' and the men on his sick list were slowly recovering.

Ann had again accused him of putting ambition ahead of their relationship.

'I cannot excuse myself now,' he replied, 'but will plead for respite until my return, when in thy dear arms I will beg for pardon and if thou canst forgive me all, will have it sealed – oh with ten thousand kisses.'[17] He scrawled: 'oh truly believe me that I pursue discovery only to be able [to] avoid the future necessity of parting from thee' – but perhaps it was because of his advanced preparations to continue his survey on the *Porpoise* that he scratched that sentence out in his first draft.[18]

He told Ann that Lieutenant Fowler was 'tolerably well', and was 'a good natured fellow', well suited to the voyage. Samuel had become more 'friendly and affectionate' with his brother as they shared the mutual loss of their father. Robert Brown was recovering from ill health and lameness, and although they were not 'altogether cordial', he thought Brown was a man of abilities and knowledge, though lacking kindness.

Matthew reassured Ann that 'Mr. Bauer, your favourite, is still polite and gentle', though he suspected that there was a 'dreadful disposition' lurking behind what he thought was a veneer. The surgeon Bell was 'misanthropic ... and pleases nobody'.[19] Westall lacked prudence, but was 'good natured', though the young artist would complain bitterly to Joseph Banks that if he had known Flinders would not be taking the ship to the 'South Sea islands', he would not have signed on.[20]

As usual, Matthew found comfort in Trim, who, like his master, 'was becoming grey'. While Matthew's hair stayed that way, he would soon report that 'to the great joy of his friends, [Trim] re-assumed his fine black robe and his accustomed portliness'.[21]

Master and cat took a little holiday. Matthew put his ulcerated feet up beside the Hawkesbury, writing with pleasure that 'the fresh air there, with a vegetable diet and medical care, soon made a great alteration in the scorbutic sores'.[22] They returned to Sydney at the beginning of July, 'nearly recovered'.

GEORGE BASS HAD LEFT TWO letters for Matthew with his business partner Charles Bishop, who was now 'wholly dispirited' that the trading partnership had foundered. Governor King would later call a jury to consider Bishop's sanity and he was declared a lunatic.

Matthew gave King a letter for Bass to collect when he returned from his Pacific voyage on the *Venus*. Matthew had heard that Bass contemplated gathering cattle and perhaps alpacas from Peru to raise in New South Wales.

He told Bass: 'My poor father has paid the last debt of nature, at a time when to my great regret, our good understanding was not complete.'[23] He filled Bass in on the adventures he'd experienced on the *Investigator*, and expressed his dismay that the fortune Bass was seeking as a trader had not yet materialised. 'This fishing and pork carrying may pay your expenses, but the only other advantage you can get by it is experience for a future voyage, and this I take to be the purport of your Peruvian expedition,' he said. If Matthew beat Bass back to England, he told him, he would call on his friend's wife and mother, along with Henry Waterhouse, who was now drinking heavily and whose 'sun seems to have passed the meridian, if they say true'.

'God bless you, my dear Bass,' Matthew concluded, 'remember me, and believe me to be your very sincere and affectionate friend, Mattw. Flinders.'[24]

Bass had sailed out of Sydney five months earlier, and none of his friends ever saw him again. For a time there was speculation that he had been captured by the Spanish in Chile and forced into slavery in their silver mines, but there were no credible sightings of his ship or crew. More than two centuries later, mystery still swirls around his fate.

Matthew also wrote to William Kent's wife, Eliza, expressing his dismay that the upper reaches of Sydney society had been rent asunder by Macarthur's wounding of Paterson in a duel. 'There is now Mrs King, Mrs Paterson and Mrs McArthur [sic] for all of whom I have the greatest regard, who can scarcely speak to each other; it is really a miserable thing to split a small society into such small parts.'[25]

On 4 July 1803, the *Porpoise* arrived back in Sydney from Van Diemen's Land and Matthew formally accepted the governor's offer of that ship and a colonial tender – most likely the *Lady Nelson* – to continue his survey work. King appointed John Aken,

Russell Mart and Thomas Moore to inspect the ship but their report sank Matthew's immediate hopes. It would take a year of work to make the *Porpoise* fit for the detailed survey, which was intended to last until the end of 1805.

The governor now proposed sending the *Porpoise* to England, with Matthew as a passenger, so that he could arrange a more suitable vessel from the Admiralty. Matthew accepted on the condition that they travel through the Torres Strait so he could double-check his charts.[26] Robert Fowler would take command of the *Porpoise*, allowing Matthew to devote himself to mapping.

Before leaving Sydney, Matthew gave some assistance to Elizabeth Macarthur, who was at Parramatta managing her farms and family while her husband was in England for his court martial. To help Mrs Macarthur with the legal aspects of the sale of cattle, Matthew consulted the *Encyclopedia Britannica* which Joseph Banks had given him.[27] He considered Mrs Macarthur his 'very sincere and affectionate friend', and prayed that the 'Almighty Power', whom she 'reverenced and adored', would 'impart such fortitude' to her so she could cope with all the responsibilities heaped upon her by her husband's absence.[28]

On 7 July 1803, the *Porpoise* was moored alongside the *Investigator* as goods and men were moved onto the new ship. The *Investigator*'s greenhouse was reconstructed on the quarterdeck of the *Porpoise*, and Matthew and twenty-one of his men from the *Investigator* boarded as passengers on 8 July; among them were Samuel Flinders, John Aken, John Allen, William Westall and Matthew's servant John Elder.

Four of the convict sailors from the *Investigator* who had been granted their freedom took the opportunity to return to England. Old habits were hard to break, though, and three of them behaved badly on the voyage. One ended up right back where he started, condemned to the prison hulks on the Thames not long after he reached England.[29]

Trim beat everyone up the gang plank onto the *Porpoise* and immediately acted as though he were master of the vessel. He shared Matthew's cabin on the upper deck, in which Matthew stored his charts and journals, along with Westall's paintings. Robert Brown and Ferdinand Bauer elected to stay in Sydney to continue their work until Matthew returned with a new ship.

Brown told Joseph Banks that overall he was 'disappointed' with the voyage so far, and 'our expectations of getting into the country by means of navigable rivers or inlets have been completely frustrated. We have seen no mountains of great height, and only once have had an opportunity of ascending one of moderate elevation … The interior of New Holland, therefore, is as completely unknown as ever.' Brown said John Allen, the mining expert, might as well have stayed home.[30] Surgeon Hugh Bell, still at loggerheads with Matthew, elected to sail home on the huge 750-ton East Indiaman trading ship the *Bridgewater*.

MATTHEW SAID GOODBYE to Sydney just before noon on 10 August. The atmosphere was festive as Governor King and his party sailed on the *Porpoise* as far as Sydney Heads, then farewelled Matthew and the others and climbed back onto the *Lady Nelson*.[31] Robert Fowler turned the *Porpoise* north for Timor, in a convoy with the *Bridgewater* and the 450-ton *Cato*.

They sailed through fair weather and light breezes, staying far out to sea to make quick time, and by 17 August they were about 450 kilometres off the mainland of Keppel Bay, having passed into the tropics.

Soon after 2 p.m., the *Cato*, on the *Porpoise*'s port side, made the signal for seeing land, which was in fact a dry sandbank. Since the *Porpoise* was the fastest of the ships, Matthew directed the others to keep going while Fowler hauled up so Matthew could take a closer look at the bank. The crew of the *Porpoise* reported water at least eighty fathoms (146 metres) deep around what Matthew later called Cato's Bank,[32] and there seemed no other reefs nearby, though it was surrounded by breakers. The *Porpoise* sailed back to the head of the convoy.

While Matthew and Fowler were apprehensive about more sandbanks and rocks as the sun came down, they were at least 200 kilometres east of the 'great Barrier Reefs'[33] and had already travelled sixty kilometres north of the sandbank. They decided it was not necessary to lose a good night's run by heaving to. Matthew agreed with Fowler that it would be 'sufficient to signal for the ships to run under easy, working sail during the night and to charge one of the *Investigator*'s warrant officers with the look-out on the forecastle'.[34] With a fresh breeze and cloudy weather, they glided on north at about fifteen kilometres an hour. Checks showed the water was at least thirty-five fathoms deep and everyone breathed easier. The *Bridgewater* was about 800 metres away on the *Porpoise*'s starboard side, the *Cato* about double that on the port side.

Russell Mart was on the *Porpoise*'s forecastle and John Aken was on the quarterdeck at 9.30 p.m. Almost at the same instant, in the pale moonlight, they cried out that they saw breakers ahead 'only 50 yards away'. The coral reef showed itself in a long line of foam in the gloom of night.[35] Matthew was in the gun room, unaware of the trouble, but Fowler sprang onto deck immediately.

When Matthew finally rushed upstairs he found the sails shaking in the wind as the ship was rocked by heavy surf. Then the *Porpoise* was sucked in among the raging breakers and, crashing into a coral reef, was flung onto its side. Screams and groans echoed around the vessel and William Westall despaired that 'not a possibility appeared that anyone could be saved'.[36]

With the fierce waves smashing at the disabled vessel, the bottom caved in and the hold quickly filled with water. Fowler did his best to rally the men. 'Come, my Lads,' he bellowed, 'I have weathered worse Nights than this!'[37] But at the second or third shock of the great waves the foremast was ripped off and carried away. The crew tried to fire a cannon to warn the other ships and set warning lights, but it was impossible in the chaos, the ship heaving in all directions.

As the briny surf lashed Matthew's eyes, the *Bridgewater* and the *Cato* were no more than 200 metres away, tacking ominously towards each other. The waves continued to pound the stricken *Porpoise*, but as its crew screamed warnings towards the other ships, without effect. Matthew shuddered, expecting to hear the dreadful collision of the two vessels. At the last minute, they turned from each other, slipping side by side, somehow without touching.

Then, from 400 metres away, Matthew gasped in horror as the *Cato* slammed into the reef. The ship fell over on her broad side, and almost instantly the masts sheared off and disappeared.

In the distance, there was a light on the masthead of the *Bridgewater*. For a moment Matthew thought that its captain, E.H. Palmer, would tack, and send boats to rescue the stricken men. He quickly realised, though, that such an operation at night would mean 'certain destruction'.[38]

Matthew had no idea how long the *Porpoise*, 'being slightly built and not in a sound state', might hold together, but the men managed to free the smallest of its boats, a four-oared gig. Matthew planned to wrap the charts and logbooks from the *Investigator*'s voyage into a seal skin, take the small boat to the *Bridgewater* and then use its boats to get the people off the reef as quickly as possible.[39] He checked that Trim was safe as Fowler organised men to row.

The boat was at a little distance from the *Porpoise*, to prevent it being smashed against the reef, and Matthew and some assistants jumped overboard and swam to it. The heavy surf rolled over them but they escaped sinking. The gig had only two oars, and three of the men with him – the armourer, a cook and a marine – did not know how to row. Matthew set them to baling water with their hats and shoes, as three other men helped him row towards the shining light on the *Bridgewater*'s masthead, keeping under the lee of the breakers.

With two awkward oars and an overloaded boat, they fought the sea on the windward side of the reef but realised that the *Bridgewater* was not slowing down. The light on her masthead was growing more distant, until finally it disappeared from view. As they grimly rowed back towards the wrecked *Porpoise*, they met the ship's cutter, which was badly damaged, although the men in it had managed to slow the leak.

The darkness of the night prevented any hope of communication with the men on the *Porpoise* until morning, by which time Fowler had started his men on making a raft of the spare topmasts and yards with short ropes all round it, so the stricken men might hold on. A cask of water, with a chest containing some provisions, a sextant and the *Investigator*'s logbooks, were secured upon the raft.[40] Fowler organised for blue lights of distress to be burnt every half hour but the *Bridgewater* did not come back.[41]

And so the crew of the *Porpoise* spent this traumatic winter's night in the boats or hanging onto the raft, shivering and shaking. Their ship creaked and groaned in its death throes as each wave delivered another merciless blow. Samuel Smith wrote that in this 'mizerable situation', every heart was 'fill'd with Horor, continual Seas dashing over us'.[42]

As the tide ebbed, Matthew waded through shallow water on the reef and climbed aboard the stricken *Porpoise*, using what was left of the broken masts as his ladder. The ship had rolled in such a way that its worn copper bottom took the brunt of the heavy surf. Matthew reported to his men that the small boats – two cutters and his gig – were safe, and since the hatches had been closed, most of the provisions had survived.

In the distance, though, the *Cato* was still being shredded, the wild surf hammering it into the coral. Everything of the ship except the forecastle and bowsprit was submerged. Terrified men, including the skipper John Park, clung to the wreckage yelling for help while the surf beat them incessantly toward the swirling deep.

AS THE SUN ROSE, THE MEN SAW the faint outline of the *Bridgewater* under sail coming towards them. Matthew spotted what appeared to be a dry sandbank, not more than 800 metres away, which looked large enough to accommodate all the men and what provisions they could salvage. It seemed to be a kilometre and a half wide, and near it were several wide and apparently deep openings, where the *Bridgewater* might anchor while sending her rescue boats. Matthew took the gig to the sandbank, and the sight of many seabirds' eggs scattered over it showed that it was above the high-water mark. He sent the gig back for Fowler, and then hoisted two handkerchiefs on a tall oar to alert the *Bridgewater*.

But by 10 a.m. the *Bridgewater* had changed course and disappeared from view. Matthew and his shipwrecked men were staggered.

Fowler sent a boat to the stricken *Cato* to begin ferrying men to the shallow water around the *Porpoise*. Captain Park and his exhausted crew threw themselves into the water, using any pieces of broken spar or plank they could find to help them swim through the breakers. Several of the *Cato*'s crew had been badly injured with cuts and deep bruises when flung against the coral rocks, and three young lads had drowned. One of these poor boys had been shipwrecked every time in his three or four previous voyages. He had launched himself upon a broken spar with his captain, Matthew recalled, 'but having lost his hold in the breakers, was not seen afterwards'.[43]

Low water came about 2 p.m., and with the reef dry next to the *Porpoise*, the men frantically got together provisions, clothes and Trim the cat, and began ferrying them to the sandbank. By twilight, five casks of water, some flour, salt meat, rice and spirits were landed, along with the pigs and sheep that had escaped drowning. All the surviving men from both ships had made it ashore. Some of the *Cato*'s sailors were wearing officers' uniforms, given to them from the *Porpoise*; despite the perilous situation they faced out here in the middle of the vast sea, they could still laugh about these instant 'promotions'. Fires were lit with dried wood, and those men who had saved greatcoats or blankets shared them with the less fortunate.

Some of the wood that had been gathered for the fire included a spar and a piece of old timber, 'worm eaten and almost rotten', which was lying in the sand. John Aken judged it to have been part of the stern post of a ship of about 400 tons; Matthew thought 'it might, not improbably', have belonged to one of the two ships under the command of 'Monsieur de la Pérouse' after he left Botany Bay fifteen years earlier. Perhaps the French skipper had 'encountered in the night, as we did, some one of the several reefs which lie scattered in this sea'.[44]

The scene of wreck of the *Porpoise* and the camp made by survivors on Wreck Reef.

Beside the fire and in their mismatched uniforms, the ninety-four exhausted survivors from the two ships lay down to sleep on the sand in 'tolerable tranquillity'.[45]

Even in the eye of the storm, Matthew was watching out for his beloved pet. 'The imagination can scarcely attain to what Trim had to suffer during this dreadful night, but his courage was not beat down,' Matthew wrote later.[46] The *Porpoise*'s two cutters and the gig were hauled up to the high-water mark, but to Matthew's horror the exhausted crew had not secured the gig strongly enough, and when the night tide rose higher than expected, it was carried away.

WITH THE BLESSING OF Fowler and Park, Matthew took command of the situation. He promised food for the men who worked hard and floggings for shirkers, exempting only the eight badly injured men from the *Cato*.

Matthew ordered a topsail yard that had survived the catastrophe to be erected as a flagstaff on the highest part of the bank, and a large blue ensign was hoisted upside down as a distress signal to the *Bridgewater*. Matthew expected that if no accident had happened, Captain Palmer would come to save them as soon as the wind moderated, but he judged it 'prudent' to act as if no help were coming. That proved to be the right course, as he later wrote:

Captain Palmer had even then abandoned us to our fate, and was, at the moment, steering away for Batavia, without having made any effort to give us assistance. He saw the wrecks, as also the sand bank, on the morning after our disaster ... he neither attempted to work up in the smooth water, nor sent any of his boats to see whether some unfortunate individuals were not clinging to the wrecks, whom he might snatch from the sharks or save from a more lingering death; it was safer, in his estimation, to continue on his voyage and publish that we were all lost, as he did not fail to do on his arrival in India.[47]

Palmer would later say that he didn't believe anyone could have survived the wreckage he saw, and the danger to his own ship was too great to investigate,[48] an explanation Governor King later blasted as 'very lame'[49] – especially since the *Bridgewater*'s third mate later wrote that the two wrecked ships were 'distinctly seen', and he was convinced 'that the crews of those ships were on the reefs'.[50] After reaching Batavia, Palmer sailed the *Bridgewater* for India, where Hugh Bell disembarked. The *Bridgewater* then sailed for England, but was never seen again.

ON WHAT BECAME KNOWN as Wreck Reef, Matthew's men spent four days in the boats transporting all the food, water, wine, spirits and porter to the sandbank, where it was placed in a large tent made from rescued spars and sails. Matthew calculated there was enough food and water to last them three months. Trim was put in charge of eliminating vermin in the stores and his 'zeal in the provision tent' was admirable.[51]

A row of tents was erected, and although most of the men worked tirelessly for mutual benefit, Matthew had one of the freed convicts whipped at the flagstaff for 'disorderly conduct', and 'to correct any evil disposition' among the others.[52]

Matthew had managed to save most of his charts, logbooks and astronomical observations, but when the mizen mast fell through his cabin roof, the papers on his deck had been washed away, among them a chart of the west side of the Gulf of Carpentaria and part of the north coast of *Terra Australis*, which he had been working on just before the crash. Some of his small library shared the same fate, though the twenty volumes of the *Encyclopedia Britannica* survived, as did James Cook's account of his first voyage to the Pacific. The rare plants they had collected were destroyed by the saltwater. Fortunately, Matthew said, Brown and Bauer, who remained at Port Jackson, had put on board only a small part of their collection of specimens.[53] William Westall, who had already grumbled about not getting the chance to see Tahiti, was in an even more sullen state after many of his sketches and drawings were water-damaged as well. They were put out to dry on the sandbank, but John Franklin inadvertently let the small flock of sheep scamper over them.[54]

By 22 August, Matthew realised the *Bridgewater* was gone for good, and he lamented 'Palmer's want of energy and humanity'.[55] He called a council of his officers to decide how they might rescue themselves.

It was proposed that an officer and crew, in the largest of the cutters, should endeavour to get to Sandy Cape, 305 kilometres south-west, and from there travel along the coast to Port Jackson, a hazardous journey of more than 1100 kilometres. But at this time of year strong winds prevailed from the south, which created 'much apprehension'. Instead:

… it was resolved that two decked boats, capable of transporting every person remaining on the bank, except one officer and boat's crew, should be immediately laid down by the carpenters, to be built from what was already and might be still further saved from the wreck; and that, if the officer in the cutter did not return with assistance in two months, the boats should then, or as soon after as they could be ready to sail, proceed to Port Jackson.[56]

Matthew volunteered to sail in the cutter, and Captain Park put up his hand to be second in command. Matthew then proposed that a smaller cutter, with two officers and crew, should remain at the sandbank; should no vessel have arrived within two months, they should make for Port Jackson also. This precaution was necessary, Matthew said, in case any unforeseen occurrence should delay his return to the sandbank. In this way, the charts, journals and papers might still be found there, to be taken on to England. By leaving the all-important charts behind on the cay, Matthew was also pledging that he would return. He designated Samuel to be in charge of the last boat crew to stay behind, but Fowler, taking responsibility for the mess they were in, insisted he take that role, as 'the post of honour'.[57]

THE LARGEST CUTTER was lightly decked out and ready for the most dangerous expedition of Matthew's life. Eighty sunburnt, hopeful men and one active black cat on this tiny sandbank hundreds of kilometres from help gathered around him and his intrepid team on the morning of Friday, 26 August 1803. It was only a few weeks since Matthew had sailed the *Investigator* into Sydney Cove in triumph, and now he was embarking again under far different circumstances, with many lives dependent on his success. Matthew named the cutter the *Hope*.

The morning was fine, and the wind light from the south. He had John Park on board and a double set of rowers, twelve men, including his boatswain, Edward Charrington, to propel them all the way to Sydney when there was no wind for the sail. They had three weeks of provisions and two casks of water, meaning the *Hope* was 'loaded rather too deeply'.

At eight in the morning they pushed off, to three cheers. One of the sailors, having obtained permission, ran to the flagstaff, hauled down the inverted ensign and rehoisted it 'right way up'. Matthew saw it as a 'symbolical expression of contempt for the *Bridgewater* and of confidence in the success of our voyage'.[58]

They sailed westward in the lee of the reef before reaching two more sandbanks. They stopped at a third one to cook a meal onshore, and Matthew shot as many noddies as he could to feed the men. They then rowed into the open sea and headed south-west for Sandy Cape, on the edge of Hervey Bay. For a while the little vessel was surrounded by playful humpback whales. The wind died and the men rowed slowly through the still night, making about eight kilometres an hour.

That afternoon, the wind picked up speed alarmingly, and Matthew realised the cutter was carrying too much weight.

> [T]he increased hollowness of the waves caused the boat to labour so much, that every plunge raised an apprehension that some of the planks would start from the timbers. Having no other resource, we emptied one of the two casks of water, threw over-board the stones of our fire place and

wood for cooking, as also a bag of pease and whatever else could be best spared ...[59]

They saw land to the west, about twenty kilometres away, at sunset on the third day of the voyage, and they then turned south, parallel to the coast. By daylight on 29 August, Matthew could make out the tops of trees somewhere between Double Island Point and Moreton Bay.

The breeze died away that afternoon, and his men took to the oars before seeing Cape Moreton at sunset. Rain poured during the night, drenching them, but they were near the halfway mark of their journey.

At Point Lookout, they saw about twenty men of the Quandamooka people performing dances for them in imitation of the kangaroo. Matthew made signs for water, which they understood, and they pointed to a small stream falling into the sea. Two of the sailors leapt overboard, with some gifts for the Quandamooka and hauling one end of the lead line, to which was tied the empty water cask. The Quandamooka stayed away from these intruders, but a shark had followed the swimmers to the beach. Fearing they might be attacked, Matthew hauled up the anchor and they rowed closer to the shore. The cask of water, a bundle of wood and the two men were soon back on board.[60]

On Sunday, 4 September, they were forced to find shelter on the north side of Tacking Point, near modern-day Port Macquarie. But their *Hope* was now becoming a reality, and salvation beckoned. Two nights later, they slept onshore for the first time, having passed Port Hunter at modern-day Newcastle.

The distance to Port Jackson was little more than eighty kilometres.[61]

The next morning, 7 September 1803, they filled their water cask again and baked some cakes in their fire before setting off for Sydney. That night the wind died away but the men took to the oars with increased vigour, knowing the finish line was close and the lives of eighty men depended on them.

They sighted the north head of Broken Bay the next morning,

and at noon on Thursday, 8 September, a welcoming sea breeze set in. Matthew crowded all sail for Port Jackson and soon after 2 p.m. passed between the heads, with his men cheering the success.

'Never do I recall,' Matthew wrote in his journal, 'experiencing sensations of joy greater than those with which my breast was animated at this time.'[62]

Matthew and Park, dishevelled, bedraggled, unshaven and unwashed, and having been cramped for twelve and a half days in their boat, hobbled up the hill from Sydney Cove to Government House, where they found Philip Gidley King having dinner with his family.

Governor King was stunned at seeing these two men who were supposed to be thousands of miles away, and who had probably never looked worse in their lives. As soon as he realised who they were and what they had just survived, Matthew later wrote, an involuntary tear came to his eye.[63]

Chapter 22

*'I am ... satisfied with your account of the exertions of the officers and
men after the loss of the ships ... nor can I sufficiently commend your
voluntary services, and those who came with you, in undertaking a
voyage of 700 miles in an open boat to procure relief for
our friends now on the reef ...'*

GOVERNOR KING TO MATTHEW FLINDERS, AFTER HIS DARING VOYAGE
IN AN OPEN BOAT.[1]

MATTHEW AND GOVERNOR KING immediately
began a mission to rescue the eighty men and Trim from
Wreck Reef.

Robert Cumming was about to sail the *Rolla* to Guangzhou,
and King commissioned him to stop at the sandbank along the
way and collect everyone who wanted to accompany him to
China, where they could then find a ship for England. Already his
passengers leaving Sydney included John Park and the astronomer
John Inman. Sydney opened its heart to the stranded men,
donating wine, livestock and vegetables for them to be taken on
Cumming's ship.

Matthew gave a detailed account to *The Sydney Gazette* about
the loss of the *Porpoise* and the *Cato*, and King told him that he
had 'every reason to be assured that no precaution was omitted by
Lieutenant Fowler and yourself to avoid the accident'.[2]

King had sent the *Lady Nelson* to Risdon Cove, in Van
Diemen's Land, to form a settlement there as a bulwark against

Baudin or any of the other French explorers entertaining ideas of claiming the island for France. There was a shortage of rescue vessels in Sydney, but two colonial schooners –the *Francis* and the newly built *Cumberland* – were available to accompany the *Rolla*, to ferry home those survivors who wanted to return to Sydney.

Matthew initially planned to sail on the *Rolla*, but worried that if he did, he would not have the chance to complete his survey among the reefs and shoals of Torres Strait. Also, he knew that once in Canton, it might take him months to arrange passage on a ship to London. He ached to see Ann again, but he also wanted to press the Admiralty to provide another ship for the continued exploration of the land he was now calling 'Australia'.

Governor King offered to give Matthew the little *Cumberland*, first to complete his survey of Torres Strait and then to sail for Britain. The *Cumberland* was built by Thomas Moore and designed to carry just ten crew and a skipper.[3] Work on it had begun years earlier, under the direction of Governor Hunter, and, according to *The Sydney Gazette*, the schooner had proven itself in 'very tempestuous weather off Norfolk Island and in Bass's Straits'.[4]

Matthew recalled:

This schooner was something less than a Gravesend passage boat, being only of twenty-nine tons burthen; and therefore it required some consideration before acceding to the proposal. Her small size, when compared with the distance from Port Jackson to England, was not my greatest objection to the little *Cumberland*; it was the quickness of her motion and the want of convenience, which would prevent the charts and journal of my voyage from being prepared on the passage, and render the whole so much time lost to this important object. On the other hand, the advantage of again passing through, and collecting more information of Torres' Strait, and of arriving in England three or four months sooner to commence the outfit of another ship, were important considerations; and joined to some ambition of being the first to undertake so long a voyage in such a small vessel, and a desire to put an early stop to the account which captain Palmer [on the *Bridgewater*] would probably give of our total loss ...

Matthew accepted the governor's offer, after being told the vessel was 'a strong, good little sea boat'. No one in Sydney knew that Britain had declared war on France again just four months earlier.

At the time, the *Cumberland* was on the Hawkesbury River, while the *Francis* was lying onshore and could not be readied before the following spring tides. The *Rolla* was not quite ready to sail either, which meant Matthew had been back in Sydney for thirteen days before the rescue convoy was ready to move.

The delay caused Matthew a great deal of 'uneasiness' as he worried that the rescue mission would not arrive at the reef before his friends, 'despairing of assistance', made some unsuccessful attempt to save themselves. Every day to Matthew now seemed like a week, such was his anxiety until he and the three vessels finally put to sea at daybreak on 21 September.

Matthew quickly realised the *Cumberland* had many faults. It was a narrow vessel and 'exceedingly crank' – tilting dangerously to one side in the wind. Three days into the rescue voyage, in his notes for Governor King, Matthew wrote:

She has also been very leaky, and in one hour and half's cessation from pumping the water washes on the cabin floor ... Hitherto we have been very uncomfortable, and dry only when at anchor. I am now sitting on the lee locker with my knees up to my chin for a table to write on, and in momentary expectation of the sea coming down the companion and sky light, for they have broken me two panes of glass out of the four already.[5]

Trying to write on the rocking, bucking, unbalanced little boat was like trying to write on horseback. What was worse, Matthew continued: of all the filthy little boats he had sailed on:

... this schooner, for bugs, lice, fleas, weavels, mosquitos, cockroaches (large and small), and mice, rises superior to them all ... I have at least a hundred lumps upon my body and arms; and before this vile bug-like smell will leave me, must, I believe, as well as my clothes, undergo a good boiling in the large kettle.[6]

Matthew promised to 'set my old friend Trim to work upon the mice' when he was reunited with the cat.[7]

On 7 October, six weeks after Matthew had left Wreck Reef in the *Hope*, the lookout atop the masthead of the *Rolla* saw a British ensign flying from the flagstaff on the sandbank. A sailor who was testing a new boat Russell Mart had built from wreckage saw a white object in the distance against the blue of the sky. At first he took it for a seabird; but, looking at it more closely through narrowed eyes, he suddenly jumped up and exclaimed to Lieutenant Fowler, 'Damn my blood, what's that?'[8] It was the topgallant sail of the *Rolla*. Shouts of joy erupted, and the sailors powered their boat back to the reef to share the news that Captain Flinders had kept his promise.

Samuel was as cool as ice, though. As the commanding officer on the bank, he was in his tent calculating some lunar distances when one of the young gentlemen ran to him, calling, 'Sir, Sir! A ship and two schooners in sight!' After a little consideration, Samuel nonchalantly said he supposed it was his brother come back, and asked if the vessels were near.

'Not yet,' was the reply.

Samuel, perhaps not wishing to concede his older brother's success yet again, told the young man to let him know when the ships reached the anchorage, 'and very calmly resumed his calculations'. Matthew noted wryly that 'such are the varied effects produced by the same circumstance upon different minds'.[9]

Matthew anchored the *Cumberland* in the lee of the bank that afternoon, to the cheers of the castaways, and Samuel organised an eleven-gun salute from the salvaged carronades of the *Porpoise*. The *Porpoise* had not yet gone to pieces, Matthew noted, 'but was still lying on her beam ends, high up on the reef, a frail, but impressive monument of our misfortune'.[10] The pleasure of rejoining his companions, 'so amply provided with the means of relieving their distress', made this one of the happiest moments of Matthew's life.[11]

The men had been busy in his absence. Russell Mart's new boat had a 32-foot keel, and the men had started work on a second boat in case Matthew failed to reappear. They had kept themselves fed by catching birds and fish and collecting turtle eggs. Trim was itching to solve the *Cumberland*'s mice plague.

MATTHEW NOW ASSEMBLED all the survivors on the sandbank. Those sailors who wanted to be discharged from service could return to Sydney on the *Francis*; 'the rest would be taken on board the *Rolla* and carried to China', with the exception of ten officers and men of Matthew's choosing, who would go to England with him on the *Cumberland*.

It was a risky venture 'in so small a vessel'. Although Matthew had discovered many 'bad qualities of the schooner', he was determined to proceed, at least as far as a port where a better vessel might be found. He gave his ten men twenty-four hours to decide.[12] Anyone who so wished could travel back to Sydney on the craft Russell Mart had built, which the men had named the *Resource*. Denis Lacy was given command of it, carrying Matthew's letter to the governor.

Samuel Flinders, Robert Fowler, John Franklin and William Westall elected to sail on the *Rolla*, and Matthew handed Fowler the surviving logs and journals of the officers, and four of Matthew's charts for the Admiralty. Westall had preserved the survival on the reef in a series of sketches. Only Matthew's clerk declined the offer to spend six months or so with him on a leaky, unbalanced schooner, but he was still joined by John Aken, Edward Charrington, the trusty servant John Elder, Trim and seven of his best sailors.

Matthew went on board the *Rolla* to farewell his brother and his friends, and at noon on 11 October the four vessels left that little sandbank, sailing off on their separate ways, the *Cumberland* heading for Torres Strait. Because of the small size of his craft, Matthew would have to make many stops for provisions along the route to London.

As they sailed along Cape York, Matthew's men worked the pumps constantly and reduced the amount of sail at even the hint of strong winds, lest the *Cumberland* be overpowered. Matthew spent two days doing survey work in Torres Strait. In the end it was thirty days before he anchored in Timor, 'the indifferent sailing of the schooner' working against making a quick passage.[13]

Johannes Geisler, the Dutch governor at Kupang, had recently died, but his replacement Mr Viertzen offered Matthew every assistance, providing him with a house near Fort Concordia. He told Matthew that Baudin had arrived at Kupang six months earlier, not long after Matthew left in the *Investigator*. Because of continued sickness on the *Géographe*, Baudin had given up his survey work and was heading for Europe, with a stop planned in Mauritius.

There were no opportunities in Kupang to fix the *Cumberland*'s failing port-side pump, and as there was no pitch available to cover the caulking of its seams, the best Matthew could do was plot a course where the winds might be less damaging. He was worried about the vessel falling apart in the approaching north-west monsoon, and made haste to leave on the evening of his fourth day at Kupang, setting a course for Cape Town, some 10,700 kilometres to the south-west.

By 17 November, squalls and lightning threatened the leaky little schooner, and Matthew erected netting to guard against Malay pirates who often swarmed the area. He considered changing course for the safe harbour at Batavia, but the wind changed direction, and though the rain poured for days, a steady trade wind pushed the *Cumberland* along at seven knots (thirteen kilometres per hour) towards southern Africa.

On 4 December, three weeks after leaving Timor, the frail craft was caught in something of a whirlpool, as wind drove waves coming from the east and a groundswell came from the south-west; 'the jumble caused by these different movements' tossed the *Cumberland* about and threatened to twist apart its fragile timbers. The leaks became so severe that the starboard pump, now the only one in good order, was worked almost continually, day and night. Had the wind been on the starboard side, Matthew noted, it is doubtful whether the schooner could have stayed afloat.[14]

He now had to consider even more carefully the danger of attempting the passage around the Cape of Good Hope in heavy winds and high seas without first having the *Cumberland* caulked and the pumps freshly bored and fitted. Governor King was wary of having any dealings at a French port, and warned Matthew of the hurricanes around Mauritius at the time of year they expected he would arrive.[15]

Matthew had a passport issued by the French, who controlled Mauritius, but he had no passport from the Dutch, who were back in charge at the Cape. He wondered if he might even be able to sell the schooner in Mauritius and take another vessel home. At the very least, he could fill the *Cumberland* with enough water and food to reach the British colony at Saint Helena, a third of the way along the west African coast. Matthew and two of his crew were suffering from what was most likely malaria, and could do with some time on land to recuperate.

On 6 December, to the satisfaction of his crew but contrary to Governor King's advice, Matthew decided to change course slightly so he could make repairs at Mauritius, a speck in the Indian Ocean 800 kilometres east of Madagascar. Isle de France, as the colonists called it, was sixty-five kilometres long and fifty kilometres wide. Matthew was sure it was his safest option.

THERE WAS NO CHART FOR Mauritius aboard the little *Cumberland*, but Matthew did have Joseph Banks' set of *Encyclopedia Britannica*, which told him that its capital, Port Louis – which since the French revolution was called Port Nord Ouest – was on the north-west side of the island[16] and had first been visited by Arab traders in the tenth century. The Portuguese and Dutch had followed, and then the French turned

Mauritius into a base for exploration and piracy, with French privateers raiding British vessels. Mauritius had once been home to a bird called the dodo, but by Matthew's day it had been extinct for more than a century.

As dawn broke on 15 December, a month and a day after leaving Kupang, Matthew saw the island rising dramatically out of the Indian Ocean. As he came closer he saw the reefs, which created stunning lagoons of green water and the long beaches of golden sand. He steered south along the edge of a reef, looking for a boat that might give him information about what he still called Port Louis.

At noon, a schooner approached from a small town, before turning back. Matthew took this as a signal, following the boat as fast as he could through a narrow pass in the reefs. He anchored in a small reef harbour which was called the Baye du Cap. The crew from the island boat jumped into a canoe and raced to shore, running up a hill, where an officer with a plume in his hat and men with muskets were waiting.

Matthew now began to fear that England and France were either at war or very near it. Still, he had his passport, and after the hospitality extended to Baudin and his Frenchmen in Sydney, Matthew assured himself, surely he and his men would be well treated after such a long voyage in such a small boat.[17] He held up letters that Governor King had given him to present to the French governor of the island, General François Louis Magallon, but his entreaties had no effect. He sent Aken to the shore in the *Cumberland*'s small boat – and Aken promptly returned with the French officer and two soldiers armed with pistols and swords. They told Matthew that France and Britain were at war again, and that he was their enemy.

The officer was a handsome 38-year-old major, Marie Claude d'Unienville, who on hearing the first reports of a British schooner near the island chasing a local boat had called for reinforcements, sent women and children to safety inland, and had the cattle and sheep on nearby farms driven into the woods lest the invaders shoot them. D'Unienville gathered half a dozen of his national guardsmen and a dozen of his slaves, and told them to get ready to fight off these unwelcome Englishmen.[18]

Matthew recalled that D'Unienville 'spoke a little English':

> ... he asked if I were the captain Flinders mentioned in the passport, whether we had been shipwrecked, and to see my commission. Having perused it, he politely offered his services, inquired what were our immediate wants, and invited me to go on shore and dine with him, it being then near three o'clock.[19]

D'Unienville promised to organise a pilot for the next day to take the *Cumberland* into Port Louis and sent a canoe for the empty casks. The master of the French schooner was sent to moor the *Cumberland* in a secure place.[20]

An hour later, D'Unienville returned with his superior officer, Thomas Etienne Bolger, a local estate owner, who had none of D'Unienville's graciousness. They had with them a man who spoke good English.

Bolger demanded Matthew's passport and commission in a rough manner. Immediately he saw that the passport was not for the *Cumberland*. Matthew explained that the *Investigator* had been unfit for the voyage, but Bolger said it was necessary that both the commission and passport should be sent to the governor, and that Matthew should remain with the vessel until an answer was returned. Matthew objected, telling them as forcefully as he dared that, since war had been declared, his papers were his sole protection and could not be given up.

At length, Matthew was told he would be escorted by soldiers overland to Port Louis with his passport and commission, and that John Aken would be furnished with a pilot and should bring the schooner around after him.

Matthew was taken to D'Unienville's house, Montrose, about a mile away, and told to be ready to set off on horseback early the next morning. The interpreter told Matthew that the governor, General Magallon, was now on nearby Réunion Island, having been superseded by the imperious young general Charles Decaen.[21]

At just thirty-four, Decaen had risen rapidly through the military ranks. Orphaned at twelve, he had been a naval gunner at Brest in 1787, and just three years before Matthew arrived on Mauritius had captured Munich and commanded a division in the Battle of Hohenlinden. He was a loyal admirer of Emperor Napoleon and he hated almost everything British.

D'Unienville was far more accommodating, and at Montrose Matthew enjoyed dinner among the same women and children who earlier in the day had been told to head for the hills because the British were coming. There was a change of plan by the French captors, and the following evening, Matthew was at last allowed to sail the *Cumberland* the seventy-five kilometres to Port Louis, anchoring at 4 p.m. on 17 December.

MATTHEW HAD HOPED to see the *Géographe* at Port Louis, because its commander would vouch for his assistance when they were in *Terra Australis*, but he was told that Baudin had died aboard his ship three months earlier,[22] most likely from tuberculosis, not long after the *Géographe* had anchored there. The ship was now under the command of Pierre Bernard Milius, who had also received Governor King's hospitality in Sydney. Unfortunately for Matthew, the *Géographe* had sailed for France just a day earlier.

Matthew was determined to take his passport and commission to General Decaen immediately, and to request his assistance to have the necessary repairs made to the *Cumberland* so he and his crew could be quickly on their way.

Dressed in his frockcoat with its epaulets and his wide black bicorne hat, Matthew was conducted by an officer and an interpreter into the government house, a large building with white pillars. The officials enquired of Decaen and told Matthew that the general was dining and that he must return in an hour or two.

And so Matthew waited impatiently in a shady lounge with the French officers. There were some who spoke English, and they asked if he had really sailed all the way from Botany Bay in that tiny vessel. 'They asked about the voyage of "monsieur Flinedare",' Matthew recorded, 'of which, to their surprise, I knew nothing but afterwards found it to be my own name which they so pronounced.'[23]

After two hours of this casual interrogation, and another half-hour during which the officers went to tell Decaen all about their conversation with 'monsieur Flinedare', Matthew was shown into a room where two young officers were standing at a table. One was Decaen, a shortish, thick-set man in an ornate officer's jacket, the other his aide-de-camp Lieutenant Colonel Auguste Fulcher de Monistrol, a genteel-looking man of twenty-nine 'whose blood seemed to circulate more tranquilly'. At first glance, Matthew knew Decaen would be hard to handle.

Decaen eyed the little Englishman dispassionately, and the contempt was returned with interest. While Decaen was resplendent in his uniform, Matthew had been at sea for more than a month and had barely had time to wash. He was wearing a uniform that smelled of the ocean, but under Decaen's withering gaze he stood firm, refusing to take off his hat.

'[Decaen] fixed his eyes sternly upon me, and without salutation or preface demanded my passport, my commission!' Matthew recalled.

Decaen glanced over the papers with a scowl.

He asked in an impetuous manner, the reason for coming to the Isle of France in a small-schooner with a passport for the *Investigator?* I answered in a few words, that the *Investigator* having become rotten, the governor of New South Wales had given me the schooner to return to England; and that I had stopped at the island to repair my vessel and procure water and refreshments. He then demanded the order for embarking in the schooner and coming to the Isle of France; to which my answer was that for coming to the island I had no order; necessity had obliged me to stop in passing ... At this answer, the general lost the small share of patience of which he seemed to be possessed, and said with

General Decaen's residence at Government House in Port Louis is one of the oldest buildings on Mauritius, completed in the mid 1700s. *State Library of Victoria 1855229*

much gesture and an elevated voice – 'You are imposing on me, sir! (Vous m'en imposez, monsieur!)' It is not probable that the governor of New South Wales should send away the commander of an expedition of discovery in so small a vessel![24]

The general was not normally an ill-natured man, but his brusque manner made Matthew bristle and the tension built.[25] Decaen tossed back Matthew's passport and the interpreter said Matthew was to follow him and another officer.

Matthew was taken back to the schooner, and with night having fallen and lamps being produced in the dark of the small cabins, he was asked to surrender all his books, charts and journals. These were placed in a trunk and confiscated. Matthew was asked to sign a statement as well, but 'the preamble set forth something upon the suspicions excited by my appearance at the Isle of France', and he refused to sign that part, only certifying that all his charts, journals and papers, together with all the letters on board the schooner, had been taken.[26]

Matthew told the French officials that he was very angry with Decaen and 'the injustice of taking away the papers of a voyage protected by a passport from the French government'. He told them that Decaen's manners would have to improve significantly if the general expected Matthew to visit him again. The interpreter replied that he had orders to take Matthew to lodgings in town.

The seriousness of what was happening now began to dawn on Matthew: 'I looked at him and at the officer, who was one of the aides-de-camp. "What!" I exclaimed in the first transports of surprise and indignation, "I am then a prisoner?" They acknowledged it to be true.'[27]

They told Matthew his detention would probably last only a few days, until his papers were examined, and that in the meantime, directions had been given that he should want for nothing.

THE NEXT MORNING, Sunday, 18 December 1803, the *Cumberland*, with Charrington and the crew, was placed under guard. Matthew, John Aken and Trim were taken by boat to a large house in the middle of the town, and then escorted through a long, dark entryway, up a dirty staircase and into the room assigned to them. Lieutenant Colonel Monistrol and the interpreter wished the pair a good night.

The room contained two low beds on wheels, a small table and two flimsy cane chairs. From its grim, dirty state, Matthew judged the lodgings provided by Decaen to be 'one of the better apartments of a common prison; there were, however, no iron bars behind the lattice windows, and the frame of a looking-glass in the room had formerly been gilt.'

Telling Aken they should probably know the extent of their incarceration soon enough, Matthew stripped and got into bed. But between the mosquitoes above and the bugs below, 'and the novelty' of the situation, it was nearly daybreak before either of them fell asleep.[28]

A long nightmare lay ahead.

Chapter 23

*'Captain Flinders imagined that he would obtain his release by
arguing, by arrogance, and especially by impertinence ...'*
GENERAL CHARLES DECAEN, GOVERNOR OF MAURITIUS,
ON HIS ENGLISH PRISONER.[1]

MATTHEW'S EXHAUSTION eventually brought on
sleep. But he had hardly drifted off when he was rudely
woken at 6 a.m. by two armed guards entering the room. One
spoke to the other, pointing to Matthew and Aken, and then
left. The remaining guard kept up a constant walk between the
beds but paid little attention to his inmates. Aken had somehow
remained asleep, but Matthew shook him by the shoulder, and
later noted that his friend's startled reaction to finding a French
soldier beside his bed would have made him laugh if the situation
had not been so serious.

The French officials had promised Matthew and Aken that
they would want for nothing, but both were unimpressed by the
hospitality. They had been locked up in a tavern called the Café
Marengo, in the Rue Nationale. It was run-down and grubby,
but the breakfast at 8 a.m. and the lunch at noon was fine, 'good
bread, fresh meat, fruit, and vegetables being great rarities'.[2]

At 1 p.m. Lieutenant Colonel Monistrol escorted Matthew to
an office, where a German secretary who spoke some English put
to him questions prepared by Governor Decaen. The questions
and answers were recorded in French and English, with the

secretary's rough translation in English beginning: 'Questions made to the commanding officer of an English shooner [*sic*] anchored in Savanna Bay, at the Isle of France.'[3]

The questions were similar to those Matthew had already been asked. How was it that he had appeared at the Isle of France in so small a vessel, when his passport was for the *Investigator*? What was become of the officers and men of science who made up part of the expedition? Did Matthew have any knowledge of the war with France before arriving? Why was a vessel chased in sight of the island? Did Matthew know the contents of the two boxes of official correspondence that he carried? The orders from Governor King, relating to the *Cumberland*, were also demanded.[4]

After four hours of grilling, Matthew was surprised to receive an invitation from Monistrol to dine with Decaen. The governor noted later that the invitation actually came from Decaen's wife, though Matthew may not have been aware of this. Already he was showing the French that while he may have been a celebrated navigator, he was no diplomat.

Matthew refused the invitation, saying later that 'it was so contrary' to everything that had happened on the island so far, that 'at first [I] thought it could not be serious'. Then he suspected the invitation was an 'experiment' to ascertain whether he really was a commander in the British Navy, since no officer worthy of the rank would accept such an invitation after being so 'grossly insulted' by the confinement.

Matthew told Monistrol that he had already dined, and on being pressed to go at least to the table, he replied that 'under my present situation and treatment it was impossible ... when I should be set at liberty, if His Excellency thought proper to invite me, I should be flattered by it, and accept his invitation with pleasure'.[5]

Monistrol took Matthew's reply to Decaen, who was staggered by the 'impertinence', writing later that 'from boorishness, or rather from arrogance, [Flinders] refused that courteous invitation, which, if accepted, would indubitably have brought about a change favourable to his position, through the conversation which would have taken place'.[6]

Monistrol returned and told Matthew that the general would certainly invite Matthew to dine again when he was 'set at liberty'. He didn't mention when that would be. Matthew later lamented that Decaen was so exasperated by the rebuff 'that he determined to make me repent it'.[7]

THE INTERROGATION continued until dark. By 7 p.m. Matthew had been feeling the heat of the questions and accusations for six hours, and that, coupled with the humidity of the night, left him feeling drained. Matthew told his inquisitors that the answer to how he happened to be in Mauritius could be found in his third log for the *Investigator*: the German secretary could take whatever extracts he needed. Matthew requested to be taken back to the tavern and that the guard be taken out of the room so he could get some sleep. All the books and papers, apart from the third volume of the log, were returned to the trunk and sealed once more, but Matthew never saw that third volume of his log again.

The next morning, the tight rein was loosened a little, the sentinel leaving the room to pace up and down the hallway. Aken was allowed back on the *Cumberland*, under the watch of a French officer, but was not allowed to stay. He returned to the tavern with the timekeeper, and Matthew's sextant and artificial horizon, and the two began a series of observations in preparation for sailing for England within a few days. Or so they thought.

On the morning of 20 December, Matthew was visited by his boatswain, Edward Charrington, who told him that some of the *Cumberland* crew were becoming restless and had started to misbehave. The interpreter, whom Matthew now knew as Joseph Bonnefoy, and an officer took Matthew to the general's house. Decaen was too busy to grant them an audience.

Matthew sat down and wrote Decaen a letter with as much civility as he could muster, despite his indignation at seeing his 'liberty and time thus trifled with'.

To the captain general and governor in chief over the
French settlements to the east of the Cape of Good Hope
Sir

... I beg your Excellencys attention to the following
circumstances ...

My officer and myself being taken out of my vessel,
all subordination and regularity amongst my seamen have
ceased; they are permitted to go into my cabin and take
spirits, they commit disorders and are even permitted to go
on shore. To correct these and for the preservation of the
vessels stores, I have to request that my officer [Aken] be
permitted to remain on board.

The principal objects for which I put in at this port were
to get the upper works of my vessel caulked, and the pumps
fresh bored and fitted ... I request that these works may be
begun upon, that I may be able to sail as soon as possible
after you should be pleased to liberate me from my present
state of purgatory.

With all due respect, I am,

Your Excellencys obedient servant

Mattw. Flinders[8]

Matthew was confident that Decaen would soon see reason and
let him go. At the time, two American ships – neutral in the
war between England and France – were in port as well. On
the afternoon of 20 December, the mate of the *Hunter*, whom
Matthew had seen at Timor in April, and the commander of
the *Fanny*, which had been in Port Jackson in July 1802, called
in at the tavern. Matthew told the two officers that once he
was at liberty, he could sell the *Cumberland* in Mauritius, with
Decaen's permission, and that he and his crew could then travel as
passengers on either of their ships to Saint Helena.

In the evening, Bonnefoy told Matthew that the corporal of
the local guard on board the schooner had been punished for
neglecting his orders, and that one of Matthew's sailors, a Prussian,
who was drunk on shore, had been put into the guardhouse.

He also announced that Decaen would give an answer to Matthew's letter in the morning.

It was not what Matthew had expected.

At about 11 a.m. on 21 December, Bonnefoy and Lieutenant Colonel Monistrol arrived at the tavern and read Matthew an order from their governor.

It said that Decaen was convinced from his examination of Matthew's journal that he 'had absolutely changed the nature of the mission for which the First Consul had granted a passport', and that he 'was certainly not authorised to stop at the Isle of France to make [himself] acquainted with the periodical winds, the port, present state Of the colony, etc'. Such conduct, Decaen said, was a violation of neutrality. To him, Matthew was a spy.[9]

Matthew did not know that François Péron, the zoologist, and Louis de Freycinet had both recently met with Decaen before leaving Port Louis on the *Géographe*. They had been spying on the defences of Sydney, despite the hospitality shown to them. Although Baudin had protested that the French had no plans to establish a settlement in New Holland, King had not been so sure. He had told Lord Hobart, Britain's Secretary of State for War and the Colonies, that he suspected collecting was not the principal objective of Baudin's mission, and that he likely had designs on forming a French outpost based on the north-west coast of Van Diemen's Land.[10]

King had been so concerned about the French that, just after they left Sydney in November 1802, and while Matthew was still sailing the *Investigator*, he had sent Acting Lieutenant Charles Robbins in the *Cumberland* to chase Baudin's team. Robbins had found them at King Island, and delivered an angry letter from King about the rumours of a French settlement, before boldly marching into the middle of the French camp and hoisting the Union Jack.[11] In September 1803, King had established a settlement at Risdon Cove on the Derwent.

That was all ammunition for Péron, who, before sailing for France, had left Decaen with a report on his visit to Sydney, telling the 'Citizen Captain-General':

These southern lands and the numerous archipelagos of the Pacific were invaded by the English, who had solemnly proclaimed themselves sovereign over the whole dominion extending from Cape York to the southern extremity of New Holland. Port Jackson ... should be destroyed as soon as possible. Today we could destroy it easily; we shall not be able to do so in 25 years' time.[12]

Decaen's attitude to Matthew grew dimmer. Bolstering his belief that this young upstart was a spy was the fact that in Matthew's papers was a letter he was delivering for Governor King to Lord Hobart, in which King advised bolstering the naval and military defences at Sydney, because the Isle of France was 'less than seven weeks' away.[13] Decaen began to suspect that Matthew was making an urgent dash to London with this missive, and had stopped at Mauritius to get a better understanding of the island's preparedness as a base from which to launch French forces westward. Decaen knew that his isolated and largely unprotected island was vulnerable to attack and he did not want Matthew telling his masters in London.

Matthew later estimated that in the first sixteen months since hostilities had been renewed, French ships from Mauritius had done damage to British worth at least £1,948,000.[14]

Matthew and Aken were ordered to be confined at the tavern, with the rest of the crew remaining on the *Cumberland*. Bonnefoy and Monistrol treated Matthew 'with much politeness, apologizing for what they were obliged by their orders to execute', and Monistrol said he would make a representation to Decaen, because the governor 'doubtless lay under some mistake'.[15] Bonnefoy and Monistrol took Matthew onto his schooner and packed up all his remaining books and papers, even confiscating letters from family and friends going back years. Everything was locked in a trunk. All Matthew and Aken could take were their clothes.

Matthew fumed. He decided to write another letter to General Decaen. While he may have been tempted to carve the words into the page as though he were using a dagger rather than a quill, he tried to remain polite and respectful, reminding Decaen of the hospitality French crews had received at Port Jackson.[16]

But he was nevertheless persistent in making his point. He reminded Decaen that France had always protected and encouraged scientific voyages that benefited all of mankind:

> Now, Sir, I would beg to ask you whether it becomes the French nation, independently of all passport, to stop the progress of such a voyage, and of which the whole maritime world are to receive the benefit? ... I sought protection and assistance in your port, and I have found a prison![17]

Decaen did not reply, but later wrote that Matthew had displayed arrogance and impertinence towards him, and that Decaen's silence over Matthew's first letter 'led him to repeat the offence'.[18] Decaen did, however, let Matthew's servant John Elder leave the *Cumberland* to stay at the tavern.

The summer humidity made Matthew weak. His room at the tavern was like a sweatbox, and he now lived in a state of high anxiety, wondering when Decaen would see reason. He was locked up in the middle of a hot, sweltering town, having already spent three months confined to a tiny, leaky vessel. He could not but feel, he said, 'the personal inconveniences of such a situation in their full rigour; and the perturbation of mind, excited by such unworthy treatment'.

But the heat and want of fresh air were nothing compared with the swarms of bugs and mosquitoes around the pallet beds at the tavern. The bites of 'these noxious insects' were made on bodies 'ready to break out with scurvy', and Matthew was soon covered with inflamed spots, some of which became ulcers on his legs and feet.[19] He admitted that his illness, frustration and embarrassment might have made him take 'too high a position' against Decaen and use 'too warm a style'.

Decaen granted Matthew's request for a surgeon, and allowed him to write to the Admiralty and his family and friends –

though Matthew's letters would first be read by a French major. The surgeon, Monsieur Chapotin, found Matthew's ulcers to be scorbutic, and after dressing them prescribed him plenty of lemonade, fruit and vegetables.

Christmas Day came and went without ceremony. The shops were open and people were at work. Matthew wrote another letter to Decaen that only irritated him more.

Decaen now told him it was useless to commence a debate about his detention, and that he was willing to attribute Matthew's 'unreserved tone' to the 'ill humour produced by your present situation'. The governor said Matthew's voyage in the *Cumberland* was 'as extraordinary as it was inconsiderate' and gave 'proof of an officious zeal, more for the private interests of Great Britain than for what had induced the French government to give you a passport'.

Decaen said he had already formed 'a poor idea of Matthew's character', but the Englishman's last letter had overstepped 'all the bounds of civility'. Since Matthew knew 'so little how to preserve the rules of decorum', he was to cease all correspondence trying to 'demonstrate the justice' of his cause.[20]

Matthew's frustration bubbled, but on 27 December Monistrol and Bonnefoy came to the tavern with some books Matthew had requested from those confiscated. The lieutenant colonel told Matthew that he'd made a mistake in the tone of his letters to the general, and thought that, rather than terminating his confinement on Mauritius, the angry letters would tend to 'protract' it.

Still, Matthew ventured, 'French republicanism involved anything rather than liberty, justice, and equality, of which it had so much boasted'.[21]

On the last day of 1803, Matthew presented to a major in Port Louis an unsealed letter to the Admiralty and two personal letters, one of them to Ann's stepfather, telling him of his predicament. None would reach their destination. Matthew was unaware that while he had been battling with Decaen, a letter from Ann's mother was making its way across the ocean to Port Jackson.

Mrs Tyler told him that Ann was 'wonderfully restor'd to health, & has been for some time as well as I ever knew her, I hope this blessing will be continued to her for many years to come, for the mutual happiness of you both'.[22]

For Matthew, though, life was becoming increasingly uncomfortable. Decaen interrogated the master of the *Hunter* about his visit to the tavern, and the governor forbade him and the crew of the *Fanny* to carry any letters for his prisoner. Instead, Matthew asked the interpreter, Bonnefoy, to take copies of *The Sydney Gazette* containing an account of the *Porpoise*'s shipwreck to both vessels, with a request that they be passed on to any English ship homeward bound.[23]

THE NEW YEAR OF 1804 brought little respite to the captives, though Monistrol organised an extra room at the tavern and some mosquito nets. Matthew's ulcers improved but the surgeon's recommendation that the prisoners be allowed outdoor exercise was refused. Instead, Matthew started work on his charts of Torres Strait and the Gulf of Carpentaria, but many of the originals had been swamped or lost on Wreck Reef, and he didn't have the third volume of his log, which contained his most recent calculations. He did not tell the French that Aken had many of the details in his log, which he had been allowed to keep.

Matthew made repeated requests to see Decaen but all were ignored. At the time the governor was busy formulating a plan for France to usurp the British in India, though the course of the war meant it would never eventuate.

Decaen decided that rather than rule on Matthew's prolonged confinement himself, he would send details of the arrest to Paris and request instructions. He knew that with the time it took to send a dispatch and receive a reply, this would ensure Matthew was in prison for many months.

On 16 January 1804, he wrote to Denis Decrès,[24] the French Minister of Marine and the Colonies, highlighting his suspicions that Matthew was a spy – or, at the very least, an arrogant and impertinent Englishman – who was carrying information from the British colony at Port Jackson to London about possible conflict with Mauritius.[25]

Charrington and Matthew's crew were now being housed on a prison ship, while the *Cumberland*, left with no protection against the heat and the rain, was mouldering and falling apart. In desperation, Matthew wrote again to Decaen, on 3 February 1804, outlining three proposals for his release. Naively, the first was that Decaen might let him go if Matthew promised not to divulge any information about Mauritius 'for a limited time'. Alternatively, he could send Matthew to France, or let Aken and the crew sail home so that at least Matthew's family and friends would know his whereabouts and the state of his health.

Matthew reminded Decaen that since being wrecked on the *Porpoise* six months earlier, he had been confined either on a small sandbank, or in a small open boat, and then in a small schooner, with virtually no exercise. He was sick with scurvy, and enervated after almost a year of 'great fatigue, bad climate, and salt provisions'.[26]

Again, Decaen did not reply. He ignored another letter three weeks later, in which Matthew signed off as 'Your Excellency's prisoner'.[27]

In his state of high anxiety, Matthew still had Trim to amuse him. Sometimes Trim contrived to elude the guard at the door, and left the tavern to make temporary excursions around Port Louis. Matthew suspected that he made some new lady friends during these visits, 'for they became more frequent than was prudent'. Fearing his little friend might be lost, he and Aken were obliged to lock the cat up after supper.[28]

Matthew also took comfort from the visits of a dashing French naval captain, 32-year-old Jacques Bergeret,[29] who, despite now being at war again with the English, had formed a close friendship over the years with the British admiral Sir Edward Pellew, after initially being his prisoner. By 1804, Bergeret was using Mauritius as his base for raiding British merchant vessels in his frigate *Psyché*, and he had already captured the East Indiamen *Admiral Aplin* and *Alfred*. On board the *Admiral Aplin* were army officers travelling with their wives, and Bergeret arranged comfortable lodgings for them.

Bergeret offered to help Matthew with money to buy better food for himself and his men, and lamented that, had he been in Port Louis when Matthew was arrested, he might have been able to sort out the situation there and then.[30]

The prisoners from the *Admiral Aplin* provided Matthew with magazines, newspapers and a copy of the publication *Steel's Royal Navy List*, up to August 1803. 'In such a place, and after so long an ignorance of what was passing in England,' Matthew wrote, '[they] were highly acceptable.'[31]

But his circumstances weren't. On 1 March 1804, Matthew asked Bonnefoy how long he thought he would remain a prisoner.

The interpreter told him: 'Probably so long as the war lasted.'[32]

CAPTAIN BERGERET began lobbying Decaen for Matthew and his men to receive more humane treatment, even though he could not persuade the governor to release them.

On 23 March, Charrington and the rest of the sailors were transferred from the prison ship to a jail in the Flacq district, on the east side of the island. An officer was present as Matthew was allowed to speak to Charrington; he urged his boatswain not to make an escape attempt, because he was sure that Bonaparte, being a sponsor of scientific discoveries, would authorise their release within six or eight months.[33] Matthew's Prussian sailor – a man who had been arrested for drunkenness – had already been released and had sailed away on a Spanish frigate, while his British companions remained under arrest.

Matthew was soon able to find foreign captains and travellers, and even some French friends he made, who were willing to smuggle letters out of Mauritius for him.

He wrote a detailed letter to the Admiralty about his capture that eventually made its way to London via America that August, and was published in *The London Chronicle*. Matthew wrote several times to Joseph Banks,[34] but he was too emotional to write to Ann, hoping that her stepfather had broken the news of his imprisonment to her.[35] Matthew had no idea whether his letters were arriving or not, but Banks received at least one and dashed off a note to Ann telling her not to fret, as a prisoner exchange was being considered:

> I am as anxious to see your Gallant & Excellent husband as any of his best friends can be, he has done much since he was employd in his last expedition to increase my Regard for him … His Recent Misfortune is one of the Calamities of war which you & I must bear with as much patience as we can muster.[36]

Mauritius had become a centre for prisoner exchange. English crews from captured vessels were being swapped for Frenchmen who arrived on British ships under flags of truce. Some of the English officers awaiting their exchange were held about a kilometre and a half north-east of Port Louis in the Maison Despeaux, a large old stone plantation house with a high wall enclosing about two hectares of tropical fruit trees, neglected long grass and uneven pathways.[37] It was known as 'The Garden Prison'.

In March 1804, Bergeret's lobbying to Decaen finally had an effect, and Matthew, Aken, Elder and Trim were moved to the Garden Prison. Matthew had just passed his thirtieth birthday and was in a debilitated state from a lack of exercise and fresh air, and a persistent kidney complaint. He could walk the 1500 metres to his new jail only with Monistrol's support.

When they arrived, Matthew and his men took up three rooms that covered a corner on the upper floor. The big old house was shabby but spacious and airy. After months in confinement, Matthew thought it was 'paradise',[38] despite being asked to pay eleven Spanish dollars per month as rent – 'perhaps the first instance of men being charged for the accommodation of a prison', he noted drily.[39]

After weeks of deprivation, pain, discomfort and despair, Matthew was worried about how his French guards were mistreating Trim, and he accepted the offer of a French woman living nearby who promised to look after his 'good-natured purring animal', because Trim would make 'a companion to her little daughter'. Scarcely a fortnight had gone by when Trim went missing.

Matthew became frantic. He was still confined to the prison but he sent John Elder to look everywhere for him. Elder took out an advertisement for Matthew in the island's gazette, offering a reward of ten Spanish dollars for Trim's return.

'I would with pleasure have given fifty dollars to have had my friend and companion restored to me,' Matthew wrote.

All these efforts were to no effect, though, and Matthew finally had to admit that 'Poor Trim was effectually lost'. It was probable, he reflected, that 'this excellent unsuspecting animal' had been stewed and eaten by a hungry enslaved person who did not realise the value of the little chap to the people who loved him, but saw only a ready meal in 'his sleek body and fine furred skin'.[40]

Matthew later wrote a 'biographical tribute' to Trim, 'the most affectionate of friends, faithful of servants, and best of creatures … He made the Tour of the Globe, and a voyage to Australia, which he circumnavigated; and was ever the delight and pleasure of his fellow voyagers.'

Matthew promised to create a lasting tribute to his pet. That is, if he ever got off the island.

Chapter 24

'I am wrung with anguish. When my family are the subject of my meditation, my bonds enter deep into my soul.'

MATTHEW FLINDERS, TRYING TO COPE WITH HIS IMPRISONMENT ON MAURITIUS.[1]

THE LOSS OF TRIM LEFT Matthew shattered, but he felt even more keenly the desperate longing for his wife, and he worried over how Ann would handle the news of his disappearance. He had written twice to her stepfather, but neither letter had reached Reverend Tyler. Finally, a smuggled note made it through.

> To the Rev. W. Tyler Partney near Spilsby Lincolnshire
> Isle of France April 26. 1804
> My dear Sir
> I again write, that I am a prisoner in this island with very little prospect of being soon set at liberty. I beg you will inform my friends of the circumstance, and especially that dear friend under your protection to whom I cannot write in my present unhappy situation ...

At the end of this note to his father-in-law, emotion got the better of Matthew, and he scrawled: "'The pleasures of memory". "Forget me not!" "Oh never!"'[2]

He sent Tyler another letter a month later, asking him to tell Matthew's stepmother that he had received in Sydney 'her melancholy letter' about his father's death, and 'had hoped by this time to have been with her to have [made] such arrangements as would have been most for her comfort, but I am debarred from all those who are dear to me'.[3] He could not go into much detail about his life in Mauritius, he lamented, as 'this letter may be subject to the inspection of every curious, idle Frenchman'.[4]

Decaen had recently rejected a proposal from Captain Bergeret to send Matthew to France, and another offer from Captain Emmanuel Halgan,[5] who, taking pity on Matthew, had volunteered to take him to Paris on the French corvette *Berceau*.

Matthew had also appealed to Rear Admiral Charles-Alexandre Linois,[6] the French commander in the Indian Ocean, who was in Mauritius after an unsuccessful raid the previous February on Britain's China Fleet – sixteen giant East Indiamen and twelve other ships bringing goods from Guangzhou to London. Most of the men who had sailed with Matthew on the *Investigator* had returned to Britain and then sailed with the China Fleet, and Robert Fowler and Matthew's brother Samuel received honours for their part in the skirmish. Matthew would not learn details of that conflict until the following year, but he heard rumours that Linois' flagship, the *Marengo*, was in port being prepared for a voyage to Paris. Matthew speculated that he could hitch a ride with this 'upright and humane man'.[7]

Linois answered Matthew immediately, but said that since he was not in port at the time of Matthew's arrival, the matter was for Decaen to decide. But Decaen was immoveable, and told the admiral that the case was now with the French government. He would not do anything until he heard from Bonaparte himself.[8]

Matthew did his best to make the most of a bad situation, and he befriended other prisoners as they came and went at Maison Despeaux. The wide branches of the trees there afforded some relief from the heat, and by using small telescopes from a flat part of the prison roof, Matthew and the other inmates could see boats in the harbour and freedom beyond.

There were eight other inmates when Matthew arrived: two officers who had come from India during the peace on account of their health and had been detained; four men from the *Admiral Aplin*; and two young midshipmen, Dale and Seymour, who had been captured aboard the French frigate *Dédaigneuse*, after having taken it as a prize.

On 1 June, Captain Neufville, the officer commanding the guard at the prison, confiscated all the telescopes and threatened solitary confinement for those who didn't comply. The door to the flat roof was nailed shut, and the prisoners were forbidden to walk in the grounds after sunset. After some of the prisoners forced the door, they were told that the guards would now shoot anyone seen on the roof.

'This produced greater circumspection,' Matthew wrote, 'but the pleasure of the walk and having a view of the sea was such, that it did not wholly remedy the evil.'

Matthew kept a level head through all of these dilemmas. 'Blessed is he that expects nothing, for he shall not be disappointed,' he wrote in his diary, underlining the words for emphasis.[9]

He developed a daily routine designed around his mapping work.

Before breakfast my time is devoted to the latin language, to bring up what I formerly learned. After breakfast I am employed making out a fair copy of the *Investigators* log in lieu of my own which was spoiled at the shipwreck. When tired of writing, I apply to music, and when my fingers are tired with the flute, I write again until dinner.[10]

Before supper, Matthew would practise one of the quartets written by the Austrian composer Ignaz Pleyel, whose works were all the rage at the time.[11]

In Paris, Minister Decrès, having received Governor Decaen's brief on Matthew's arrest, passed the matter on to France's Council of State. In July 1804 the council sent formal approval of Decaen's actions, but added that 'from a pure sentiment of generosity', they chose to grant Matthew's liberty and the restitution of the *Cumberland*, pending a final decision by Bonaparte.[12]

Decaen received their dispatch but was in neither a sentimental nor a generous mood.

MATTHEW SMUGGLED OUT another letter to Joseph Banks, dated 12 July 1804. He told his patron that since his imprisonment, he had written several letters to him 'by several conveyances', and that surely Banks must have received something, though on Mauritius he had no way of knowing.

> I have now been kept in prison seven months ... my dependence, therefore, is upon the Admiralty demanding me to be given up, by virtue of the French passport, in which even here it is acknowledged there has been no infringement on my part ...[13]

He told Banks that the charts which he had sent back to England had been completed to the best of his ability, despite 'the rottenness of the *Investigator*' and their having reached the north coast of *Terra Australis* late in the season.

'I think our labour will not lose on a comparison with what was done by the *Géographe* and *Naturaliste*,' he said, unaware that his charts were in an unopened box at the Hydrographic Office and Banks had not seen them.

Despite his predicament, Matthew's ambition was palpable. As he told Banks, he could not 'rest in the unnoticed middle order of mankind'. He had not been born into a famous or wealthy family, but his actions would ultimately 'speak to the world'. Though he might never surpass Cook in achievement, he believed he could certainly 'secure the second place'.[14] With that in mind, Matthew declared:

> No part of the unfortunate circumstances that have since occurred, can, I believe, be attributed to my neglects or mistakes; and therefore I am not without hope, that when the Admiralty know I am suffering an unjust imprisonment, they will think me worthy to be put upon the post captains list ...[15]

Banks was doubtful that the Royal Navy would promote an imprisoned officer, but not knowing when Matthew would receive his assurances, he wrote: 'From the moment that I heard of your detention I have used every effort in my power towards Effecting your release.' He said he was hoping for the 'Gracious Condescention' of Emperor Napoleon.[16] Banks wrote to France's esteemed scientific academy, the Institut de France, of which he was also an honoured member. He related how Matthew, 'for want of water and provisions', was now detained as a prisoner of war, and 'in consequence of some misunderstanding of the Governor, accused of having come in as a spy and treated with some harshness'.[17]

When Banks wrote to Governor King, he pointed out that he had some sway with the institute:

> Our Government have no communication with the French; but I have some with their literary men, and have written, with the permission of Government, to solicit his release, and have sent in my letter a copy of the very handsome one M. Baudin left with you. If this should effect Flinders's liberation, which I think it will, we shall both rejoice.[18]

Matthew continued a letter for Governor King, which he had started in Kupang.

> This Account will not a little Surprize you, my Dear Sir, who have so lately shewn every attention to the Geographe and Naturaliste, but a Military Tyrant knows no Law or principle ... Passports, reciprocal kindness, and National Faith are baits to catch Children and Fools with ....[19]

King was incensed. He realised Decaen had read his secret correspondence to Hobart, which had clearly fuelled Decaen's anger. King had Matthew's letter reprinted in *The Sydney Gazette*,[20] and it eventually made its way into newspapers in London and India. He also fired off a letter of protest to Decaen,[21] enclosing Baudin's letter of recommendation, which King had forgotten to give Matthew before he sailed. He also informed Admiral Pellew,[22] who by this stage had recaptured his old friend Jacques Bergeret once more.

King also sent dispatches about Matthew's imprisonment to the esteemed Irish orientalist William Marsden, the First Secretary to the Admiralty, and to John Pratt, the Secretary of State for War and the Colonies. The letters left Sydney on 23 May 1805 under the stewardship of Matthew's old friend William Kent, who was returning two members of Matthew's scientific team, Robert Brown and Ferdinand Bauer, and their collections to Liverpool. It gave Matthew no comfort to learn that the ship they sailed home in was his old vessel the *Investigator*, which had been patched up sufficiently to make the voyage.[23]

THIRTEEN THOUSAND kilometres away, in the fenlands of Lincolnshire, Ann Flinders remained in the dark about her husband's welfare. She had assurances from her stepfather and from Joseph Banks that he would be freed soon, but her mind must have returned often to the dark days of her childhood, when she and her mother realised they would never see John Chappelle again.

Well aware of this, and realising that his ambition had put his wife in this turmoil, Matthew began pouring his heart out to her. He had started a letter to Ann in Sydney, and had thought he might as well deliver it by hand. Now he intended to finish it and put his wife's mind at ease. One of his fellow inmates, a British surgeon named Walter Robertson, was being sent home to England via the United States in a prisoner swap. He said he would take Matthew's letter with him.

Matthew told his wife that he was still 'closely confined … although in 7 months nothing has been found to confirm their villainous suspicions against me'.

Comfort thyself my dearest Ann, a few months more after thou shalt receive this will restore me to thy arms, and in our happiness the memory of this misery shall be buried, as a frightful dream ... To my dear Belle, to Thomas Franklin and the charming family, and to my cousin Henny and her family, pray my dear, give my best love and remembrance, and my respects to my uncle at Bolingbroke.[24]

He asked Ann to give some money to help his aged aunt and uncle in Spalding, and arranged with his agent in London to forward the money. He suggested that Ann and Belle might find some good exercise riding on horseback in the fine weather.[25]

On 25 August, Matthew also wrote to his brother Samuel, congratulating him on the accolades he and Fowler had received in the battle with Admiral Linois' raiders.

They formed erroneous opinions of me on my arrival, – they imprisoned me, – I remonstrated, – they were enraged that a prisoner should accuse them of injustice, and determined to punish me I was too obstinate to sacrifice one tittle to them either of the honour of my country or of myself, and therefore prepared myself to suffer ...[26]

Matthew said if the Admiralty was inclined to fit out another ship to finish their mapmaking voyage, Samuel could perhaps arrange another twelve-month assignment 'on the home station', and then if Matthew could arrange two vessels for Sydney, Samuel could command one of them.

In failure of this however, I have many plans in my eye for both [of] us; you know I am a notable builder of aerial castles ... I assure you that what with reading, writing and chart making, I am almost as busy in this prison as ever you knew me.[27]

He had almost finished the hydrographical part of recording their voyage, and along with Samuel's letter he was sending Joseph Banks a general chart of New Holland – though Matthew was now inclined to 'call the whole island Australia or Terra Australis'.[28]

He sent Banks his chart[29] and marked the top of it 'Australia'. Matthew was popularising the name, an abbreviation from *Terra Australis*, Latin for 'southern land', but in fact it had first been recorded on maps in the fifteenth century, although only to mark a theory that there was a landmass far away that would 'balance' the continents of the Northern Hemisphere.

'Australia' was specifically applied to the large landmass for the first time in 1794, when botanists George Shaw and Sir James Smith wrote of 'the vast island, or rather continent, of Australia, Australasia or New Holland' in their 1793 *Zoology and Botany of New Holland*.[30] James Wilson included it on a 1799 chart in his book *A Missionary Voyage to the Southern Pacific Ocean*.[31]

Matthew told Banks that 'the propriety of the name, Australia or Terra Australis, which I have applied to the whole body of what has generally been called New Holland must be submitted to the approbation of the Admiralty and the learned in geography ... the whole body should have one general name, since it is now known ... that it is certainly all one land'.[32]

Two days after he wrapped up his chart of 'Australia' for Banks, along with his letter for Samuel, four British warships under Commodore John Osborn appeared off Mauritius, in an attempt to blockade Linois' fleet. The French were thrown into a panic. But if Matthew thought his deliverance had arrived, he was mistaken. He was closely confined in the Garden Prison – but Charrington was not so closely watched at Flacq, and together with six other inmates who were officers from merchant ships, he commandeered a boat and reached the safety and freedom of one of the British vessels.

Decaen flew into what Matthew called a 'paroxysm of rage'. He sacked the officer in charge at Flacq and ordered every Englishman on the island into close confinement. Matthew's sword, his symbol of rank and respect, was confiscated, and he and Aken began to hope that they might be classified as prisoners of war and exchanged.

Late the next month, Captain George Cockburn,[33] who would later supervise the burning of Washington in another war with the Americans, arrived off Port Louis in HMS *Phaeton* under a flag of truce, proposing to see Decaen for a swap of prisoners, especially the young captain Matthew Flinders. Cockburn was blindfolded, taken onto the prison ship, and told that swapping Flinders was not an option, since the governor did not consider him a prisoner of war but a spy. Cockburn was eventually released, feeling insulted and indignant.

After Charrington's escape, the six remaining sailors from the *Cumberland* were moved to an even more austere prison, leaving most of their clothes behind. The blockade failed, and to Matthew's great disappointment, the British flotilla sailed away.

FINALLY, ON 14 NOVEMBER 1804, there were fleeting rays of spring sunshine in Matthew's life when his friend the surgeon Walter Robertson was finally sent from Mauritius on the British ship *Bellisaurus*. Matthew was dejected that he and his men could not go too.

The prolonged anxiety and disappointment was taking a toll on the health of Matthew and his men. By the end of the year, John Aken had been seriously ill for three months, and George Adler, one of Matthew's crewmen, had been brought to the Garden Prison to care for him. Matthew became more pale and gaunt,[34] suffering from what he called 'a depression of spirits'.[35] He had pain in his legs and swellings in the glands of his neck. Years of making maps and charts by dim lanternlight or looking into the bright sun as he did his survey work had caused 'dimness of sight, and head ach[e]', which made further work difficult.[36] Matthew even lost contact with his friend Bonnefoy, the interpreter, after Decaen sacked him, and Bonnefoy became too afraid to ask permission to visit Matthew again.[37]

On 17 December 1804, Matthew wrote again to the governor, reminding him that he had now spent a full wasted year of his life in confinement, when he could have been producing charts of 'Australia' to benefit all mankind. He tried to make his letter forceful without offending his thin-skinned captor, and rewrote his plea four times before sending it. There was no reply. He wrote again. Dead silence.

Christmas 1804 came and went without celebration, though after weeks of pleading with officials, Matthew's near-naked crew were finally given extra clothes. But all Matthew received was a return of what he called his 'gravelly complaint' – the kidney disorder that had troubled him for years. It became excruciatingly painful, as he passed small crystallised stones in his urine.[38]

He was only thirty but had the appearance and stiffness of a man twice his age. Still he looked for the positives, telling Ann:

> I shall learn patience in this island, which will perhaps counteract the insolence acquired by having had unlimited command over my fellow men. You know, my dearest, that I always dreaded the effect that the possession of great authority would have upon my temper and disposition ... My brother will tell you that I am proud, unindulgent, and hasty to take offence ... In this island, these malignant qualities are ostentatiously displayed and I am made to feel their sting most poignantly.[39]

WHILE JOSEPH BANKS LOBBIED all the French friends he could think of to obtain Matthew's release, he also kept a close eye on the administration of Governor King in New South Wales. King was wearying of life on the frontier, and of the frequent clashes with John Macarthur and the Rum Corps, who still controlled much of the economy through their liquor trade.

King had made enemies among the establishment in Sydney by giving former convicts positions of responsibility, declaring that they should not remain in disgrace forever. He even allowed the celebration of mass for Catholics – at least until Irish convicts staged the Castle Hill rebellion in March 1804, which involved at least 230 prisoners.

These conflicts all took a toll on King's health, and he now wanted to return to London. And the president of the Royal

Society knew the perfect replacement: William Bligh, the master mariner who had taught Matthew so much about seamanship and navigation.

Matthew saw other prisoners come and go on Mauritius, and he was visited by many French friends, including the young merchants Thomi Pitot[40] and his brother Edouard. Thomi told Matthew he was doing himself no favours with the tone of his letters to Decaen, but he elected to write some himself to French men of science and influence, including the explorer Bougainville. Matthew, in turn, wrote to Lord Bentinck, Governor of Madras and son of the Duke of Portland, seeking the release of two prisoners there who were Pitot's relatives. Bentinck not only complied, but wrote to Decaen on Matthew's behalf – though once again all appeals drew a blank.

John Aken remained so ill that by May 1805, Decaen consented to send him and his carer, George Adler, home to London by an American ship bound for New York. Matthew pleaded with Decaen for permission to go as well, sending his captor 'as humble a letter as I could bring myself to write',[41] but again he received no answer.

India's governor-general, Richard Wellesley, wrote to Decaen offering to swap the recently captured Jacques Bergeret for Matthew, but there was no deal. Instead, Bergeret was returned to Mauritius and began to visit Matthew again as a free man, having given his word not to fight against the British anymore.

On the night of 25 July 1805, two French officers invited all the inmates in the Garden Prison to the theatre, telling Matthew that Decaen had given his permission. The theatre was so full that, being late, Matthew scarcely got a sight of the stage during 'a Danse Anglaise'. Matthew was stunned to see 'so many handsomely dressed women in the audience'.

> … the necks of almost all, and the shoulders, and bosoms, and nearly half the breasts were uncovered, as well as the arms nearly up to the shoulders. An equal number of women, equally dressed, would I think raise an uproar in

one of our English theatres. The modest would be offended, the prudes would break their fans, the aged would cry shame! the libertines would exult and clap, and the old lechers would apply to their opera glasses ...[42]

The officer in charge of the outing, Monsieur Reignier, would not be denied in buying tickets for all three of the prisoners under his watch. Reignier behaved with great politeness, Matthew said, but two other officers with them 'were too fond of carousing and libertine conversation for my taste'.[43]

By August, most of the prisoners at the Garden Prison had been exchanged, including Dale and Seymour, the two midshipmen. Matthew, John Elder, who refused to leave his captain, and a crewman named William Smith, an Englishman recovering from a broken leg, were the only inmates who remained. The rest of Matthew's crew had also been freed.

Matthew sometimes played billiards with the 'old serjeant, who had behaved kindly to all the prisoners', but he would much rather have been home playing house with Ann.

He thought often about making a daring escape.

Chapter 25

*'After the time of my premeditated escape, I remained some time
in a state of sullen tranquillity.'*

MATTHEW FLINDERS, DEPRESSED AT HAVING LET SLIP AN OPPORTUNITY
TO FLEE HIS ISLAND PRISON.[1]

MATTHEW'S STRENGTH and appearance were now
so changed after twenty months of mental torture and
physical deprivation that he was scarcely recognisable as the same
person 'who had supported so much fatigue in exploring the
coasts of Terra Australis'.[2]

On 18 August 1805, he made himself as comfortable as he
could in his jail and prepared a hearty breakfast with plenty of
milk in his tea. He then snorted three pinches of snuff – finely
ground tobacco – while he sat drinking his hot brew and thinking
of his wife and friends in England. He wrote himself a note:

Mem. Must not take so much snuff when I return, for it
make me spit about the rooms ... Find myself better this
morning than usual, and less head ach[e]. Took up my flute
and played the 1st. and 5th. Duo of Playels opera 9 ... Must
have all Playels musick when I return to England, that is
set for the flute, and Mozarts, and Haydens, and some of
Hoffmiesters and Deviennes, but the whole will be too
expensive, musick is so very dear in England; and indeed
so is almost everything else. Hope Mrs. F. will have got the

better of the inflamation in her eyes ... Must take a house in the country when I return, and enjoy myself two or three months before I engage in any service; but, God knows, it is now three years since I heard from anybody at home; and what may have happened it is impossible to say ...[3]

He was determined to keep to his plan of eating more puddings and vegetables, and less meat, because it seemed to lessen the severity of his headaches. He had so much time to think on Mauritius because there was little else to do. He contemplated the possibility of 'some river or large opening upon the north-west coast of Australia ... Would I go out as governor of a settlement there, should it be proposed to me? I can't tell, it would depend on many circumstances. Wish to finish the examination of the whole coast of Australia before I do anything else.'[4]

But how would his health hold up?. His diary entries that afternoon showed an exhausted, dispirited, sick man on the cusp of oblivion, isolated, lonely and downhearted. But one who still had a sense of humour.

Must have some kind of trap set for that rat, which comes disturbing me every night ... Dropped asleep soon after ten. Waked about one by the noise of the soldiers in the guard house, who are playing about and running after each other like children. Wish the loud-voiced fellow had taken a dose of opium.[5]

Decaen finally showed Matthew some mercy after considerable lobbying by Bergeret, Pitot and other friends. Just a day after his sad diary entries, Matthew received a letter from Monistrol saying he, Smith and Elder were being moved to a more comfortable house in the country.[6]

MATTHEW'S NEW HOME WOULD be 'Le Refuge', a large plantation owned by a widow named Louise D'Arifat at Wilhems Plains, in the southern interior of the island, far from the sea.

Matthew noted that some of the officers, including Monistrol, seemed extremely courteous to him, perhaps because they feared an imminent British invasion of their isolated posting.

Before being escorted to Wilhems Plains, Matthew was allowed to spend four days in Port Louis without restriction. He stayed at Thomi Pitot's house and did the social rounds, attending a musical recital and a supper party at the home of Pitot's relative Monsieur Deglos.[7]

Recreating his adventure climbing Mount Beerburrum in Australia, and trying to restore his vitality, Matthew and Edouard Pitot climbed an 800-metre mountain called Le Pouce, carrying umbrellas against the driving rain.

At 4 p.m. on 24 August, Matthew set off for his new home, accompanied by nearly all of Pitot's family riding donkeys. At sunset they reached Pitot's large and airy plantation house beside the 60-metre-deep chasm of the Grande Riviere, about eight kilometres from the town, and stayed the night.

Pitot's sons Robert and Frederick taught Matthew the card game bouillotte, and the next morning, with some 'small bandy-legged hounds',[8] took him shooting for hares, though they had little success. They rode along roads that were 'generally bad and very circuitous on account of the chasms which fall into the Grande Riviere'.[9] Most people in the area travelled by horse and donkey, though some of the women were carried in palanquins by enslaved people. They passed through orchards and coffee plantations, fields of sugarcane and maize on a plateau 300 metres above sea level. The surrounding forests were full of monkeys, deer and wild pigs.

At Le Refuge, Matthew was offered the full use of the estate, which included a main house, two outbuildings, which he called pavilions, and a row of huts for the plantation slaves. He chose the two pavilions, one for himself and one for Elder and Smith. Madame D'Arifat was the widow of a French nobleman, a captain in France's royal regiment stationed on the island. She was living in Port Louis but would be moving back to her estate in the summer with her three young daughters and her two youngest sons. She had three older sons in their twenties who had homes of their own.

The change in scenery and circumstance was a tonic for Matthew, who immediately began to experience the benefits of walking as far as he might within his mandated two-league boundary.

Within a few weeks, the D'Arifats arrived to take up residence on the estate. They immediately treated Matthew as one of the family, demanding he live with them without cost, though out of pride he insisted on contributing to their expenses.[10] Along with Madame D'Arifat were her daughters Delphine, twenty-one, Sophie, seventeen, and Lise, who was nearly thirteen. There were two smaller boys, Marc and Aristide, who Matthew thought were about ten and seven.[11] He found the widow and her family 'so very agreeable and interesting' that he became 'desirous of being as much with them as possible'.[12]

I cannot enough be grateful to them for such kindness, to a stranger, to a foreigner, to an enemy of their country for such they have a right to consider me if they will, though I am an enemy to no country in fact … My employments and inclinations lead to the extension of happiness and of science, and not to the destruction of mankind.[13]

Still, it was always in Matthew's mind that he had a beloved wife in England who was expecting his return 'in sickness and in tears', and he still wanted to get off Mauritius as soon as possible.

ON 22 OCTOBER, a packet of letters from Matthew's family and friends finally reached him on an American brig via Walter Robertson and Thomi Pitot. After three years without any word from his loved ones and from Banks, he tore the letters open and read and re-read them alone, savouring all the love they brought from Ann, his half-sister Hannah and his cousin Henny.

Matthew at last could smile at some good news from home, though there was sadness too. His cousin William had become bankrupt, and his uncle Samuel Ward had died.[14] The tone of Banks' missive and his appeals to the French men of science gave Matthew hope that deliverance was near, though he told Ann he had 'been so often disappointed that I am almost afraid now to permit myself to expect with much anxiety'.[15]

It was hard to know how to react when mistreated. 'Should the same circumstance happen to me again,' he said, 'I fear I should follow nearly the same steps.'[16]

Banks told him that 'Mrs Flinders ... was full of anxiety for your Return ... I have heard many times from her on the subject and always done my utmost to quiet her mind & sooth her apprehensions.'[17]

Matthew tried to stay positive as he awaited freedom. He bathed each morning in a stream behind the house, and took two-hour morning walks, exploring the different roads, streams and plantations in his neighbourhood, sometimes going out to the Grand Bassin, a lake at altitude. He searched through ancient caves and walked across fields once owned by Lapérouse. Sometimes he would go shooting with Edouard Pitot to the deep cataracts around the Tamarind River. He befriended a young French officer, Charles Baudin,[18] who had sailed on the *Géographe* to New Holland, though he was unrelated to its late commander. He drank tea and coffee, and ate lashings of eggs and bread and butter, sometimes with radishes and salad.[19]

Sometimes 'a cold or an approaching fit of the gravel' would send him to bed 'in a cold shaking fit', but he was much happier than he had been at Port Louis.

Matthew practised his French with as many people as he could, but especially with the two elder D'Arifat girls, who were just as eager to practise their English. He thought they would quickly find husbands if they were in Port Jackson, but on Mauritius young women far outnumbered young men. Matthew sometimes escorted them both to balls at the big plantation homes, and he taught their little brothers mathematics and navigation. He befriended Irish carpenter Thomas Druce, who had lived on Mauritius for twenty years, and Jean Barrow, a free black man, who was kind and generous, and sold Matthew fowls and eggs, and would not hear of taking money for vegetables.

Ann wrote him again, suggesting she sail to Mauritius to join him, but he told her that was dangerous and impractical. He told her all about Madame D'Arifat's beautiful daughter Delphine, but quickly added that she was very desirous of one day sailing to England to meet the Mrs Flinders she had heard so much about.

On the first day of 1806, Matthew wrote a rough draft of a very affectionate letter to Delphine, asking her to think of him and his loneliness often. But he never sent it.

BACK IN PORT JACKSON William Bligh's command of the colony would soon prove as unpopular as his time in charge of the *Bounty*, as John Macarthur whipped up trouble against him.

On Mauritius, two cyclones ravaged the countryside in February 1806. At the same time, news arrived at Port Louis of Napoleon's successes in Europe at Ulm, Vienna and Austerlitz, but there was no mention of him ordering Matthew's release.

'My hopes are, therefore, now more feeble than ever,' Matthew lamented in a note to Samuel, 'it appears that amidst the great events with which the French ministers are occupied, they have either forgotten or have not time to think of me here.'[20]

Despite his predicament, and the wasted years, he longed for another chance to map Australia. 'Should a peace speedily arrive and their Lordships of the Admiralty wish to have the N.W. coast of Australia examined immediately,' he wrote, 'I will be ready to embark in any ship provided for the service that they may chuse to send out. My misfortunes have not abated my ardour in the service of science.'[21]

As that letter was on its way to Banks' mansion in Soho Square, Ann wrote to a friend to complain: 'The Navy Board have thought proper to curtail my husband's pay, so it behoves me to be as careful as I can; and I mean to be very economical, being determined to do with as little as possible, that he may not deem me an extravagant wife.'[22]

Toussaint Chazal, a talented artist who owned a plantation near Le Refuge, painted the portrait in oil that is on the cover of this book and also made a sketch of Matthew in military uniform. *State Library of South Australia B 16400*

In May 1807, Captain Gamaliel Matthew Ward, of the American ship *Recovery*, visited Matthew at Le Refuge and offered to smuggle him off the island.[23] Matthew replied that he could not go. He had so many important papers from his voyage to Australia that he did not want to risk having to leave them behind or see them destroyed if he were captured. He had also given his word as an officer and a gentleman that he wouldn't make a bid for freedom.

The opportunity passed, and Matthew soon regretted it. He wrote to John Aken, now happily at home in London, to tell him that he could 'have made my escape long ago', if not for his promise to Decaen not to.[24]

In August, Thomi Pitot left Mauritius for the neighbouring island of Réunion, and Matthew felt the wrench of losing a great support. That same month, Delphine left Le Refuge to stay in town with a friend, and Matthew sank into what he called 'a state of melancholy and weakness of mind which destroys my happiness and renders me unfit for and miserable in society'.

In 'a depression of spirits inconceivable', he retired to his couch. Then Elder arrived with some letters from home, and Matthew snapped out of his downward spiral. Later, he said the 'dejection of spirits ... might have proved fatal' had he not 'sought by constant occupation to force my mind from a subject so destructive'.[25]

But he could never rest easy. He was shocked and confused to read that the *Investigator* was sailing again, and perhaps with a touch of paranoia he feared someone else would command it on another expedition and reap what he had sown. When he heard that Bligh was the new governor of New South Wales, Matthew told Banks that he did not wish to be placed under his old commander, lest Bligh 'monopolise' all the credit for Matthew's hard work.[26]

Later, Matthew would write to Bligh congratulating him on his appointment as governor, and including some advice on the growing of maize which he'd observed in Mauritius.[27] He also pointed out that, despite the 'unfortunate antipathy' he claimed Bligh took against him 'in the latter part of our voyage on the *Providence*', he would always be grateful that he had gained his first knowledge of nautical science from a commander who, in turn, had learned from the master, James Cook.[28]

Bligh's time as governor proved a disaster. He was in command for just sixteen months when John Macarthur and his Rum Corps led a military coup[29] that placed Bligh under house arrest, throwing the colony into more turmoil than it had seen in its thirty years.

BY THE END OF 1806, Matthew had been fighting serious depression. His friend Toussaint Chazal,[30] a talented artist who owned a plantation near Le Refuge, gave him a helpful distraction when he asked him to pose for a portrait in oil on canvas. Chazal finished the painting, sixty-five centimetres high by sixty centimetres wide,[31] in January 1807, by which time Matthew was nearly thirty-three.

It is a striking work, showing a thin-faced young man with thin lips, a narrow chin, piercing dark eyes and short, close-cropped curly hair. Chazal may have taken some licence with the hair, since John Elder later wrote to Ann Flinders to say Matthew's hair had become 'very white'. Matthew is dressed in an officer's black jacket with gold trim, his jaw set firmly against all the hardships he had faced.

He had also occupied his mind with a whimsical tribute to Trim, written in a breezy style at odds with the darkness that often visited his mind. Before long Matthew was translating the adventures of his cat into French for the amusement of Madame D'Arifat's young sons.[32]

THE FRENCH COUNCIL of state had approved Matthew's release from Mauritius way back on 14 July 1804. It had been ratified by Napoleon Bonaparte on 11 March 1806. But Decaen had still done nothing about giving Matthew his liberty.

Perhaps, in a time when France was trying to win a scientific as well as military war against the British, supporters of François Péron wanted to see his *Voyage de découvertes aux Terres Australes* in print before Matthew's findings from the *Investigator* proved Australia was an island continent.[33] In any case, on 30 December 1806, Secretary William Marsden of the Admiralty wrote to Admiral Pellew, who was based in Madras as commander of the British fleet in the Indian Ocean. He included a copy of Bonaparte's order.

The diplomatic pressure mounted, and three separate French ships were sent to Mauritius carrying Bonaparte's order to free the esteemed English cartographer. Two were attacked by English vessels, and all their documents were thrown into the sea. The third copy reached Decaen in August 1807, just as Pellew ordered the frigate *Greyhound* to sail to Port Louis under a flag of truce with a fourth copy of the order. It left Calcutta on 19 June 1807.

Unaware of these developments, Matthew had become increasingly fearful about the mental health of his servant John Elder. While Elder had constantly refused opportunities to leave

Matthew behind and sail to freedom, he had been suffering for a long time in silence. Matthew had hoped to see Decaen in April 1807 to again appeal for his release, but the audience was refused. Elder then fell into a deep despair.[34]

Matthew's next petition did succeed, and despite initially having fears that Elder was in no state to be entrusted with crucial documents, he gave him a trunk full of important papers, letters and gifts to take to England. For Ann there was a necklace made by the D'Arifat girls, and some Indian silver coins.

Matthew was allowed into Port Louis on 6 July 1807 to pay for Elder's voyage home via Baltimore on an American ship, the *Phoebe*, but Elder elected to sail as part of the crew. He arrived in London on 31 January 1808. With his charts in safe hands back in England at last, Matthew now felt unencumbered to make his escape.

'Pray heaven,' he asked Ann, 'my project may succeed.'[35]

THE *GREYHOUND*, CARRYING Napoleon's directive for Matthew's release, arrived in Port Louis on 20 July 1807. The message was handed over but Decaen ordered the ship to get out the following day. Then, on 24 July, Monistrol delivered Matthew a letter from Admiral Pellew in Madras, telling him of Napoleon's decision and the fact the French government had recommended his release three years earlier. Matthew was elated and furious in the one turbulent moment. As soon as his head stopped spinning, he fired off a demand to Decaen to let him go.

A week later the governor informed Matthew, through another written despatch, that he had received the directive and, 'so soon as circumstances will permit, you will fully enjoy the favour which has been granted you by his Majesty the Emperor and King'.[36]

For Decaen, though, circumstances did *not* permit. He wrote to Minister Decrès to say he awaited 'a more propitious time for putting into execution the intentions of His Majesty'.[37]

It was costing Decaen 5400 francs a year to keep Matthew confined, and although he was a thrifty administrator, he was still not prepared to let his prisoner go. Mauritius was small, remote, isolated and easy pickings for the British if they decided to storm it. Decaen was sending his brother René to Paris with his plan to invade the British positions in India from Persia and Mauritius, and the general knew that Matthew was well aware that only 'a moderate force' was needed to capture the island and its imperious governor.[38]

Matthew was indeed 'always dangerous'. The French had suffered a crushing setback against Horatio Nelson's forces at the Battle of Trafalgar, off Spain, and perhaps Decaen genuinely feared that Matthew's knowledge of Mauritius might lead to a Union Jack flying over his government house.

Life continued on around Matthew, even as he felt his was ebbing away. In January 1808 he attended the Roman Catholic wedding of Madame D'Arifat's youngest daughter, fifteen-year-old Lise, to Matthew's friend Charles Desbassayns, a wealthy young plantation owner from Réunion. Matthew signed their marriage certificate. The next month, Matthew began teaching Lise's little brothers algebra, all the time calculating that he should have made his escape years earlier.

EARLY IN MARCH 1808, Madame D'Arifat's oldest son, Labauve,[39] told Matthew that British warships had been sighted off his plantation facing Tamarind Bay. The British had in fact increased their blockade activities in the surrounding waters. After more than four years of imprisonment, Matthew began to think the British ships had actually come to rescue him. He packed up his belongings and sent them to Labauve's plantation house, Les Tamarinds.

Matthew's eyesight had been failing for some time, but at dusk on 15 March he saw a vessel very near the coast, manoeuvring as if the skipper wished to remain there all night. Matthew suspected it was an English frigate. He arranged with friends for a canoe to collect him before daybreak the next day.

On the rainy morning of his thirty-fourth birthday, 16 March 1808, Matthew was on the beach at dawn looking at the same ship about three kilometres away. His canoe had not arrived, but three boats, with six or eight men in each, steered towards an opening in the reef where Matthew had planned to embark, and he bolted towards them until he was gasping for breath. He almost dropped dead with disappointment when he saw that the rowers were Africans. To his heartbreak, the ship proved to be the French vessel *La Perle*. Matthew's canoe finally arrived on the beach at 7.30 a.m. after being detained by a soldier guarding the point.[40]

Matthew's mood sank further when the French frigate *Semillante* arrived in Port Louis – now called Port Napoleon – with his young friend Charles Baudin, who had lost his right arm in a battle with an English vessel, *Terpischore*. Matthew was allowed to visit Baudin, taking fifty-three of the finest oranges he could find.[41]

Three months later, Baudin made his first attempt at a letter using his left hand, writing to thank Matthew for his kindness.[42] Matthew replied that he wished mankind would pursue the arts and sciences rather than war.[43]

Back in Lincolnshire, Ann had her own heartbreaks. Her stepfather, Reverend Tyler, died on 14 July 1808, and with a new minister moving to Partney, Ann, her mother and her half-sister Belle had to vacate the home they had known for a quarter of a century. They found a small house at Walkergate in the Yorkshire market town of Beverley.

Matthew received another cruel blow as his captivity continued on into 1809. He read in a French newspaper, *Le Moniteur Universel*, that François Péron had published his *Voyage de découvertes aux Terres Australes*, and that part of the southern coast of Australia was to be called Terre Napoleon, Kangaroo Island was to be called Isle Decres, and 'my two gulphs are to be named *Golphe Bonaparte* and *Golphe Josephine*'.[44] Matthew told Banks:

Thus, whilst General De Caen keeps me prisoner here, they search at Paris to deprive me of the little honour with the scientific world which my labours might have procured me … I have much scruple to believe that my imprisonment is connected with this invasion of the maritime reputation of England …[45]

His anger intensified with the arrival at Port Louis of the new commander of the French fleet in the Indian Ocean, Captain Jacques Hamelin, who had commanded the *Naturaliste* under Baudin and knew Matthew well. While he was a veteran of war on the water, Hamelin did not wish to make waves with Decaen over Matthew's situation and did not visit the prisoner.

Matthew wrote that adding to the 'bitterness of my situation' was the knowledge that 'no means were spared by the French government to enhance the merit' of Baudin's voyage to New Holland.

All the officers employed in it had received promotion; but the *Investigator*'s voyage seemed to obtain as little public notice in England as in France, no one of my officers had been advanced on their arrival, and in addition to so many years of imprisonment my own promotion was suspended.[46]

BRITISH FORCES BEGAN to close in on Mauritius by the middle of 1809, by which time Matthew was thirty-five. Decaen quickly realised Matthew was not the only person trapped on the tiny island outpost. In September, the British attacked Réunion, destroying shore batteries and sparking a rebellion by the island's slaves. Decaen had 1600 troops on Mauritius, and though he wanted to use slaves as troops against the British, the rebellion on Réunion made him think otherwise.

On 12 December, the British East India Company ship *Harriet* was allowed through the British blockade off Mauritius to negotiate another exchange of prisoners. The ship was commanded by John Ramsden, who had been a fellow inmate of Matthew's at the Garden Prison, and it was carrying captured French naval officers. A young civil servant on board named Hugh Hope[47] had been specifically tasked by Lord Minto,[48]

Britain's new governor-General of India, with bringing Matthew home.

Hope wrote to Matthew to say: 'although I should be sorry to raise expectations which might afterwards be disappointed, I confess, my hopes are great'.[49]

As 1810 dawned, Hope told Matthew that he had received 'a very favourable reception' from Decaen, and that he had 'hopes of a final success'. Matthew's long experience with the governor, though, 'prevented any faith in the success of his application for my release'.[50]

In January, other British ships left Port Louis, having retrieved 200 of their prisoners, and Matthew despaired that he had, again, literally missed the boat. But as British warships hovered off the coast of Mauritius like hungry sharks about to attack, Hope shone through. He kept lobbying Decaen, who knew that he and the French were fighting a losing battle.

On 13 March, while Matthew was revising some of his notes on the magnetism of the Earth and on ships from the *Investigator*'s voyage, he received another note to say that Hugh Hope had secured 'an indirect promise' from Decaen for Matthew's release. Hope wrote:

> I [asked] to have it fixed before that any communication should be made to you, as I am told the great sometimes change their minds here ... Sir I now consider it fact ... that I shall have the pleasure of your company to India, and to express at the same time the satisfaction which I feel in being the means of restoring to his friends, his family and his country, a man so worthy of their esteem and admiration.[51]

Matthew still wasn't convinced, and his friends, wishing to prevent another crushing disappointment, said it was unlikely Decaen would let him go with an invasion imminent.[52]

Then, on 28 March 1810, twelve days after Matthew turned thirty-six, a messenger rushed to Matthew with a missive from Monistrol:

> His Excellency the captain-general charges me to have the
> honour of informing you, that he authorises you to return
> to your country in the cartel *Harriet*, on condition of not
> serving in a hostile manner against France or its allies
> during the course of the present war ...
>
> P. S. The cartel is to sail on Saturday next (31st.)[53]

With his head spinning like never before, Matthew rushed about
the island telling his 'incredulous' friends of the good news: he
had just three days remaining on the island. Most were joyful
and sad at the same time, as they had grown fond of the plucky
little Englishman. After dark, Matthew spent hours by candlelight
writing to other friends he would never see again. His servant
and two of the D'Arifats' slaves carried his luggage into town,
and after a farewell dinner Matthew bade a tearful goodbye to
the kindly widow and her children, who had likely rescued him
from an early death four years before. In Port Louis the next day,
Matthew and Hope met with Monistrol, who told them that
while Matthew was free to go, Decaen would not be returning
the *Cumberland*, his telescopes, or the third volume of Matthew's
logbook.

Matthew was not home yet. The British blockade of Mauritius
prevented the *Harriet* from sailing, and Matthew spent the next
few weeks doing the social whirl, attending parties, dinners and
dances in his honour and meeting with other British prisoners
who were to travel with him. After so long with only French
company, he had difficulty for a time in speaking English again.[54]

All the time he feared his release might be cancelled. On 8
May 1810, though, he boarded the *Harriet*, along with other
British prisoners to be expatriated. He was only one of five
prisoners to have his own cabin, small as it was. He signed a
parole document promising not to fight against the French for the
duration of the present war. The *Harriet* was still prevented from
leaving for several more weeks, but rather than sink back into
depression, Matthew occupied his racing mind by practising love
songs he had written for Ann, and teaching himself the Malay

language, using a borrowed dictionary. Despite the fact that he had not been on the sea for more than six years, and had not seen his wife in almost a decade, Matthew told himself: 'This language may be useful to me in exploring the islands between Timor and New Guinea which I propose to do in my future voyage.'[55]

In England, Ann, unaware of the latest developments, was beside herself with worry. Joseph Banks wrote to her in Beverley to say that he had all but failed in his attempts to force Matthew's release, even though he had gone as high as Bonaparte himself.[56]

But in Port Louis, at daybreak on 13 June 1810, the pilot boarded the *Harriet*, and Matthew's sword was returned to him. Finally, after several anxious hours, at 3 p.m. Captain Ramsden gave the command of 'Anchors aweigh'. Matthew waved furiously to his dear friends on shore as the *Harriet* sailed out of the harbour.

Being back on the vast sea at last caused a flood of emotions. 'After a captivity of six years, five months, and twenty-seven days,' Matthew wrote, 'I at length had the inexpressible pleasure of being out of the reach of general De Caen.'[57]

Matthew breathed in the salty air and thought of home. He was free at last.

Chapter 26

'For more than twenty years, He was the Idol of my heart, the centre of my earthly happiness, & altho deprived of his dear society above nine years absence, never for one moment weaned my affection.'

ANN FLINDERS, ON THE UNDYING LOVE SHE HAD FOR HER HUSBAND DESPITE THEIR LONG SEPARATION.[1]

MATTHEW FINALLY FELT English soil under his feet for the first time in nine years and three months when he stepped ashore at Portsmouth on 24 October 1810.[2] There was no one to meet him, as no one knew when he was coming back, though Banks, holidaying on his estate at Revesby Abbey, had delightedly announced to Ann a month earlier:

> Madam
> I have infinite satisfaction in informing you that Capt
> Flinders has at Last attaind his Release & is expected in
> England in a few weeks & that on his arrival he will be
> immediately made a Post Captain.[3]

Banks' letter was actually confirmation of 'news from the Cape' that Ann, her mother and Belle Tyler had read in a newspaper about a ship carrying the famous Captain Flinders home. They were getting ready to go to church at the time, and the news item was like a thunderbolt from heaven. They read the item again and again, asking each other in anxious tones, 'could it be true?'[4]

Matthew had left Mauritius on the *Harriet* bound for India. Not far off the coast, though, he learned that one of the ships in the British blockade, the *Otter*, a sloop of war, was returning to England via the shorter route around the Cape of Good Hope. The next day, as all aboard the *Harriet* cheered this remarkable man, Matthew transferred to the *Otter* for what he hoped would be an express ride back to Ann.

Once again, his passage was delayed. He arrived at the Cape on 11 July, but Britain's Vice-Admiral Thomas Bertie ordered Matthew to divulge everything he knew about the defences of Port Louis, as an invasion was imminent.[5] It was six weeks before Matthew could head for England.

He left Simon's Bay on 28 August aboard the small cutter *Olympia*, an 'indifferent sailing vessel, very leaky, and excessively ill found'.[6] Fourteen days later they reached Saint Helena, and a broad smile creased Matthew's face as he gave the *Olympia*'s skipper, Lieutenant Henry Taylor, the benefit of his experience in these waters, telling him not to travel 'so far west as ships usually do in returning to England'.

On 21 October, as the *Olympia* neared the English Channel, Matthew was remined of all the death-defying battles he'd fought with the sea, as a savage gale assaulted the little ship. Two days later, Matthew had passed the chalk outcrops called the Needles and entered the Solent, and the ship was anchoring at Spithead.

The roads and pathways of Portsmouth must have felt like air under Matthew's feet, such was his elation at coming home, but he followed Navy protocol and he reported to Admiral Sir Roger Curtis, who had been in charge at the Cape when Matthew first sailed there in the *Investigator*. Matthew named a port in Australia after him. There being no ready transport to London, Matthew spent much of the day reminiscing with James Park, the master attendant at the dockyard, who knew Matthew well from a decade earlier when the *Investigator* was being readied for its voyage.

Then Matthew travelled on the mail coach through the night to London, 130 kilometres to the north-east, eagerly expecting to take another coach ride to his wife in Lincolnshire. He arrived in the great city at 7 a.m. on 25 October and went to Fleet Street to see an old friend, Charles Bonner, who told him that Ann had come to town to meet her beloved. The news shocked Matthew: because of the uncertainties of shipping, he had not heard from Ann for four and a half years.

He took lodgings at the Norfolk Hotel on the Strand, and then continued on to the Admiralty, opposite Whitehall, to report his return. There he was met with flattering attention by the First and Second Secretaries, John Wilson Croker and John Barrow.

AT NOON THAT DAY, 25 October 1810, Ann Flinders, escorted by one of her relatives, Mrs Penelope Proctor, in case she became overwhelmed by the situation, alighted from a carriage and walked between the grand columns of Admiralty House and into the imposing entrance hall, all the time looking for the dashing young man she had farewelled in 1801. She saw him talking with John Franklin, who had also come to welcome Matthew home.

Matthew was no longer the young tearaway who had stolen her heart. The years in Mauritius had left him white-haired and sickly. Stress, grief and ill health had aged them both, but there was no stopping their passion for each other. Matthew and Ann embraced tearfully and shared many kisses.

Franklin didn't know where to look as husband and wife abandoned all restraint. He sidestepped the passionate meeting, and would later apologise to Matthew for both the intrusion and the hasty exit.[7]

Matthew later recalled his emotion on seeing his wife again: 'I need not describe to you our meeting after an absence of nearly ten years. Suffice it to say I have been gaining flesh ever since.'[8]

Love was grand, but it didn't pay the bills, and while at the naval headquarters, Matthew was determined to squeeze every penny he could from the Royal Navy for the work he had done in mapping Australia, and the years he had spent incarcerated. He met with Charles Phillip Yorke, now First Lord of the Admiralty, who received Matthew with 'urbanity' and seemed to feel for his suffering in Mauritius.[9] Matthew had named a peninsula in Australia after Yorke.

The returned seafarer now asked that his promotion to post-captain be backdated to 1804, as he'd been promised that pay rise by Earl Spencer, a previous Lord of the Admiralty. Yorke told him that was against Navy regulations, since Matthew had been a prisoner for all those years. King George could make an exception for Matthew, Yorke said, but he was virtually blind, in severe pain from rheumatism and suffering a relapse into a mental disorder.

'His Majesty was then incapable of exercising his royal functions,' Matthew wrote, 'thus the injustice of the French governor of Mauritius, besides all its other consequences, was attended with the loss of six years post rank in His Majesty's naval service.'[10]

As Matthew and Ann left the Admiralty and headed for the Norfolk Hotel, they ran into Ferdinand Bauer, who had travelled home on the *Investigator* five years before. Bauer had brought home 2000 drawings from Australia and was about to have his favourites published.

That evening, Robert Brown called on Matthew. He had just published a book about his botanical work in Australia, *Prodromus Florae Novae Hollandiae*, and would soon start work as Joseph Banks' librarian. While there had been a frostiness between Brown and Matthew aboard the *Investigator*, the pair now embraced as old comrades.

Matthew wrote a quick letter to Banks telling him he was home, and then, after all his professional duties had been performed for the day, Mr and Mrs Flinders resumed their married life, picking up from where they had left off in 1801.

THE NEXT MORNING, Matthew had his hair cut and had a tailor measure him for a smart new uniform, complete with gold braid and epaulets. He visited the Transport Office to pass on letters and money for French prisoners from families and friends on Mauritius. Eventually, he would write that 'I had the gratification of sending five young men back to the island, to families who had shown kindness to English prisoners'.[11]

He visited his old skipper from the *Reliance*, Henry Waterhouse, and found him in poor shape from heavy drinking. They talked about Waterhouse's brother-in-law George Bass, and the tragic mystery of his disappearance. William Kent, their mutual friend, would write to Matthew from his station at Lisbon, asking if Waterhouse was still alive or if the 'pernicious Grog [had] washed him to death'.[12] Kent asked Matthew to visit the monument to his late wife Eliza in a church at Paddington Green.[13]

Matthew also bumped into the former governor of New South Wales, John Hunter. In contrast to Waterhouse, Hunter was, at seventy-three, in rude health, with a mane of wavy white hair complementing his admiral's uniform. Matthew saw Anna King, who was now a widow living with her three young daughters on a small pension, Philip Gidley King having died in London two years earlier.

Colonel William Paterson had seen a new governor, Lachlan Macquarie, arrive in Sydney in 1809, and had sailed from Sydney with his wife, Elizabeth, as Matthew was leaving Mauritius. Paterson died on the voyage and Elizabeth was now struggling to make ends meet. She would soon marry the former lieutenant-governor Francis Grose.

Matthew was saddened to hear that Denis Lacy, the *Investigator*'s eager young Irish midshipman, had been lost at sea in 1804. Robert Fowler was already a commander, and would eventually become an admiral, while Matthew's faithful servant John Elder was now the master-at-arms on HMS *George*, a huge 110-gun warship.

On 5 November 1810, Matthew settled his account at the Norfolk Hotel, paying £17 10s for eleven days. Having now been married for nearly ten years, Matthew and Ann at last set up home together close to Banks' mansion, though at the far more modest accommodation of Mrs Major's boarding house at 16 King Street, Soho, where Matthew had briefly lived as a lieutenant.

On their first night there, Wallingham Franklin and Matthew's brother Samuel called in. Samuel had fallen foul of the Navy. After he and Fowler had been exonerated over the wreck of the

Porpoise, Samuel had gone on to command the twelve-gun brig *Bloodhound*, but he had undergone a court martial two years later, charged with disobedience to orders and falsifying logbook records. He had his seniority as a lieutenant downgraded by three years, and retired on half pay to lay about in Devon. Matthew wrote to Samuel's former commander and had him put back on the Navy List.

On 7 November, Matthew dined with some of the most powerful businessmen in the world, the directors of the British East India Company, at the City of London tavern, where he 'received the same obliging attention' as he had found everywhere since his homecoming.[14] He told the directors that they could save a fortune in time and money by following his track through Torres Strait. Matthew also pressed them to pay him and his officers the remaining £600 in 'table money' Banks had promised from them before the *Investigator* sailed. They paid up two months later, with Matthew receiving £300, as agreed.

Banks wrote exuberantly from Revesby Abbey to say he was coming south posthaste and looked forward to 'Shaking you by the hand & congratulating you',[15] and on 8 November he did just that after Matthew called in at Banks' Soho mansion. Banks was now sixty-seven, crippled by gout, and often cantankerous because of the pain. But he was only too willing to help Matthew get his promotion backdated, and to enquire about payments for writing his account of the voyage to *Terra Australis* – or Australia, as Matthew preferred to call it.

Eight days later, Matthew called on Banks again after Banks had prepared 'the skeleton' of a memorial – an appeal to the King in Council over the backdated promotion. Banks had a guest with him, his great friend William Bligh. Matthew and his former commander were certainly not friends, but they talked at length about how Macarthur had organised a coup in Sydney in 1808, toppling the monarch's government and briefly establishing a military junta.[16]

Matthew kept pressing Charles Yorke for the backdated promotion and submitted his memorial, but Yorke told him that 'the Lords Commissioners had not judged it expedient' to give Matthew any compensation, 'on account of the precedent it might establish', but that they had approved £500 for his services. Yorke told Matthew he looked upon the matter favourably, but that in his appeals to have his promotion backdated Matthew should not flatter himself too much.[17]

On 23 November, Matthew and Ann took the coach to Cambridge, a pleasant ride of just under 100 kilometres. They spent two days at the Blue Boar Tavern there and visited relatives at Wisbech and Tydd St Mary. Matthew deposited the £600 plus interest left to him by his father, which had been in the keeping of his cousin Charles Hursthouse. Matthew and Ann then travelled in the Hursthouses' two-wheeled curricle to Spalding. They revelled in the crisp autumn day, with the wind coming off the sea across the flat, cool fenlands. They collected Matthew's 73-year-old uncle John Flinders, Flinders Sr's brother, and took a post-chaise to Donington.

Matthew's stepmother Elizabeth had moved out of the family home on the market square and was living nearby with her youngest daughter, Henrietta, Matthew's half-sister. Matthew and Ann spent a pleasant, joyous week visiting relatives, and telling the gobsmacking tale of his arrest and survival at the hands of the French tyrants. He visited his other half-sister, Hannah, and her husband, Joseph Dodd, as well as his sister Susanna and her husband, George Pearson, who now had six children, and finally his old teacher, Reverend John Shinglar. Matthew also played with the two children of his sister Betsey, who had died eleven years earlier.

On 3 December, Matthew went with his uncle to get extracts from the parish register relating to his ancestors. He made provisions in his will for monuments in the Church of St Mary and the Holy Rood to honour his father, grandfather and great-grandfather.

In Boston, Matthew was visited by John Allen, the miner from the *Investigator*, and by William Bowles, who had bought Matthew's 120 hectares at Banks Town.

Matthew and Ann moved on to the village of Enderby, where some of the Franklins lived, and Matthew and Willingham walked through the snow to Spilsby to settle the affairs of Matthew's uncle Samuel Ward, who had left him £200 in his will. They walked home at dusk, and Matthew was proud that he did not feel the cold as badly as the others.

There were more family visits in Partney and Louth, and then in Hull, where the couple visited Matthew's cousin Henny and her husband, John Newbald, as well as Ann's mother and Belle in nearby Beverley.

Three days after Christmas, friends took Matthew to the outskirts of Hull to see a jaw-dropping piece of scientific wizardry, 'the steam engine, made by Watt', which raised the water level to supply the town.[18] Matthew and Ann returned to London on 3 January 1811, their six-week second honeymoon ending with a boat ride across the Humber River from Hull to Barton and then a gruelling thirty-four-hour ride in a swaying, rocking, bouncing coach to the capital.[19]

MATTHEW RETURNED to London to find Banks with a copy of Péron's book about the Baudin expedition, in which French names were applied to many of the places Matthew had mapped. No charts of the French voyage had been published yet, and Matthew ventured that the French were waiting to see his before 'pilfering' them too.[20]

He took immense satisfaction in the news that reached London in February 1811 of Decaen's capitulation on Mauritius. The island had fallen to British forces on 3 December 1810. Rather than becoming a prisoner of the British, Decaen had sailed home to France in style, and was soon made the head of the Army of Catalonia in Napoleon's battles with the Spanish. Decaen took Matthew's third volume of his logbook with him; it would not be returned to the British ambassador in Paris until 1825.

At Réunion, Lise and Charles named their baby Henry Flinders Desbassayns in Matthew's honour. Matthew invested in the family's plantation, and in a cattle ranch on Mauritius.

On 13 March 1811, Matthew went to meet Banks and John Barrow from the Admiralty to make arrangements for writing *A Voyage to Terra Australis*. The Admiralty would pay the expenses of reducing and engraving the charts and illustrations, while paper, printing, binding and the fees for Matthew's words and the artwork of Bauer and Westall would be paid out of the profits from the book. Matthew demanded that he be on full pay as a post-captain while writing the two volumes, because he had to live in London, and it was twice as expensive as if he was on half-pay and semi-retired in the country.

Matthew was therefore in a bad mood that afternoon when he 'dined ... stupidly' with his landlady, 'Mrs. Major and a small party of Goths', uncouth characters who spoiled his day.[21] The next day he found a bigger and better rental home around the corner at 7 Nassau Street, Soho, for the same money, and he and Ann moved in immediately.[22]

Matthew called on all his stubbornness to pursue the backdating of his promotion to 1804. Henny's husband, John Newbald, wrote on his behalf to the local member of parliament at Hull, William Wilberforce,[23] best known for his crusade against the slave trade. On 21 May, Matthew received 'a polite note from Mr. Wilberforce', asking him to meet that morning, and the bold MP said he would fight for Matthew's money.[24] A month later, Wilberforce said he had found Yorke 'strongly bent against anti-dating [Matthew's] rank'. Wilberforce now proposed to bring the complaint before the House of Commons.[25]

But Matthew knew that Wilberforce attacking the Admiralty in parliament could only spell disaster for any future naval career he might have, and any voyages of discovery he planned. He therefore politely declined Wilberforce's offer of help.[26] Matthew never received the backdating of the promotion, and he always felt cheated by the Admiralty. Nevertheless, he recognised Wilberforce – 'the worthy representative of Yorkshire' – by naming a cape in Arnhem Land after him on his map of Australia.[27]

MATTHEW WAS NOW a regular guest of Joseph Banks at meetings of the Royal Society, and Yorke appointed Matthew, Banks and John Barrow to oversee the publication of Matthew's account of the *Investigator*'s travels.

Matthew went ahead with gusto, even though he was unhappy to still be only on half-pay from the Navy. The money he'd invested from his father and uncle gave him about £120 a year, and his half-pay brought in another £250. He estimated it would take almost double his income for him and Ann to live 'economically but decently' in the city, and he estimated that *A Voyage to Terra Australis* would take at least two years to produce.[28] He asked the Admiralty to compensate him for the extra astronomical work he'd done on the *Investigator* after John Crosley disembarked at Cape Town, but that claim was rejected too.

Matthew now started to suspect that Banks was not offering to help because he wanted to keep him poor and at his service, and he began meditating 'upon the general conduct of Sir J.B. towards me; in which I find many things not easy to be explained'.[29] Eventually, the Admiralty gave Matthew another £200 after 'Sir J.B.'s Kind interference'.

At one meeting with the cartographer charged with preparing the charts for publication, Aaron Arrowsmith, both Banks and Bligh were present, and Matthew was chuffed that they praised the quality of his work.[30] Matthew became a regular at the breakfast gatherings Banks hosted at his mansion, and Matthew would discuss the work with George Nicol,[31] publisher to the King, and a man who had previously published the writings of Captain Cook.

Samuel moved from Devon to live in London to help his brother with recalculations on the charts because of errors in the lunar tables published during the *Investigator*'s voyage, and the Board of Longitude agreed to pay Samuel and the original astronomer on the *Investigator*, John Crosley, to do that work. Anna King and Elizabeth Paterson loaned Matthew charts that their husbands had kept so he had some more interior detail for his maps. He interviewed John Hunter about his travels and met with the botanist George Caley, newly returned to England after eight years collecting plants for Banks in New South Wales.

Matthew honoured with place names all those who had helped him on the voyage.

IN NASSAU STREET, Matthew and Ann lived a quiet life that revolved around his work. Ann never enjoyed good health and was often bedridden with headaches, but when she was well they liked to walk together in Soho Square after their afternoon dinner, and would then return home to play chess. Ann was now forty, and a small, frail woman, but she and Matthew still thought about having a family.

When Belle Tyler travelled down from Beverley to stay, Matthew had another chess partner, and he treated her to the sights of town, including William Westall's art gallery and the Kew Gardens, where plants were growing from seeds Brown had brought from Australia on the *Investigator*. Matthew had hoped that Belle and his brother Samuel might became a couple, but eventually had to concede that there was 'no probability' of a union: 'she is a very decisive lady, and seems to have taken a dislike to him'.[32]

In August 1811, Matthew and Ann took a stagecoach across town to Hackney to dine with the Hippius family, her relatives. Matthew's eyesight had worsened. He had trouble seeing anything in the distance and had taken to 'wearing spectacles when in the street'.[33] Together, the group watched the flight of 'aeronauts' in a balloon; Matthew thought they had done it 'in a good stile'. 'I went into the gardens to see the balloon filled,' he wrote, 'and there were about 2,000 people.' With his wry humour, he noted that there were reports of 200,000. 'We heard at night that the aeronauts had descended at Tilbury fort, and in returning to town we met a coach with the balloon and car on the top of it.'[34]

These distractions and entertainments could not stop Matthew powering through his writing of the two volumes of *A Voyage to Terra Australis*. The book comprised a history of early European exploration of Australia, Matthew's early voyages, the full history of the *Investigator*'s travels, the shipwreck of the *Porpoise* and Matthew's years of confinement. It also included Matthew's views on the effect of magnetism on compass readings and Robert Brown's observations on Australian botany.

Matthew was concerned about the accuracy of Westall's drawings, which pictured the Australian landscape in places as being far lusher than he remembered, but they were the type of drawings popular with audiences at the time. Matthew consulted books and charts in Banks' extensive library and also in the British Museum for his narrative of early voyages. The Admiralty provided him with Cook's logbook and his own from the *Providence*. Hunter had his records from the *Tom Thumbs*, the *Francis* and the *Norfolk*.

At the end of September 1811, Matthew and Ann moved house again, this time to Mary Street, off Hampstead Road, then at the edge of the city. It was a more peaceful abode, though like all of the Flinders' homes in London, it was a furnished terrace house on a narrow lane, and this one had other lodgers in the building.

The couple owned very little but it cost Matthew twenty-five shillings for removalists to take his trunks, boxes and new bookcase to the new house. The rent was £90 a year, £19 less than in Nassau Street.[35]

Money was not the issue, though. Ann had been plagued again by headaches, which had required a doctor to visit on four days at Nassau Street. The new property was 'cleaner and better arranged', and gave them a rural aspect, looking towards Highgate and Hampstead. Ann's health improved, and before long Matthew could write that he 'walked for an hour with Mrs. F. in the fields at the back of the street'.[36]

Her health was his major priority now. Just shy of her forty-first birthday, Ann was also pregnant, and the couple – who had once wondered if they would ever see each other again – now looked forward to the birth of a child who would reflect their deep love.

Matthew's chart of Van Diemen's Land, with all its little inlets and bays, was particularly difficult to make. But he still had time to walk with Ann to St James's Church in Hampstead Road, although Matthew himself was never overtly religious. At other times he practised his flute at home with new sheet music, expensive as it was. Sometimes he read parts of his book to his wife. His kidneys still gave him grief, and he decided to make a will.

BANKS AND JOHN BARROW favoured the term *Terra Australis* for the island continent, and in his introduction to his book Matthew said that 'with the concurrence of opinions', he had adopted it as the term for New Holland and New South Wales 'in a collective sense', along with 'the adjacent isles, including that of Van Diemen's', although he added: 'Had I permitted myself any innovation upon the original term, it would have been to convert it into AUSTRALIA; as being more agreeable to the ear, and an assimilation to the names of the other great portions of the earth.'[37]

Samuel again proved difficult, and the Board of Longitude confirmed only John Crosley's permanent appointment to recalculate Matthew's astronomical observations. Samuel had worked on the project for months and was furious, and against Matthew's advice fired off written broadsides to the senior officials. On 14 December 1811, Matthew was called to Soho Square, where a big, lumbering, red-faced Banks told him he would not take 'any further steps in the business of [Samuel] being employed as a recalculator'.[38] The brothers then fell out when Samuel refused to give up not only his own record books, but also any of Matthew's in which he had made calculations.[39]

Matthew composed a withering letter for Samuel, writing and rewriting it several times until his blows were deadly accurate:

> The books in question were not originally yours ... You now retain them to the detriment of that voyage; and therein commit, in my opinion, an act of injustice and a breach of confidence. The Board of Longitude will not suffer by this violent proceeding ... But it seems, that

however much I may suffer, it does not prevent you from
following the baneful influence of the passions that possess
you; justice, confidence, gratitude are all trampled under
foot. After having disobliged all other of your relations by
your overbearing pride, intolerance, and want of feeling;
you now ... injure the protector of your youth, and best
remaining friend.[40]

If Samuel did not hand over the books, Matthew promised, he
would never to speak to him again.[41]

Samuel eventually conceded, grudgingly, and Matthew
collected the books from his brother's lodgings. Samuel was,
however, 'hurt that I did not chuse to shake hands with him, at
parting'.[42] Samuel's mood worsened a few weeks later when he
was badly beaten up, suffering a black eye and 'face much swelled'
after an attempted robbery at Temple Bar.[43]

Matthew, angry but still loyal to his troublesome sibling,
ensured that Samuel was eventually paid £250 for the
recalculation work he had done. Despite the troubles with his
brother and the Admiralty, frequent changes of address, and a
limited income, Matthew had become a much more patient man
following his incarceration on Mauritius. Ann told a friend that
'Day after day, month after month passes, and I neither experience
an angry look nor a dissatisfied word. Our domestic life is an
unvaried line of peace and comfort.'[44]

Matthew finished writing the introduction to his book in
mid-January 1812, and on the twenty-third of that month started
the narrative of the *Investigator*'s voyage. Two weeks later, Bligh
arranged for Matthew to show his chart of Australia to the Duke
of Clarence, who would go on to become Lord High Admiral and
then King William IV. The Duke was impressed, and Matthew
slipped into the conversation that he needed royal help in securing
his back-dated promotion. Matthew felt the Duke was keen to
help but nothing came of it.

By 13 March 1812, Ann was almost due. Matthew hired
a servant and Belle came down from Beverley, in order, as

Matthew whimsically put it, to take 'upon herself office of regent of my domestic affairs, till Mrs. F. is able to re-assume the reins of government'. Belle already had a nickname for the baby: 'Timothy'.

On 25 March, with fresh snow on the rooftops around the house, Matthew, now thirty-eight, appeared before a committee at the House of Commons looking into convict transportation. John Hunter and Henry Waterhouse both ventured their opinions, as did Reverend Richard Johnson, who had conducted the first divine service in Sydney in 1788. The committee was looking into which areas of Australia were best suited for colonisation. Matthew told them the area he preferred was the south of Van Diemen's Land around the Derwent.[45]

Matthew's house was awash with visitors as Ann's labour drew nearer. Elizabeth Paterson and Anna King stopped by as did the Hippius and Proctor families.

Ann was scared of the ordeal ahead, and of the prospects for her and the baby. Infant mortality rates were high even for young mothers, let alone for this frail and often sickly woman of forty-one. She wrote Matthew a note, urging him to find another wife if she did not survive, and to read the Bible and remember her in his prayers. And to always love and take care of their child.

THE NIGHT OF 31 MARCH 1812 was a boisterous one of high winds that ended in a quiet, foggy morning. On the afternoon of 1 April, after Matthew had visited Samuel to see how he was recovering from the beating he'd received, he ran into Belle's room at the Mary Street house and implored her to jump up immediately. 'Timothy is coming,' he said.

'Nonsense,' Belle replied, 'You will not make an April fool out of me.'[46]

But Matthew wasn't joking. No one named 'Timothy' had arrived, but Matthew wrote in his diary: 'This afternoon Mrs. Flinders was happily delivered of a daughter; to her great joy and to mine.'[47] They named the tiny baby Anne, after her mother, adding an 'e' in keeping with family tradition.

Matthew wrote to his stepmother that his wife:

... bore the trial with heroic fortitude, and thus far is doing remarkably well ... The child is a little black-eyed girl, without blemish ... and has a decent appearance enough; so that I hope you will have no occasion to be ashamed of it as a grand daughter.[48]

Three days after his daughter's arrival, Matthew suffered another attack of 'the gravel' – which he took to be stones in the bladder. Despite his poor health, he left London on 19 April and travelled by coach to Sheerness, eighty kilometres away, to conduct three days of experiments on magnetic variations for compasses. He undertook more experiments at Portsmouth the following month, and journeyed to Gosport to see Nathaniel Portlock, who had sailed the *Assistant* on the voyage to Tahiti twenty years earlier.

Matthew's booklet *Magnetism of Ships* became required reading for British naval officers, and he wrote about the subject in detail in *Voyage to Terra Australis*. By the late nineteenth century, the Flinders Bar was in common use on ships: a soft vertical iron bar placed in a tube on the foreside of a compass to counteract the magnetism within a ship.

Anne Flinders was baptised on 7 May 1812 at St Giles in the Fields, a grand church with a towering Gothic spire. Her parents and Belle would dance about with her in their arms, and the little bundle of joy 'began to crow and laugh & enjoy it all amazingly'.[49] Matthew called Anne a 'lively girl' who added 'indescribably' to his happiness. As was the custom at the time, Anne was placed with a wet nurse. Matthew wanted nothing more out of life now to make it complete – except to see his income match his expenses.[50]

LIKE ISAAC NEWTON and James Cook, Matthew wanted to build up knowledge for the world rather than tear down his fellow humans as a fighting man. He knew that opportunities for another voyage of discovery were slim now that Bonaparte had invaded Russia and hostilities had resumed between Britain and the United States. He contemplated retiring to live in the country with his wife and baby.[51]

Following his father's example of financial prudence, Matthew noted with alarm on the first day of 1813 that his expenses for the previous year were £422 3s, 'being £156 19[s] more than my income'.[52]

On 17 January, Ann suffered a miscarriage. She still had not recovered five days later. Their once pleasant home with its rural outlook had become an unhappy place. Matthew was writing the long account of his imprisonment by Decaen for his book, and that no doubt contributed to a bad mood, though he complained also that 'the conduct of the hostess here, and some of her lodgers [were] what we cannot approve'. He and Ann went looking for a new home.[53] They settled on 45 Upper John Street, agreeing to rent the first and second floors with kitchen for £95 per annum. The lodgings were newly furnished, and the owners, the Thompsons, appeared respectable.[54]

Two months later, the Flinders family moved to 7 Upper Fitzroy Street at £100 a year. It was a difficult time for Matthew. His friends Henry Waterhouse and William Kent had both died recently, and Labauve wrote from Mauritius to say that Madame D'Arifat had also passed away. But there was still the unbridled joy of little Anne, who came home to stay for good on 27 July 1813 after fifteen months with the wet nurse. She showed her parents that she could run stoutly and speak a few words.

There was some tension between Matthew and Banks, who wanted 'New Holland' used throughout Matthew's book instead of *Terra Australis*. Having already given ground on using the word 'Australia', Matthew now stood firm.

THE NEW YEAR OF 1814 arrived with a bitter winter in London that covered the streets in snow, though it was nowhere near as chilling as the scenes in Russia, where Bonaparte's army was being slaughtered. Matthew became ill again with kidney pain, and there was a sense of urgency as he tried to hurry along the printing of his book. He moved house yet again, this time to 14 London Street, Fitzroy Square, just two kilometres from the mansion that would become Buckingham Palace. Something about the way he felt told him he might not be moving again.

On 27 February 1814, he noted in his diary: 'Communicated to Mr. Hayes our surgeon all the symptoms of the complaint which alone ever troubles me, and which appears to be either stone or gravel in the bladder. It [has] troubled me more within some months and become painful.'[55]

He felt better for a while and spent the next few days correcting proofs and examining engraved charts, but he soon had to summon Hayes again. The surgeon became a regular at the house, and by 9 March Matthew was complaining that he was 'suffering much pain all day'.[56]

A week later, Matthew marked his fortieth birthday, but he seemed much older, a little white-haired man with glasses. The next evening he 'passed some rough pieces of gravel, which appear to have separated from a stone in the bladder'.[57] He documented his condition with the same meticulous care that his father showed in the delivery of babies. The next day Matthew passed 'gravelly sand', and it became a regular occurrence. The 'gravel' was mostly 'oblong small crystals, some bright, others discoloured with blood' and usually 'enveloped in mucus forming a pulpy mass'. Hayes used a thin, flexible bougie to explore passages in Matthew's body but it revealed little and irritated much.

He and Hayes experimented with treatments: distilled water, onion water, leek water, citric and muriatic acid, cups of tea which a servant delivered at all hours, and calcined magnesia – a substance which did more harm than good.

On 24 April, Matthew noted he had to rise twelve times during the night to pass some 'imperfect crystals with mucus, and some concreted broken pieces'. In the twenty-four hours of 3 May, he calculated, he had to urinate thirty times. All the while he also faced the enervating demands of the book, proofreading pages, amending his comments, checking maps and charts.

Visitors arrived constantly with the approach of summer – his honeymooning half-sister Henrietta Chambers and her husband,

James, other relatives, as well as Samuel and former shipmates Brown and Fowler. They found a man in great pain, who could only sit using 'a cushion made hollow'. The 'gravel' was now rough and sharp and he was obliged to lie down for long periods on account of pain, complaining that his urethra became blocked with mucus.[58] He was moved to a downstairs bedroom to make him more comfortable. Little Anne had the measles but Matthew looked worse.

By late June, Belle sadly recorded that 'Dear Matthew' was 'wrapped in a long flannel Gown, with grey hair and sunken cheeks'. He spoke rarely and 'his pleasant cheerful manner was gone'.[59] Ann was horrified that her husband now looked '70 years of age, & was worn to a skeliton'.[60]

On 29 June, Matthew wrote that, with Belle nursing him, he had at last experienced a tranquil night, and was able to rise before noon and shave. Arrowsmith brought him a set of proofs of all the charts in the book's atlas, and he wrote a note for Banks saying he approved. Soon after, the Hippius family offered Matthew their more comfortable home for his recovery, and with that end in sight he became more animated and talked cheerfully about the things he would do when he was well again.[61] But he was too sick to take up the offer.

Matthew rallied with all his strength. Ann said he lived to know that the work on which he'd spent his whole life would at last be shown to the world, and to see the two volumes of the finished book.[62]

Matthew kept scratching in his diary about his condition and treatments. And his pain. When the *Naval Chronicle* advertised a new edition of *Robinson Crusoe*, his mind drifted back to his carefree youth, and he dashed off a note asking to be put on the list of subscribers.[63]

On 10 July, with a shaky hand, he wrote into his daily journal: 'Did not rise before two being I think, weaker than before.'

Friends and relatives came to encourage him. One writer claimed that Matthew shook Samuel's hand and said his last goodbye as though he would see him the next day. Others said that George Nicol arrived on 18 July with copies of the two volumes of *A Voyage to Terra Australis* only to find Matthew unconscious – and that Ann placed them on his bed and put his hand atop them. He was said to have called for 'my papers'[64] with his last breath, though other stories had him harking back to his days on the *Investigator* and whispering hoarsely, 'But it grows late, boys – let us dismiss.'[65]

Belle Tyler, who was there, recorded none of those occurrences. She said Matthew's fretting wife had been persuaded to leave his bedside and retire to her bedroom upstairs to get some much-needed sleep. Belle wrote:

> I was awake by my Sister arising, she was going to the sick room. I begged her to let me go first – the sun shone brightly on me as I went down the stairs, – all seemed so still! What could it mean? – I entered the drawing room – his bedroom opened into it, the door was ajar, I went in – there lay the corpse – the spirit had flown, the countenance placid & at rest. Dear Matthew.[66]

Belle stood at the foot of the bed stunned, looking at Matthew in quiet repose, his pain, and his astonishing life, gone for good. She then ran upstairs to break the news Ann suspected, and then to console her.

'She was soon in the room of death & pressed his cold lips to hers – it was a heart breaking effort. All was now over.'[67]

Captain Matthew Flinders RN had died on 19 July 1814.

Ann always blamed her beloved husband's time on Mauritius for the sickness that killed[68] her idol for twenty years. She said that an autopsy showed that his bladder was 'in the most dreadful state of decay, the inner membrane was torn to shreds by an incalculable number of small crystals, which were found sticking in every part'. 'On being opened,' Ann wrote, Matthew's bladder more resembled 'a mass of fibrous strings & honeycombed sponge than anything like the human organ'.

In modern terms, Matthew died from renal failure. He was

probably suffering from cystitis, an inflammation of the bladder, and nephritis, an inflammation of the kidneys. His kidney complaint, which had first attacked him in childhood, was no doubt worsened by the effects of scurvy and dehydration during his long voyages. Medical knowledge at the time was not sufficient to stop the progress of Matthew's condition.

At the house on London Street, little Anne, just two years old, knew something dreadful had happened and begged her mother not to cry.[69] Mrs Tyler, who had lost two husbands, came down from Beverley to console her daughters, and the three women and Matthew's infant child made a home together.

Matthew was buried at St James's Church in Hampstead Road before a small gathering of mourners. Ann erected a fitting tombstone over her husband's grave and promised that he would never be forgotten,[70] but in time the cemetery was redeveloped. Though his memory lives on in the name of Australia, and in the many placenames that now honour him, Matthew Flinders' tombstone and body were apparently lost forever.

Epilogue

'Matthew Flinders had an Indigenous Australian on board with whom
he clearly had a close rapport. His name was Bungaree, a person
Flinders described as "worthy and brave". "Worthy and brave" is a
description that is just as apt for Captain Flinders himself.'
PRINCE WILLIAM, DUKE OF CAMBRIDGE, AT THE UNVEILING
OF A STATUE HONOURING MATTHEW IN 2014.[1]

WHILE HE WAS IMPRISONED on Mauritius, Matthew had written to Joseph Banks to venture that '[a]fter a misfortune has taken place we all see very well the proper steps that ought to have been taken to avoid it: to be endued with a never-failing foresight is not within the power of man'.[2]

Matthew was rarely blessed with such foresight during his life. From the time he defied his father's wishes to go to sea, to his attempts to smuggle Ann aboard the *Investigator*, to his hurried and aborted trip home on the *Cumberland*, Matthew's driving ambition, his arrogance and his impetuousness caused most of his problems in life. He pursued his dreams regardless of the consequences, and in many ways was his own worst enemy.[3] Samuel Flinders regarded his brother as something of an upstart, 'a fatally-clever man, who, confiding in his own discernment, too easily believed that no branch of knowledge was beyond his reach'.[4]

Until he learnt patience as a prisoner of the French, Matthew displayed little prudence or caution, and so it was when it came to providing for his wife and daughter after his death.

Ann wrote that Matthew died as a 'martyr to his zeal for his country's service', and that he had been so preoccupied with his work – with telling the world about Australia through his book – that he had not properly attended to leaving his family a legacy. 'Thus in the prime of life, fell an officer unexampled in his devotions to the service of his country – from the age of 16 to 40, a period of 24 years, he served it faithfully & unremittingly.'[5] Matthew had barely taken a break from his work all that time, and Ann lamented that he had died an agonising death while working tirelessly to ensure the accuracy of his charts and descriptions in *A Voyage to Terra Australis.*

Ann complained to Matthew's cousin Charles Hursthouse that had Matthew seen the danger of his illness, and had he not been so absorbed with work, 'he would never have left me in such a situation, with neither a bed nor a chair to call my own'. She said he had been overly generous to others 'out of his little property', and that if 'anything happen to the child [Anne], before she is of age', his friends and brother Samuel would receive even more.[6]

Ann Flinders' pain was compounded by battles with Samuel over his will. Matthew had initially bequeathed his investments in Mauritius to his brother, but amended the will after the birth of little Anne. Still, his generosity to other relatives and friends left his widow and daughter struggling financially, even though Matthew did not die a poor man. The sale of his English investments raised £3498 16s 1d, but more than £1000 went to pay off debts, and Matthew's wife and child did not own a home of their own. Samuel was to receive all the books and papers relating to the *Investigator's* voyage but Matthew's grieving widow kept them.

Samuel married at age thirty-seven and raised a family of three daughters and a son, whom he named Matthew. He died at age fifty-two with a reputation for being clever but 'idle, selfish, quarrelsome – an unkind brother, husband and father'.[7]

Ann paid £88 17s 6d[8] for four plaques honouring Matthew and his ancestors on the north wall of the chancel Donington's Church of St Mary and the Holy Rood. She also paid £14 for mourning rings to be sent to Matthew's closest friends, including Thomi Pitot. Ann wrote to Pitot to say she had outlaid £1 12s 6d for three months of *The Times*, newspapers which she would send on to him in Mauritius, but that she would have to charge him for the full amount, as she was 'now too poor' to make them a gift.[9]

THE BRITISH PUBLIC WAS not yet ready for Matthew's two volumes about Australia, a land of which they knew almost nothing. His detailed maps and charts intrigued only other navigators. George Nicol printed 1000 copies but the work did not sell well; Ann would receive just £190 in royalties, spread over many years. She sent copies of the book to Pitot, Labauve D'Arifat and Charles Desbassayns, as well as twenty-six additional copies to Mauritius for sale, though she was crestfallen to learn that most of the books sent there perished in a fire.

Despite poor sales, *A Voyage to Terra Australis* was monumentally influential, though. Lachlan Macquarie, the new governor of New South Wales, adopted the use of the name 'Australia', and two years after Matthew's death wrote to Henry Goulburn, the Under-Secretary of State for War and the Colonies, to declare that Australia should 'be the name given to this country in future, instead of the very erroneous and misapplied name, hitherto given it, of "New Holland"'.[10] The name stuck.

Ann and her daughter survived on a £90 naval widow's pension, combined with a £55 annuity provided by Matthew's legacy. Joseph Banks and William Wilberforce both lobbied for Ann to receive a pension in line with the £300 a year being paid to Captain Cook's widow, Elizabeth, but their attempts failed. In 1815, Ann and her daughter, her half-sister Belle and their mother, Anne Tyler, moved to the vicarage in Chobham in Surrey, one of many rented homes Matthew's family would occupy over the next three decades.

Almost a year to the day after Matthew's funeral, Napoleon Bonaparte led the procession at his own demise. Following his defeat at Waterloo, Napoleon handed himself over to Captain Frederick Maitland, commander of Matthew's old ship

HMS *Bellerophon*, on 15 July 1815, and wrote a letter of surrender, throwing himself on the mercy of Britain's Prince Regent.

That same year, Governor Macquarie dubbed Bungaree 'Chief of the Broken Bay Tribe'[11] and presented him with fifteen acres (six hectares) of land on George's Head, in what is now Mosman. Bungaree had remained on good terms with the colonists after Matthew had left Sydney. Way back in 1804 he had assisted in capturing runaway convicts from Newcastle but the escapees later took revenge by murdering Bungaree's father. Still, he appeared to enjoy the company of the British authorities and the feeling seemed mutual. Bungaree became known to the Europeans in Sydney as 'The King of Port Jackson', and his main wife, Cora Gooseberry,[12] was called his queen.

In 1817 Bungaree sailed with Captain Phillip Parker King, the former governor's son, around Australia on the Indian-built cutter *Mermaid*, as the Royal Navy undertook more survey work of the Australian coastline, looking for an inner route through the Great Barrier Reef.

For the rest of his life, and dressed in the uniform of a naval officer, complete with bicorne hat, Bungaree acted as an official greeter for visiting dignitaries, welcoming, among others, the crew of the Russian exploration ship *Vostok* to Sydney in 1820. He would talk to visitors about Aboriginal culture and give displays on boomerang throwing, as well as taking part in corroborees.

In 1828, when Bungaree was thought to be fifty-three and said to be 'in the last stages of human infirmity',[13] he and his extended family moved to the Governor's Domain and were supplied with rations. *The Sydney Gazette* reported on 2 July the following year that the 'poor fellow was seen in the forenoon of Thursday last in a state of perfect nudity, with the exception of the old cocked hat, graced with a red feather, supporting his trembling frame with a large staff. [Bungaree] is identified with Sydney, and something ought to be done to make his few remaining days easy.'[14]

Bungaree died at Garden Island on 24 November 1830 and was buried at Rose Bay. Sydney's major newspapers, *The Sydney Gazette* and *The Australian*, both wrote glowing obituaries for 'his

Aboriginal Majesty'.[15] Eventually, he would have many localities in Australia named in his honour.

AFTER A DISTINGUISHED naval career that included victories against the French and three gruelling expeditions to the Arctic, John Franklin was appointed lieutenant governor of Van Diemen's Land in 1837. Before embarking on his final, fatal voyage to the Arctic, Franklin wrote to the now elderly Ann Flinders, informing her that he had erected a great monument 'to the memory of your deeply lamented husband – and my earliest friend' over Port Lincoln in the colony of South Australia.[16]

Franklin's obelisk was one of many monuments celebrating Matthew's extraordinary life that would be erected around the world, including in Donington, in Mauritius, in London and throughout Australia. Many of the monuments feature Matthew with Trim, and while Matthew's name is remembered in such places as Flinders University, the Flinders Ranges, Flinders Street Station and Flinders Island, there is also the Café Trim at Sydney's State Library of New South Wales.

By 1851, Ann Flinders was living at Woolwich on the Thames, in a home she shared with her daughter and Belle, her half-sister and close companion for almost all of her life. Anne Jr, a scholar at a time when women were scarce in academia, had a fascination with Egyptology, and in 1845, under the pseudonym Philomathes, had published a book, *The Connexion between Revelation and Mythology Illustrated and Vindicated*.[17] On 2 August 1851 at the nearby St Thomas Church in Charlton, Ann watched on as her daughter, then aged thirty-nine, married thirty-year-old William Petrie,[18] a brilliant young electrical engineer who had helped to develop carbon arc lighting.

Six months later, and having mourned her late husband for almost four decades, Ann Flinders died at Woolwich aged eighty-one. She was buried at the same church where her daughter had become Mrs Petrie.

Belle Tyler died fifteen years later, and just as she had shared her whole life with her older half-sister, she was buried beside her.[19]

All Matthew's siblings had died by then, but his sister Susanna's daughters, Susanna Newsham and Eliza Jackson, migrated with their families to Australia. Their descendants soon made lives for themselves around Sydney, Adelaide and Geelong and, later, in western Queensland. Another relative, Charles Wilson Hursthouse, became a prominent New Zealand surveyor and politician.

In 1810 a decision had been made in London to sell the *Investigator* for breaking up but the dilapidated old vessel was sold to a trader and rebuilt as a commercial sailing vessel. Remarkably, Matthew's ship sailed extensively around the globe for six more decades until 1872.

On 3 June 1853, Matthew Flinders' only grandchild, William Matthew Flinders Petrie,[20] began his voyage through life at Charlton. That year the governments of New South Wales and the newly named colony of Victoria announced that they had each intended to pay Matthew's widow a yearly pension of £100, but following her death would now send the money to his daughter instead. Anne Petrie wrote back to say she would use the money to educate her son at home. The boy would eventually become a renowned explorer in his own right as Sir Flinders Petrie, often called 'the Father of Egyptian Archaeology'. One of his students, Howard Carter, uncovered the tomb of Tutankhamun in 1922.

That same year, Flinders Petrie offered all of his grandfather's papers to the first Australian state to erect a statue in Matthew's honour, and New South Wales won the race.[21] The papers are now housed in the Mitchell Library, as part of the State Library of New South Wales. Outside the building, Matthew's likeness, overlooking Macquarie Street, is cast in bronze and depicted in full naval uniform clutching a raised sextant in his right hand. A rendition of Trim watches on from a windowsill.

More than a century after Flinders Petrie donated those papers, philanthropist Barbara Mason donated to the National Archives of Australia the three volumes of James Cook's *First Voyage* that Matthew had carried in his library on the *Investigator*. The books had been bought from the Flinders family at a Christie's auction. They were estimated to be worth about $150,000 but sold for five

Like his grandfather, Sir Flinders Petrie became a world-renowned explorer. *The Petrie Museum of Egyptian Archaeology*

times that amount. Experts said the volumes were made more valuable because of the handwritten notes Matthew made on some pages, including inserting Indigenous words and correcting Cook's coordinates.[22]

A letter from Matthew to Belle Tyler sold at the same auction for $129,000,[23] and the twenty-volume set of *Encyclopaedia Britannica* that Joseph Banks gave Matthew for the voyage sold for $637,000.

IN 2014, AT AUSTRALIA HOUSE in London, Prince William unveiled a statue of Matthew and Trim that would eventually grace the main concourse at Euston Station, just down the road from London's Petrie Museum of Egyptian Archaeology. The sculpture depicts Matthew bent over a map of Australia, holding his surveying instruments, alongside his beloved cat. The inscription reads: 'In commemoration of CAPTAIN MATTHEW FLINDERS RN 1774–1814 who named Australia and charted its unknown coast with the help of Bungaree and the crew of HM Sloop *Investigator*'.

Prince William told his audience:

> Australia is a very dear country to me and Catherine, and so I am particularly honoured to have been invited today to celebrate a man who did far more than anyone to place Australia quite literally on the map. Thank you for inviting me to commemorate this great man with you all.[24]

Prince William was not to know that Matthew still had one great adventure left.

Matthew had been buried at St James's Church, near what became Euston Station. His body was thought to have been lost after the cemetery became a public park. But 205 years after Matthew died, he became central to a new age of discovery as part of the $100 billion HS2 rail project, providing a high-speed train service between London and England's north-west.

As part of the biggest infrastructure project in Europe, a team of archaeologists worked underneath a white tent the size of a football field. Using techniques pioneered by Flinders Petrie to preserve bones and artefacts, they gently brushed away at the North London soil to explore the old graveyard, which had once contained 60,000 bodies, ensuring that any remains were treated with care and dignity.

Most of the recovered coffins could not be identified, as the gravestones were long gone and the caskets lacked nameplates. Some had tin plates, which had eroded.

But on 15 January 2019, one worker's trowel tapped down onto a solid surface: a coffin plate that was made from lead. The inscription was faint and shrouded in dirt, but as the lead was carefully cleaned, the cursive inscription on the ornately decorated plate called down through the ages.

It said simply:

Capt Matthew
Flinders, R.N.
Died 19 July 1814
Aged 40 Years

Matthew had been found at last. The lead in the plate had also helped preserve parts of the wooden coffin, though the rest had crumbled away under the weight and moisture of the soil.[25]

Matthew's complete skeleton was a little further below in the earth, though it was not well preserved. His old bones were gently removed from the grave and washed and cleaned, as relatives prepared to bury him again.

In July 2019, at a ceremony inside Donington's St Mary and the Holy Rood Church, three generations of Matthew's descendants – Lisette Flinders Petrie, her daughter Rachel Flinders Lewis, and Rachel's infant son Jacob Flinders-Simmons – gave thanks for his remarkable life and the recovery of his remains.[26]

Matthew's descendants, along with local Donington residents and politicians, formed a Matthew Flinders Bring Him Home (MFBHH) group,[27] and plans were made for Matthew to be reburied in his hometown, in the church where he was baptised, and where Samuel Flinders, Matthew Flinders Sr and other ancestors rest in peace.

The reburial at the church of St Mary and the Holy Rood was set for 16 March 2024, the 250th anniversary of Matthew's birth in the little town on the Lincolnshire Fens where the marshland merges with the sea, and where the sea disappears into the horizon and the great, wide world beyond. It was the town where Matthew nurtured a dream of exploration, inspired by the adventures of Robinson Crusoe and Captain Cook.

After a life spent at sea, in which he put Australia on the map and gave the vast island continent its name, the last voyage of Matthew Flinders would see him finally return home.

Acknowledgments

Matthew Flinders surrounded himself with the best team possible when he set sail on his great voyages of discovery. I felt that I did exactly the same when I began my voyage around the life of one of Australian history's most fascinating characters.

My wife Colleen and my friends Bruno Rizzo and Everald Compton offered invaluable support as I charted the life of the man who put Australia on the map.

This book would not have been possible without the backing of my wonderful publishers HarperCollins and ABC Books, especially Roberta Ivers, Lachlan McLaine, Brigitta Doyle, Helen Littleton, Hannah Lynch, Nicolette Houben and Jude McGee.

I owe a great debt of gratitude to Louisa Maggio who designed the sublime cover for this book, so wonderfully encapsulating the beauty of Flinders' life and work.

Thanks to my tireless and diligent editors Julian Welch and Kevin McDonald, and to the writings of such acclaimed authors and Flinders experts as Miriam Estensen, Rob Mundle, Ernest Scott and Dr Gillian Dooley.

Dr Meg Foster, Research Fellow of Newnham College at Cambridge University, provided invaluable advice on colonial historical accounts and insights into the attitudes that shaped European views of the First Peoples of Australia.

Thanks also to Matt Lee from the Australian National Maritime Museum and the staff of the State Library of New South Wales, the State Library of Queensland, the State Library of Victoria,

Bibliography

ACRONYMS USED IN THE BIBLIOGRAPHY AND ENDNOTES

HRA   Historical Records of Australia
HRNSW   Historical Records of New South Wales
NLA   National Library of Australia
NMM   National Maritime Museum (UK)
SLNSW   State Library of New South Wales

BOOKS

Sidney John Baker, *My Own Destroyer*, Currawong Publishing, 1962.

Martyn Beardsley and Nicholas Bennett (eds), *Gratefull to Providence: The Diary and Accounts of Matthew Flinders, Surgeon, Apothecary, and Man-Midwife*, two volumes, Lincoln Record Society, 2007.

F.M. Bladen (ed.), *Historical Records of New South Wales*, Government Printer, 1893.

William Bligh, *A Voyage to the South Sea*, Lords Commissioners of the Admiralty, 1792.

William Bligh (edited by Owen Rutter), *The Log of the Bounty*, two volumes, Cockerel Press, 1937.

Paul-Gabriel Boucé, *Sexuality in Eighteenth-century Britain*, Manchester University Press, 1982.

James Stanier Clarke, *Naufragia: Or Historical Memoirs of Shipwrecks and of the Providential Deliverance of Vessels*, Vol. 1, I. Gold, 1805.

David Collins, *An Account of the English Colony of New South Wales*, Vol. 2, T. Cadell Jr & W. Davies, 1802.

Adam Courtenay, *Three Sheets to the Wind*, ABC Books, 2022.

Warren R. Dawson, *The Banks Letters: A Calendar of the Manuscript Correspondence of Sir Joseph Banks*, Trustees of the British Museum, 1958

Miriam Estensen, *The Life of Matthew Flinders*, Allen & Unwin, 2002.

Matthew Flinders, *A Voyage to Terra Australis; Undertaken for the Purpose of Completing the Discovery of That Vast Country*, two volumes, G. & W. Nicol, 1814.

Paul Gaffarel, *La Politique coloniale en France*, Librairies Félix Alcan Et Guillaumin Réunie, 1908.

Francis Galton, *Inquiries into Human Faculty and Its Development*, Macmillan, 1883.

Peter Goodwin, *The Ships of Trafalgar: The British, French and Spanish Fleets October 1805*, Conway Maritime Press, 2005.

Holinshed's Chronicles of England, Scotland and Ireland, Vol. II, J. Johnson, 1807.

Geoffrey C. Ingleton, *Matthew Flinders: Navigator and Chartmaker*, Genesis Publications, 1986.

William James, *The Naval History of Great Britain*, Vol. 1, Richard Bentley, 1837.

Ida Lee, *Captain Bligh's Second Voyage*, Longmans, Green and Co., 1920.

Pieter Arend Leupe, *The Voyages of the Dutch: To the Zuidland or New Holland in the 17th and 18th Centuries*, G. Hulst van Keulen, 1868.

George Mackaness, 'Fresh Light on Bligh: Some Unpublished Correspondence', *Australian Historical Monographs*, Vol. 5, Review Publications, 1949.

W.G. McDonald, *The First-Footers: Bass and Flinders in Illawarra*, Illawarra Historical Society, 1975.

Joseph Henry Maiden, *Sir Joseph Banks: The Father of Australia*, William Applegate Gullick, Government Printer, 1909.

John Masefield, *Sea Life in Nelson's Time*, Methuen & Co, 1905.

Rob Mundle, *Flinders: The Man Who Mapped Australia*, Hachette, 2012.

Arthur Phillip, *The Voyage of Governor Phillip to Botany Bay*, John Stockdale, 1789.

Catharine Retter & Shirley Sinclair, *Letters to Ann: The Love Story of Matthew Flinders and Ann Chappelle*, HarperCollins, 1999.

Anne Salmond, *Bligh: William Bligh in the South Seas*, Penguin Random House, 2011.

Ernest Scott, *The Life of Captain Matthew Flinders, R.N.*, Angus & Robertson, 1914.

William Smellie, *A Treatise on the Theory and Practice of Midwifery*, Vol. 1, D. Wilson & T. Durham, 1766.

Edward Smith, *The Life of Sir Joseph Banks*, John Lane, 1911.

Lyn Stewart, *Blood Revenge: Murder on the Hawkesbury 1799*, Rosenberg Publishing, 2015.

John St Pierre, *Matthew Flinders: The First English Explorer in Moreton Bay 1799*, Redcliffe Historical Society, 1997.

George Suttor (ed.), *Joseph Banks: Memoirs Historical and Scientific of the Right Hon. Sir Joseph Banks*, E. Mason, 1855.

Watkin Tench, *A Complete Account of the Settlement at Port Jackson, in New South Wales, Including an Accurate Description of the Colony; of the Natives; and of Its Natural Productions*, G. Nicol and J. Sewell, 1793.

Pishey Thompson, *Collections for a Topographical and Historical Account of Boston and the Hundred of Skirbeck in the County of Lincoln*, J. Noble, 1820.

Frederick Watson (ed.), *Historical Records of Australia [HRA]*, Library Committee of the Commonwealth Parliament, 1914.

James Wilson, *A Missionary Voyage to the Southern Pacific Ocean*, London Missionary Society, 1799.

Rif Winfield, *British Warships in the Age of Sail 1793–1817*, Seaforth, 2008.

MAGAZINE ARTICLES

'Biographical Memoir of Captain Matthew Flinders, RN', The Naval Chronicle for 1814, *General and Biographical History of the Royal Navy of the United Kingdom*, Vol. 32, Joyce Gould, London.

Arthur Boylston, 'Daniel Sutton, a Forgotten 18th Century Clinician Scientist', *Journal of the Royal Society of Medicine*, February 2012.

J.G.L. Burnby, 'The Flinders Family of Donington: Medical Practice and Family Life in an Eighteenth Century Fenland Town', *Society for Lincolnshire History and Archaeology Journal*, no. 23, 1988.

Kate Fullagar, 'Bennelong in Britain', *Aboriginal History*, Vol. 33, ANU Press, 2009.

'Memoirs of the Circumnavigator', *The Town and Country Magazine or Universal Repository of Knowledge, Instruction and Entertainment*, Vol. 5, 1773.

INTERNET

N. Barquet & P. Domingo, 'Smallpox: The Triumph over the Most Terrible of the Ministers of Death', *Annals of Internal Medicine*, Vol. 127, No. 8, part 1, 1997, pp. 635–42.

Rebecca Brice, 'Sacre Bleu! French Invasion Plan for Sydney', abc.net.au, 10 December 2012.

Gillian Dooley, 'Matthew Flinders: The Man Behind the Map of Australia', transactionsvic.blogspot.com, 14 October 2015.

Gillian Dooley, 'The Limits of Empathy: Matthew Flinders' Encounters with Indigenous Australians', theconversation.com, 6 December 2016.

Edward Duyker, 'Mary Beckwith', dictionaryofsydney.org.

Jean Fornasiero & John West-Sooby, 'A Cordial Encounter? The Meeting of Matthew Flinders and Nicolas Baudin (8–9 April, 1802)', News and Papers from the George Rudé Seminar, h-france.net.

Olivier Goossens, 'The Glorious First of June 1794', historic-uk.com.

Richard Hiscocks, 'Captain Bligh's Second Breadfruit Mission – August 1791–August 1793', morethannelson.com, 9 December 2018.

William Michael Hunt, 'The Role of Sir Joseph Banks, K.B., P.R.S., in the Promotion and Development of Lincolnshire Canals and Navigations', PhD thesis, The Open University, 1986, oro.open.ac.uk.

'In Search of the Lost Coal Mines of Newcastle', hunterlivinghistories.com. Diana S. Jones, 'The Baudin Expedition in Australian Waters (1801–1803): The Faunal Legacy', museum.wa.gov.au.

Stefan Riedel, 'Edward Jenner and the History of Smallpox and Vaccination', *Baylor University Medical Center Proceedings*, Vol. 18, no. 1, 2005, pp. 21–25.

Keith Vincent Smith, 'The Mystery of Yemmerrawanne's Resting Place', State Library of New South Wales [SLNSW], www2.sl.nsw.gov.au, 2014.

Keith Vincent Smith, 'Woollarawarre Bennelong', dictionaryofsydney.org

JOURNALS

William Bligh, *A Log of the Proceedings of His Majesty's Ship Providence on a Second Voyage to the South Sea Under the Command of Captain William Bligh, to Carry the Breadfruit Plant from the Society Islands to the West Indies, written by himself*, SLNSW.

'A Memoir of Captain Flinders, Apparently in His Wife's Handwriting', Papers of Matthew Flinders (as filmed by the AJCP), 1698-1852: [M444, M3033-M3037]/Series FLI 101 - 110/File FLI 103-110/Item FLI 104, NLA.

William Bradley, Journal titled 'A Voyage to New South Wales', December 1786–May 1792, SLNSW.

James Cook, *Journal of H.M.S. Endeavour*, 1768–1771, NLA.

Matthew Flinders' Biographical Tribute to His Cat Trim, 1809, flinders.rmg.co.uk.

Matthew Flinders, *Journal in the Norfolk Sloop*, 1799, SLNSW.\

Matthew Flinders, *Journal Kept in the Bellerophon*, September 1793–July 1794, NLA.

Matthew Flinders, *Journal on HMS 'Investigator'*, SLNSW.

Matthew Flinders, *Narrative of Voyages in the Tom Thumb*, NLA.

Matthew Flinders, *Portion of Journal, Relating to Voyage on Providence with Bligh*, circa 1791, NLA.

Matthew Flinders, *Private Journal*, 17 December 1803–8 July 1814, SLNSW.

John Hunter, *Journal Kept on Board the Sirius During a Voyage to New South Wales*, May 1787–March 1791, SLNSW.

Journal of Richard Atkins, NLA.

Log of Reliance (Commander: H. Waterhouse), NLA.*Memoirs of George Suttor, F.L.S., 1774–1859*, SLNSW.

Daniel Paine, *Diary as Kept in a Voyage to Port Jackson N.S.W*, NLA.

Samuel Smith, *Journal on Board Investigator and Porpoise under Captain Matthew Flinders*, SLNSW.

William Henry Smyth, 'A Brief Memoir of Captain Matthew Flinders, RN', State Library of Victoria.

'The Journal of Peter Good, Gardener on Matthew Flinders Voyage to Terra Australis 1801–03', *Bulletin of the British Museum (Natural History)*, Historical series, Vol. 9, 1981.

George Tobin, *Journal on HMS Providence 1791–1793*, SLNSW.

Isabella Tyler, 'Biographical Outline of Capt. & Mrs Flinders 1852', NLA, NMM. [Incorrectly attributed to Elizabeth Tyler]

Henry Waterhouse papers, SLNSW.

Endnotes

Prologue

1   'A Memoir of Captain Flinders,
    Apparently in His Wife's Handwriting',
    Papers of Matthew Flinders (as filmed
    by the AJCP), 1698-1852: [M444,
    M3033-M3037]/Series FLI 101 - 110/
    File FLI 103-110/Item FLI 104, NLA.

Chapter One

1   Martyn Beardsley & Nicholas Bennett
    (eds), *Gratefull to Providence: The Diary
    and Accounts of Matthew Flinders, Surgeon,
    Apothecary, and Man-Midwife, 1775–1802*,
    Vol. 1, Lincoln Record Society, 2007,
    p. 10.

2   William Flinders, b. 1520 (Carlton,
    Nottinghamshire), d. April 1596
    (Gedling, Nottinghamshire).

3   *Holinshed's Chronicles of England,
    Scotland and Ireland*, Vol. 2, J. Johnson,
    1807, p. 58.

4   Pishey Thompson, *Collections for a
    Topographical and Historical Account of
    Boston and the Hundred of Skirbeck in the
    County of Lincoln*, J. Noble, 1820, p. 31

5   Matthew Flinders, Private Journal,
    17 December 1803 – 8 July 1814,
    State Library of NSW, Safe 1/58,
    3 August 1810.

6   John Flinders, b. 17 December 1682
    (Gedling, Nottinghamshire), d. 13 April
    1741 (Donington, Lincolnshire).

7   From marble tablet in Donington Parish
    Church, National Maritime Museum,
    Greenwich, England, (NMM) FLI06.

8   John Flinders, b. 3 January 1713
    (Donington, Lincolnshire), d. 26
    December 1776 (Donington).

9   Matthew Flinders Sr, b. 17 February
    1750 (Donington), d. 1 May 1802
    (Donington).

10 *Gratefull to Providence*, Vol. 1.

11 J.G.L. Burnby, 'The Flinders Family
    of Donington: Medical Practice and
    Family Life in an Eighteenth Century
    Fenland Town', *Society for Lincolnshire
    History and Archaeology Journal*, no. 23,
    1988, p. 52.

12 *Gratefull to Providence*, Vol. 1, p. xvi.

13 *Ibid.*, p. 53.

14 *Ibid.*, p. xvi.

15 *Ibid.*, p. 136.

16 Burnby, 'The Flinders Family of
    Donington', p. 51.

17 Neil Wright, 'Lincolnshire Towns
    and Industry 1700–1914', *History of
    Lincolnshire XI*, 1982, p. 3.

18 Aaron O'Neill, 'Child Mortality in the
    United Kingdom 1800–2020', statista.
    com, 21 June 2022.

19 'The Kongouro from New Holland' is a
    1772 oil painting by George Stubbs.

20 *Gratefull to Providence*, Vol. 1, p. 14,
    13 May 1775.

21 *Ibid.*, p. 36, 4 June 1776.

22 Elizabeth Hursthouse, b. 1715 (Moulton
    near Spalding, Lincolnshire), d. 27
    March 1768 (Donington).

23 *Gratefull to Providence*, Vol. 1, p. 12,
    22 April 1775.

24 Furneaux to Secretary Stephens, 5 April
    1774, from F.M. Bladen (ed.), *Historical
    Records of New South Wales* [*HRNSW*],
    Vol. 1, Part 1 (1762–1780), Government
    Printer, 1893, p. 375.

25 Sarah Banks, 'Memoranda on Various Subjects', Papers of Sir Joseph Banks, National Library of Australia, (NLA), MS 9, Series 2/Item 32, 32k.

26 'Captain Cook to Mr Banks', 24 May 1776, *HRNSW*, Vol. 1, p. 396.

27 *Gratefull to Providence*, Vol. 1, p. 3, 5 January 1775.

28 Jonathan Gleed, b. 1748, d. 17 March 1820 (Donington, Lincolnshire).

29 Thomas Arnall Gleed, b. December 1774 (Donington, Lincolnshire), d. 13 February 1814 (Donington).

30 *Gratefull to Providence*, Vol. 1, p. 4.

31 *Ibid.*, p. 7, 9 February 1775.

32 *Ibid.*, p. 10, 17 March 1775.

33 *Ibid.*, p. 7, 21 February 1775.

34 *Ibid.*, p. 10, 16 March 1775.

35 *Ibid.*, p. 9, 17 March 1775.

36 William Smellie, *A Treatise on the Theory and Practice of Midwifery*, Vol. 1, D. Wilson & T. Durham, 1766, p. xiv.

37 *Gratefull to Providence*, Vol. 1, p. 9, 17 March 1775.

38 *Ibid.*

39 *Ibid.*, p. 11, 18 April 1775.

40 *Ibid.*, p. 14, 1 May 1775.

41 *Ibid.*, p. 89, 8 December 1779.

42 *Ibid*, p. 20, August 1775.

43 Elizabeth 'Betsey' Flinders, b. 24 September 1775 (Donington), d. 27 October 1799.

44 *Gratefull to Providence.*, Vol. 1, p. 136.

45 *Ibid.*, p. 24, 20 October 1775.

46 *Ibid.*, p. 10, 16 March 1775.

47 *Ibid.*, p. 34, 15 May 1776.

48 *Ibid.*, p. 37, 30 July 1776.

49 *Ibid.*

50 *Ibid.*, p. 41, 13 November 1776.

51 *Ibid.*, p. 42, December 1776.

52 *Ibid.*

53 *Ibid.*, p. 49, 12 August 1777.

54 N. Barquet & P. Domingo, 'Smallpox: The Triumph over the Most Terrible of the Ministers of Death', *Annals of Internal Medicine*, Vol. 127, no. 8, part 1, 1997, pp. 635–42.

55 Stefan Riedel, 'Edward Jenner and the History of Smallpox and Vaccination', *Baylor University Medical Center Proceedings*, Vol. 18, no. 1, 2005, pp. 21–25.

56 Arthur Boylston, 'Daniel Sutton, a Forgotten 18th Century Clinician Scientist', *Journal of the Royal Society of Medicine*, Vol. 105, no. 2, 2012, pp. 85–87.

Chapter Two

1 Paul Gaffarel, *La Politique coloniale en France*, Librairies Félix Alcan Et Guillaumin Réunie, 1908, p. 34.

2 *Gratefull to Providence*, Vol. 1, p. 65, 18 May 1778.

3 *Ibid.*, p. 68, 30 May 1778.

4 *Ibid.*, p. 72, 21 September 1778.

5 *Ibid.*, p. 87, 30 July 1779.

6 Captain Clerke to Secretary Stephens, 8 June 1779, *HRNSW*, Vol. 1, part 1, p. 415.

7 Later Vice-Admiral William Bligh, b. 9 September 1754 (Plymouth), d. 7 December 1817 (London).

8 *Gratefull to Providence*, Vol. 1, pp. 101–02, August 1780.

9 John Flinders, b. January 1767 (Spalding, Lincolnshire), d. 13 August 1793 (of yellow fever aboard the ship *Cygnet* travelling to Britain from the West Indies).

10 Philemon Pownoll, b. circa 1734, d. 15 June 1780.

11 *Gratefull to Providence*, Vol. 1, p. 106, 8 January 1781.

12 Sir Thomas Charles Bunbury, 6th Baronet, b. May 1740, d. 31 March 1821.

13 *Journals of the House of Commons (JHC)*, Vol. 37, 1803, pp. 311.

14 *Ibid.*

15 Henrietta Flinders, later Newbald, b. 1765, d. 1842.

16 *Gratefull to Providence*, Vol. 1, p. 108, 1 March 1781.

17 John Flinders, b. 5 April 1781 (Donington), d. March 1834 (Yorkshire).

18 Samuel Ward Flinders, b. 3 November 1782 (Donington), d. 30 December 1834 (Hadleigh, Suffolk).

19 *Gratefull to Providence*, Vol. 1, p. 132, 31 December 1782.

20 *Ibid.*, p. 136, 31 March 1783.

21 *Ibid.*, p. 138, 3 June 1783.

22 Elizabeth Ellis (nee Weekes), b. 1753 (Donington), d. 20 July 1841 (Donington).

23   *Gratefull to Providence,* Vol. 1, p. 143,
     9 August 1783.

24   *Ibid.,* p. 146, 11 October 1783.

25   *Ibid.,* p. 150, 1 January 1784.

26   Thomas Adams Franklin, b. 1773,
     d. 1807.

27   'Captain Matthew Flinders R.N.',
     Launceston Examiner, 30 April 1892, p. 2.

28   *Gratefull to Providence,* Vol. 1, pp. 164–65,
     3 January 1785.

29   Mary 'Polly' Franklin, b. 1775, d. 1799.

30   *Gratefull to Providence,* Vol. 2, p. 12,
     22 July 1785.

31   William Lygon, 1st Earl Beauchamp,
     b. 25 July 1747, d. 21 October 1816.

32   Sir Evan Nepean, 1st Baronet, b. 9 July
     1752, d. 2 October 1822.

33   Thomas Townshend, 1st Viscount
     Sydney, b. 24 February 1733, d. 30 June
     1800.

34   Beauchamp Committee Report, 9 May
     1785, *JHC,* Vol. 40, p. 955.

35   National Archives Kew, Home Office,
     7/1, ff. 71–76.

36   *Ibid.*

37   *Gratefull to Providence,* Vol. 2, p. 30,
     14 August 1786.

38   *Ibid.,* p. 32, 3 October 1786.

39   *Ibid.,* pp. 17–18.

40   Later Admiral Arthur Phillip, b. 11
     October 1738 (Cheapside, London),
     d. 31 August 1814 (Bath).

41   Mary Carr, nee Ward, sister of
     Matthew's late mother, Susanna.

42   Two 500-ton storeships were reclassified
     as the frigates *Boussole* (under
     Lapérouse) and *Astrolabe* (under skipper
     Fleuriot de Langle).

43   Jean-François de Galaup, Comte de
     Lapérouse, b. 1741, d. 1788.

44   Louis XVI of France, b. 23 August 1754,
     d. 21 January 1793.

45   Phillip to Sydney, 15 May 1788, *HRA,*
     Series I, Vol. 1, 1788–1796, p. 18.

46   *Ibid.*

47   Arthur Phillip, *The Voyage of Governor
     Phillip to Botany Bay,* John Stockdale,
     1789, p. 58.

48   King sailed for his new post on
     14 February 1788 and arrived on
     6 March.

49   Later Vice Admiral John Hunter, b. 29
     August 1737 (Leith, Scotland), d. 13
     March 1821 (London).

50   William Bradley, *A Voyage to New
     South Wales, December 1786–May 1792;*
     compiled 1802, State Library of New
     South Wales (SLNSW), Safe 1/14, p. 76,
     5 February 1788.

51   John Hunter, *Journal Kept on Board the
     Sirius During a Voyage to New South Wales,
     May 1787–March 1791,* SLNSW, SAFE/
     DLMS 164, p. 42, 5 February 1788.

52   *Ibid,* p. 40.

53   Both ships were wrecked on the reefs
     of Vanikoro, in Solomon Islands. Some
     of the survivors were believed to have
     been massacred by the local inhabitants,
     while others survived for nine months
     before perishing on a two-masted craft
     from the wreckage of the *Astrolabe.*

54   Ernest Scott, *The Life of Captain Matthew
     Flinders, R.N.,* Angus & Robertson,
     1914, p. 29.

55   John Masefield, *Sea Life in Nelson's Time,*
     Methuen & Co, 1905, p. 69.

56   Muster Book, *HMS Alert,* ADM
     36/10793, The National Archives, Kew.

57   Later Admiral Sir Thomas Pasley, b. 2
     March 1734, d. 29 November 1808.

58   Scott, *The Life of Captain Matthew
     Flinders,* p. 18. Matthew's father makes
     no mention of the visit.

59   *Gratefull to Providence,* Vol. 2, p. 79,
     26 April 1790.

60   Geoffrey C. Ingleton, *Matthew Flinders:
     Navigator and Chartmaker,* Genesis
     Publications, 1986, p. 425.

61   *Gratefull to Providence,* Vol. 2, p. 79,
     26 April 1790.

62   'Biographical Memoir of Captain
     Matthew Flinders, RN', *The Naval
     Chronicle for 1814, General and
     Biographical History of the Royal Navy
     of the United Kingdom,* Vol. xxxii, Joyce
     Gould, London, p. 178.

63   *Ibid.*

64   *Gratefull to Providence,* Vol. 2, p. 81,
     31 May 1790.

Chapter Three

1   *Gratefull to Providence*, Vol. 2, p. 94, 20 May 1791.

2   Arthur Phillip to Joseph Banks, 17 November 1791, SLNSW, SAFE/Banks Papers/Series 37.18.

3   Isabella Tyler, 'Biographical Outline of Capt. & Mrs Flinders 1852', Flinders Papers 60/017, FLI 107, NMM. [Incorrectly attributed to Elizabeth Tyler].

4   *Gratefull to Providence*, Vol. 2, p. 86, 10 August 1790.

5   Admiral of the Fleet Richard Howe, 1st Earl Howe, b. 8 March 1726, d. 5 August 1799.

6   Anne Salmond, *Bligh: William Bligh in the South Seas*, Penguin Random House, 2011.

7   Bligh to Banks, 4 August 1787, SLNSW, SAFE/Banks Papers/Series 46.02.

8   Fletcher Christian, b. 25 September 1764, d. 20 September 1793.

9   William Bligh, *A Voyage to the South Sea*, Lords Commissioners of the Admiralty, 1792, p. 162.

10   'Memoirs of the Circumnavigator', *The Town and Country Magazine or Universal Repository of Knowledge, Instruction and Entertainment*, Vol. 5, 1773, p. 458.

11   William Bligh (ed. Owen Rutter), *The Log of the Bounty*, Vol. 2, Cockerel Press, 1937, pp. 16–17.

12   William Eden, 1st Baron Auckland, FRS, b. 3 April 1745, d. 28 May 1814.

13   'Capt. Bligh Hints for an Outfit, Being Suggestions Made by William Bligh for the Second Breadfruit Voyage', circa March 1791, SLNSW, Banks Papers, Series 49.05.

14   Phillip Stephens to the Lords of the Admiralty, 'Additional Guns for *Providence*', 13 June 1791, PRO, Admiralty records, 1673-1957 [microform], Records of the Navy Board and the Board of Admiralty File 2216. AJCP Reel No: 6119, NLA.

15   Nathaniel Portlock, b. circa 1748, d. 12 September 1817.

16   Later Rear-Admiral Francis Godolphin Bond, b. 1765, d. 1839.

17   William Bligh to Banks, 23 February 1791, SLNSW, Banks Papers, Series 50.02.

18   Admiral Louis-Antoine, Comte de Bougainville, b. 12 November 1729, d. August 1811.

19   *Gratefull to Providence*, Vol. 2, p. 94, 20 May 1791.

20   *Ibid.*

21   Thomas Pasley to Matthew Flinders, 3 June 1791, NMM, FLI01.

22   'A Memoir of Captain Flinders, Apparently in His Wife's Handwriting'.

23   Bligh to Banks, 7 July 1791, SLNSW, Banks Papers, Series 50.04.

24   James Wiles, b. 16 January 1768 (Holywell, near Stamford, Lincolnshire), d. 9 October 1851 (Monmouth Mount, Jamaica).

25   Joseph Banks to James Wiles, 16 July 1791, SLNSW, Banks Papers, Series 52.05.

26   Ida Lee, *Captain Bligh's Second Voyage*, Longmans, Green, and Co., 1920.

27   *Ibid.*

28   George Tobin, *Journal on HMS Providence 1791–1793*, 2 August 1791, SLNSW, Safe 1/406 (A 562).

29   *Ibid.*, 17 August 1791.

30   *Ibid.*

31   *Ibid.*, September 1791.

32   Bligh to Banks, 30 August 1791, SLNSW, Banks Papers, Series 50.13.

33   Tobin, *Journal on HMS Providence 1791–1793*, September 1791.

34   Portion of journal, relating to voyage on *Providence* with Bligh, circa 1791 (File FLI 8a), Series FLI 8-14. *Journals and Narratives of Captain Flinders*, circa 1791–1809, NLA.

35   *Ibid.*

36   *Ibid.*

37   *Ibid.*

38   *Gratefull to Providence*, Vol. 2, p. 101. Letter from Matthew dated 28 August 1791.

39   Bligh to Banks, 30 August 1791, SLNSW, Banks Papers, Series 50.13.

40   *Ibid.*

41   *Ibid.*, 13 September 1791, Series 50.14.

42   Later Rear-Admiral George Tobin, b. 13 December 1768, d. 10 April 1838.

43 Tobin, *Journal on HMS Providence 1791–1793*, September 1791.
44 Tobin, *Journal on HMS Providence 1791–1793*, September 1791.
45 Flinders, journal on *Providence* with Bligh.
46 He gave it the scientific name *Malacopterygii*.
47 Tobin, *Journal on HMS Providence 1791–1793*, 22 September 1791.
48 Flinders, journal on *Providence* with Bligh.
49 Tobin, *Journal on HMS Providence 1791–1793*, 15 October 1791.
50 Flinders, journal, 15 October 1791.
51 Tobin, *Journal on HMS Providence 1791–1793*, 30 October 1791.
52 Bligh to Banks, 7 November 1791, SLNSW, Banks Papers, Series 50.13.
53 Tobin, *Journal on HMS Providence 1791–1793*, 6 November 1791.
54 *Gratefull to Providence*, Vol. 2, p. 102. Letter dated 8 November 1791, received in Donington on 29 January 1792.

Chapter Four
1 Matthew Flinders, *A Voyage to Terra Australis; Undertaken for the Purpose of Completing the Discovery of that Vast Country*, Vol. 1, G&W Nicol, 1814, p. lxxv.
2 'A Log of the Proceedings of His Majesty's Ship Providence on a Second Voyage to the South Sea Under the Command of Captain William Bligh, to Carry the Breadfruit Plant from the Society Islands to the West Indies, Written by Himself', Vol. 1, 17 July 1791–19 July 1792, SLNSW, SAFE/A 564/1, 7 November 1791.
3 William Bligh to Joseph Banks, 17 December 1791, SLNSW, Banks Papers, Series 50.17.
4 Tobin, *Journal on HMS Providence 1791–1793*, 22 December 1791.
5 Flinders, journal on *Providence* with Bligh.
6 *Ibid*.
7 *Ibid*.
8 Tobin, *Journal on HMS Providence 1791–1793*, November 1791.

9 Bligh to Banks, 24, 26 November 1791, SLNSW, Banks Papers, Series 50.16.
10 Later Lieutenant-General Francis Grose, b. 1758 (Greenford, Middlesex), d. 8 May 1814 (Croydon, Surrey).
11 Bligh to Banks, 17 December 1791, SLNSW, Banks Papers, Series 50.17.
12 Ida Lee, *Bligh's Second Voyage*.
13 Bligh to Banks, 17 December 1791, SLNSW, Banks Papers, Series 50:17.
14 Bligh log, 22 December 1791.
15 Bligh to Banks, 18 December 1791, SLNSW, Banks Papers, Series 50:18.
16 Tobin, *Journal on HMS Providence*, 24 December 1791.
17 Bligh log, 13 January 1792.
18 *Ibid.*, 10 February 1792.
19 Tobin, *Journal on HMS Providence*, 23 February 1792.
20 Bligh log, 9 February 1792.
21 *Ibid*.
22 Tobin, *Journal on HMS Providence*, 23 February 1792.
23 *Ibid*.
24 *Ibid*.
25 Flinders, *A Voyage to Terra Australis*, Vol. 1, p. xc.
26 Tobin, *Journal on HMS Providence*, 23 February 1792.
27 *Ibid*.
28 Bligh log, 20 February 1792.
29 James Wiles and Christopher Smith to Joseph Banks, 17 December 1792, SLNSW, Banks Papers, Series 52.09.
30 When Jacques Labillardière, the French naturalist, saw the inscription a year later on the voyage of Bruni d'Entrecasteaux, he scoffed at Bligh's arrogance and 'the despotism which condemned men of science to initials and gave a sea captain a monopoly of fame'.
31 Bligh log, 20 February 1792.
32 *Ibid*.
33 Tobin, *Journal on HMS Providence*, 24 March 1792.
34 Flinders journal, 26 March 1792.
35 Tobin, *Journal on HMS Providence*, 1 March 1792.
36 *Ibid.*, 4 March 1792.
37 Bligh log, 14 March 1792.
38 *Ibid.*, 22 March 1792.

39 *Ibid.*, 31 March 1792.

40 Bligh log, 25 March 1792.

41 Flinders journal, 25 March 1792.

42 *Ibid.*, 2 April 1792.

43 *Ibid.*, 27 March 1792.

44 *Ibid.*, 5 April 1792.

45 Tobin, *Journal on HMS Providence*,
    2 April 1792.

46 Paul-Gabriel Boucé, *Sexuality in
    Eighteenth-century Britain*, Manchester
    University Press, 1982, p. 9.

47 Bligh log, 8 April 1792.

48 Tobin, *Journal on HMS Providence*,
    7 April 1792.

49 Flinders journal, 7 April 1792.

50 *Ibid.*, 9 April 1792.

51 *Ibid.*

52 Tobin, *Journal on HMS Providence*,
    7 April 1792.

53 The *Pandora* was wrecked on the Great
    Barrier Reef, with the loss of thirty-five
    men, four of them mutineers.

54 *Star* (London), 2 March, 1793.

55 Tobin, *Journal on HMS Providence*,
    10 April 1792.

56 Bligh log, 11 May 1792.

57 *Ibid.*, 8 July 1782.

58 '*Providence* pay book', Records of the
    Admiralty, Naval Forces, Royal Marines,
    Coastguard, and related bodies, ADM
    35/1361, The National Archives, Kew.

59 Tobin, *Journal on HMS Providence*, 1792,
    Notes on 'Otahytey'.

60 *Ibid.*

61 Flinders journal, 22 June 1792.

62 Bligh log, 3 May 1792.

63 *Ibid.*, 19 June 1792.

Chapter Five

1 Flinders journal, 6 September 1792.

2 James Wiles and Christopher Smith
   to Joseph Banks, 17 December 1792,
   SLNSW, Banks Papers, Series 52.09.

3 Tobin, *Journal on HMS Providence*,
   18 July 1792.

4 *Ibid.*

5 Bligh log, 14 July 1792.

6 Bligh log, 9 July 1792. Bligh called him
   'Omai', a common misspelling of the
   Tahitian's name.

7 James Wiles, 'Postcript to the Royal
   Gazette', 26 October–2 November
   1793, p. 3, SLNSW, Banks Papers,
   Series 89.02.

8 Bligh log, 19 July 1792.

9 Flinders journal, 20 July 1792.

10 *Ibid.*, 9 August 1792.

11 Scott, *The Life of Captain Matthew
    Flinders R.N.*, p. 33.

12 Flinders journal, 2 September 1792.

13 *Ibid.*

14 *Ibid.*, 6 September 1792.

15 Flinders, *A Voyage to Terra Australis*, Vol. 1,
    p. xxi.

16 *Ibid.*, p. xxiii.

17 Flinders journal, 6 September 1792.

18 *Ibid.*

19 *Ibid.*, p. xxii.

20 Bligh log, 7 September 1792.

21 Flinders, *Terra Australis*, Vol. 1, p. xxiv.

22 Flinders journal, 10 September 1792.

23 Bligh journal, 11 September 1792.

24 Flinders, *Terra Australis*, Vol. 1, p. xxv.

25 *Ibid.*

26 Flinders journal, 11 September 1792.

27 The island where they landed is now
    called North Possession Island.

28 Flinders, *Terra Australis*, Vol. 1,
    p. xxix.

29 Miriam Estensen, *The Life of Matthew
    Flinders*, Allen & Unwin, 2003, p. 477.

30 Francis Godolphin Bond to Thomas
    Bond, in George Mackaness, 'Fresh
    Light on Bligh: Some Unpublished
    Correspondence', *Australian Historical
    Monographs*, Vol. 5, Review Publications,
    1949, p. 69.

31 Bligh to Banks, 2 October 1792,
    SLNSW, Banks Papers, Series 50.19.

32 Flinders journal, 24 September 1792.

33 William Henry Smyth, 'A Brief Memoir
    of Captain Matthew Flinders, RN', Box
    81/3(d) La Trobe Collections LTC, State
    Library of Victoria.

34 Flinders journal, 5 February 1793.

35 Bligh log, 10 October 1792.

36 *Ibid.*, 6 November 1792.

37 Flinders journal, 20 November 1792.

38 Bligh log, 17 December 1792.

39 Flinders journal, 18 December 1792.

the Queensland Museum, the Redcliffe Museum and Britain's National Maritime Museum at Greenwich.

I also owe a great deal of thanks to Flinders descendant Rachel Flinders Lewis for her kind permission to reprint family artwork and to Jane Pearson from the Matthew Flinders Bring Him Home group for facilitating this contact.

Matthew Flinders was one of the great figures in the history of Australia, and almost 250 years after his birth in rural England it was an honour and privilege to tell the story of his astonishing life and achievements.

40 Tobin, *Journal*, December 1792.

41 Bligh log, 3 March 1793.

42 James Wiles, 'Death of Pappo', *Postscript to the Royal Gazette* (Kingston, Jamaica), 26 October–2 November 1793, p. 3.

43 James Wiles to Banks, 16 March 1793, SLNSW, Banks Papers, Series 52.14.

44 Peter Goodwin, *The Ships of Trafalgar: The British, French and Spanish Fleets October 1805*, Conway Maritime Press, 2005, p. 66.

45 Thomas Pasley to Matthew Flinders, 7 August 1793, Papers of Matthew Flinders (as filmed by the AJCP), 1698–1852: [M444, M3033-M3037]/Series NMM FLI, NLA.

46 Tobin, *Journal on HMS Providence*, August–September 1793.

47 *Gratefull to Providence*, Vol. 2, p. 127, 12 September 1793.

48 *The Naval Chronicle* for 1814, Vol. XXXII (from July to December), p. 180.

Chapter Six

1 Henry Waterhouse, 'Account of the Battle of the 1st of June 1794', Henry Waterhouse papers, MLMSS 6544/3 (Safe 1/187), SLNSW.

2 Banks at Revesby Abbey to Joseph Huddart, 11 August 1793, Natural History Museum Library and Archives (London) DTC 8.240-241.

3 'Sir David Attenborough on Joseph Banks' (video), British Library, bl.uk.

4 James Wiles to Banks, 16 March 1793, SLNSW, Banks Papers, Series 52.14.

5 *Ibid.*

6 *Gratefull to Providence*, Vol. 2, p. 127, 12 September 1793.

7 *Ibid.*, p. 130.

8 Catharine Retter & Shirley Sinclair, *Letters to Ann: The Love Story of Matthew Flinders and Ann Chappelle*, HarperCollins (Kindle edition), 1999.

9 Ann Chappelle, b. 21 November 1770 (Hull, Yorkshire), d. January 1852 (Lewisham, London). Baptised 24 December 1770 Holy Trinity, Hull. The spelling of Ann's surname was sometimes rendered as Chappell, to Anglicise it in a time of wars with France.

10 Flinders to Misses Chappelle and M. Franklin, 10 March 1795, Papers of Matthew Flinders (as filmed by the AJCP), 1698-1852: [M444, M3033-M3037]/Series NMM FLI/File FLI 4, NLA.

11 Retter & Sinclair, *Letters to Ann*.

12 At Sulawesi Tengah, in modern-day Indonesia.

13 Retter & Sinclair, *Letters to Ann*.

14 Isabella 'Belle' Tyler, baptised 26 January 1785 (Partney, Lincolnshire), d. October 1867 (Blackheath, Kent).

15 Retter & Sinclair, *Letters to Ann*.

16 Henry Waterhouse, b. 13 December 1770 (Westminster, London), d. 27 July 1812 (Westminster, London).

17 Maria Waterhouse, b. 1791, d. 1875.

18 Journal kept in the *Bellerophon*, September 1793–July 1794, 11 September 1793, Papers of Matthew Flinders (as filmed by the AJCP), 1698-1852: [M444, M3033-M3037]/Series FLI 8 - 14/File FLI 8b, NLA.

19 *Ibid.*, 18 November 1793.

20 *Ibid.*, 19 November 1793.

21 *Ibid.*, 27 November 1793.

22 *The London Gazette*, 16 September 1794, No. 13704, p. 946. *La Blonde* was renamed the *Princess* and became a British whaling boat until the French recaptured it off Mozambique two years later.

23 Later Vice Admiral Sir William Johnstone Hope, b. 16 August 1766, d. 2 May 1831.

24 Flinders to Banks, 20 March 1794, SLNSW, Banks Papers, Series 53.20.

25 *Ibid.*

26 Some estimates say as many as 350 merchant ships.

27 Later Admiral Louis-Thomas Villaret de Joyeuse, b. 29 May 1747, d. 24 July 1812.

28 Flinders, *Bellerophon* journal, 23 May 1794.

29 *Ibid.*, 28 May 1794.

30 Built in Brest in 1766 as the *Bretagne* but renamed the *Révolutionnaire* during the French Revolution.

31 Flinders, *Bellerophon* journal, 29 May 1794.

32 *Ibid.*

33 *Ibid.*, 30 May 1794.

34 *Ibid.*

35 Scott, *The Life of Captain Matthew Flinders, R.N.*, p. 45.

36 Henry Waterhouse to William Waterhouse, June 1794, William Waterhouse papers, 1782–1803, SLNSW, MLMSS 6544.

37 The Naval Chronicle for 1814, Vol. XXXII (from July to December), p. 180–81.

38 Flinders, *Bellerophon* journal, 1 June 1794.

39 William James, *The Naval History of Great Britain*, Vol. 1, Richard Bentley, 1837, pp. 154–55.

40 Henry Waterhouse, 'Account of the Battle of the 1st of June 1794', Henry Waterhouse papers, SLNSW, MLMSS 6544/3 (Safe 1/187).

41 Henry Waterhouse to William Waterhouse, 2 June 1794, William Waterhouse papers, 1782–1803, SLNSW, MLMSS 6544.

42 Flinders *Bellerophon* journal, 1 June 1794.

43 *Ibid.*

44 Olivier Goossens, 'The Glorious First of June 1794', historic-uk.com, 18 January 2022.

45 Flinders, *Bellerophon* journal, 5 June 1794.

46 Estensen, *The Life of Matthew Flinders*, p. 38.

47 *Gratefull to Providence*, Vol. 2, p. 135–36, 30 June 1794.

Chapter Seven

1 Flinders, *Terra Australis*, Vol. 1, p. xcvii.

2 *Gratefull to Providence*, Vol. 2, p. 141–42, 25 September 1794.

3 Flinders, *Terra Australis*, Vol. 1, p. xcvi.

4 Hunter to Home Secretary Henry Dundas, 13 February 1794, *HRNSW*, Vol. 2, p. 117.

5 Woollarawarre Bennelong, b. circa 1764, d. 3 January 1813.

6 Yemmerrawanne, b. circa 1775, d. 18 May 1794.

7 Phillip to Banks, 3 December 1791, SLNSW, Banks Papers, Series 37.20.

8 Watkin Tench, *A Complete Account of the Settlement at Port Jackson, in New South Wales*, G. Nicol and J. Sewell, 1793, p. 178.

9 *Ibid.*, p. 86.

10 William Waterhouse, b. circa 1752, d. 1822.

11 Keith Vincent Smith, 'The Mystery of Yemmerrawanne's Resting Place', SLNSW, www2.sl.nsw.gov.au, 2014.

12 Ibid., 'Woollarawarre Bennelong', 2013, dictionaryofsydney.org.

13 *London Observer*, 29 September 1793.

14 Kate Fullagar, 'Bennelong in Britain', *Aboriginal History*, Vol. 33, ANU Press, 2009, p. 40.

15 Captain Paterson to Henry Dundas, *HRA*, Series 1, Vol. 1 (1788–1796), 15 June 1795, pp. 501–02.

16 *Gratefull to Providence*, Vol. 2, p. 141–42, 25 September 1794.

17 *Ibid.*

18 *Ibid.*, p. 133, 'Conclusion of the Year 1793'.

19 *Ibid.*, p. 112, 17 July 1792.

20 *Ibid.*, p. 141–42, 25 September 1794.

21 The Admiralty sold the ship in July 1792 and the new owners renamed it *Thomas and Nancy*. The ship then carried coal around London until 1806.

22 William Kent, b. 1760, d. 1812.

23 *Gratefull to Providence*, Vol. 2, p. 141–42, 25 September 1794.

24 Flinders to Mary Franklin, 23 January 1795, Papers of Matthew Flinders (as filmed by the AJCP), 1698–1852: [M444, M3033-M3037]/Series NMM FLI/File FLI 4, NLA.

25 *Ibid.*

26 *Ibid.*

27 *Ibid.*

28 Hunter to Under Secretary John King, 25 January 1795, *HRNSW*, Vol. 2, p. 281.

29 *Ibid.*

30 George Bass, b. 30 January 1771 (Aswarby, Lincolnshire); disappeared 5 February 1803 after leaving Port Jackson, New South Wales.

31 'The Discovery of Bass Strait', *HRNSW*, Vol. 3, p. 312.

32 Flinders to Bass, 15 February 1800, SLNSW, MLMSS 7046.

33 Scott, *The Life of Captain Matthew Flinders R.N.*, p. 84.

34 With an eight-foot keel, the boat may have been ten feet or so in length.

35 W.G. McDonald, *The First-Footers: Bass and Flinders in Illawarra*, Illawarra Historical Society, 1975, p. 5.

36 Hunter to Portland, 1 March 1798, *HRNSW*, Vol. 3, p. 363.

37 Hunter to Nepean, 3 September 1798, *HRNSW*, Vol. 3, p. 474.

38 Flinders, *Terra Australis*, Vol. 1, p. xcvii.

39 John Shortland, b. 1769, d. 1810.

40 Log of *Reliance* (Commander: H. Waterhouse), mfm PRO 412-7269-Public Record Office, Admiralty records, 1673-1957 [microform]/Fonds. Records of HM Ships/Series ADM 51/File 1121. AJCP Reel No: 5736, NLA.

41 *The Naval Chronicle* for 1814, Vol. XXXII (from July to December), p. 180–81.

42 Flinders to Misses Chappelle and M. Franklin, 10 March 1795, Papers of Matthew Flinders (as filmed by the AJCP), 1698-1852: [M444, M3033-M3037]/Series NMM FLI/File FLI 4, NLA.

43 *Ibid.*

44 Retter & Sinclair, *Letters to Ann.*

45 Flinders to Misses Chappelle and M. Franklin, 10 March 1795, NLA.

46 Hunter to Under Secretary John King, 20 May 1795, *HRNSW*, Vol. 2, p. 297.

47 Hunter to Portland, *HRNSW*, Vol. 2, p. 297.

48 *Ibid.*

49 Waterhouse to Arthur Phillip, 24 October 1795, Banks Papers, SLNSW, Series 37.28.

50 Hunter to the Navy Board, 19 March 1794, ADM 106/1353, PRO.

51 Christopher Smith to Flinders, 30 June 1799, Flinders Papers 60/017, FLI/1, NMM.

52 Log of *Reliance*, NLA.

53 Journal of Richard Atkins, NLA, MS 4039, 7 September 1795.

54 Hunter to Portland, 11 September 1795, *HRA*, Vol. 1, p. 528.

55 Henry Waterhouse to William Waterhouse, 24 October 1795, SLNSW, MLMSS 6544.

56 Waterhouse to Arthur Phillip, 24 October 1795, SLNSW, Banks Papers, Series 37.28.

57 *Ibid.*

58 Hunter to Banks, 12 October 1795, SLNSW, Banks Papers, Series: 38:01.

59 Hunter to Portland, 12 November, 1796, *HRA*, Series 1, Vol. 1, p. 676.

60 Ibid., 10 August 1796, *ibid.*, p. 574.

61 *Ibid.*

62 'Condition of the Settlements', 20 August 1796, *HRNSW*, Vol. 3, p. 81.

63 David Collins, *An Account of the English Colony of New South Wales*, Vol. 2, T. Cadell Jr and W. Davies, 1802, p. 63.

64 The wedding took place on 5 April 1796 (*Gratefull to Providence*, Vol. 2, p. 162–63, 7 June 1796).

65 Ibid., p. 169, 14 October 1796.

Chapter Eight

1 Matthew Flinders, *Narrative of Voyages in the Tom Thumb*, Papers of Matthew Flinders (as filmed by the AJCP), 1698-1852: [M444, M3033-M3037]/Series FLI 8 - 14/File FLI 9a, NLA, p. 2.

2 Flinders, *Terra Australis*, Vol. 1, p. xcv.

3 *Ibid.*, p. xcvi.

4 *Memoirs of George Suttor, F.L.S., 1774–1859*, SLNSW, A 3072.

5 Daniel Paine, *Diary as Kept in a Voyage to Port Jackson N.S.W*, p. 35, NLA, Collections held by the NMM (as filmed by the AJCP)/Fonds JOD/Series JOD/172/.

6 Flinders, *Terra Australis*, Vol. 1, p. xcvii.

7 Flinders, *Tom Thumb*, p. 2.

8 Held in February 1795.

9 Collins, *An Account of the English Colony of New South Wales*, p. 365.

10 Flinders, *Terra Australis*, Vol. 1, p. xcv.

11 *Ibid.*

12 'Condition of Norfolk Island', *HRNSW*, Vol. 3, 18 October 1796, p. 146–150.

13 Henry Hacking, b. circa 1750
(Blackburn, Lancashire), d. 21 July 1831
(Hobart, Tasmania).

14 David Collins, *An Account of the English
Colony*, p. 70.

15 *Ibid.*

16 Ibid., p. 311.

17 Collins, *An Account of the English Colony*,
p. 311.

18 Later Lieutenant-Colonel George
Johnston, b. 19 March 1764, d. 5 January
1823.

19 Daniel Paine, diary, p. 39.

20 Flinders, *Tom Thumb*, p. 3.

21 'Government and General Order',
HRNSW, Vol. 3, 9 October 1797, p. 304.

22 Adam Courtenay, *Three Sheets to the
Wind*, ABC Books, 2022, p. 189.

23 Flinders, *Tom Thumb*, p. 4.

24 *Ibid*, p. 5.

25 *Ibid.*

26 *Ibid.*

27 *Ibid.*, p. 7.

28 Flinders, *Terra Australis*, Vol. 1, p. xcvii.

29 Flinders, *Tom Thumb*, p. 9.

30 *Ibid.*, p. 11.

31 Flinders, *Terra Australis*, Vol. 1, p. c.

32 Flinders, *Tom Thumb*, pp. 12–13.

33 *Ibid.*, pp. 13–14

34 *Ibid.*, p. 15.

35 *Ibid.*

36 *Ibid.*, p. 16.

37 *Ibid.*, p. 17.

38 *Ibid.*, p. 18

39 Flinders, *Terra Australis*, Vol. 1, p. ci.

40 Flinders, *Tom Thumb*, p. 19.

41 *Ibid.*, p. 20.

42 Flinders' original log of the voyage
lists the date as 1 April, but his book
A Voyage to Terra Australis, written two
decades later, skips a day to 2 April.

Chapter Nine

1 *Matthew Flinders' Biographical Tribute to
His Cat Trim*, 1809, p. 1, flinders.rmg.
co.uk, FLI/11.

2 'Public Works', *HRNSW*, Vol. 3, p. 337.

3 Hunter to Under Secretary Nepean, 31
August 1796, *HRNSW*, Vol. 3, p. 91.

4 Flinders, *Terra Australis*, Vol. 1, p. cv.

5 *Ibid.*

6 Collins, *An Account of the English Colony*,
Vol. 1, p. 493.

7 Portland to Hunter, August 1796,
HRNSW, Vol. 3, p. 96.

8 Flinders' journal entry and certificate
of seamanship, 19 January 1797, flinders.
rmg.co.uk, FLI05.

9 *Ibid.*

10 It was confirmed by the Admiralty the
following year, on 21 January 1798.

11 Flinders' journal entry and certificate of
seamanship, flinders.rmg.co.uk, FLI05.

12 *Gratefull to Providence*, Vol. 2, p. 160, 20
August 1796.

13 *Ibid.*, p. 175, 1 February 1797.

14 *Ibid.*, p. 174, 12 January 1797.

15 Banks to Henry Waterhouse, 8 July
1806, NLA, MS9, Item 24.

16 Henry Waterhouse to William
Waterhouse, 28 August 1797, SLNSW,
MLMSS 6544.

17 Hunter to Portland, 25 June 1797,
HRNSW, p. 236.

18 *Ibid.*, p. 237.

19 *Ibid.*, 6 July 1797, p. 238.

20 Henry Waterhouse to William
Waterhouse, 28 August 1797, SLNSW,
MLMSS 6544.

21 Hunter to Portland, 6 July 1797,
HRNSW, Vol. 3, p. 236.

22 Lieutenant Kent to Secretary Nepean,
16 May 1797, *HRNSW*, Vol. 3, p. 286.

23 *Ibid.*

24 Captain Waterhouse to his father, 20
August 1797, *HRNSW*, Vol. 3, p. 287.

25 Henry Waterhouse to William
Waterhouse, 28 August 1797, SLNSW,
MLMSS 6544.

26 *Flinders' Biographical Tribute to Trim*.

27 *Ibid.*

28 Hunter to Portland, 25 June 1797,
HRNSW, Vol. 3, p. 236.

29 Hunter to Nepean, 10 July 1797,
HRNSW, Vol. 3, p. 278.

30 The *Supply* was eventually broken up in
1806.

31 Hunter to Portland, 25 June 1797,
HRNSW, Vol. 3, p. 237.

32 Henry Waterhouse to William
Waterhouse, 28 August 1797, SLNSW,
MLMSS 6544.

33 *Flinders' Biographical Tribute to Trim.*
34 Appendix A, 'Narrative of the Shipwreck of Captain Hamilton and the Crew of the *Sydney Cove*', *HRNSW*, Vol. 3, p. 757.
35 *Ibid.*
36 *Ibid.*, p. 758.
37 *Ibid.*
38 Courtenay, *Three Sheets to the Wind*, p. 119.
39 *Ibid.*, p. 127.
40 Appendix A, 'Narrative of the Shipwreck of Captain Hamilton', *HRNSW*, Vol. 3, p. 761.
41 Ibid., p. 762.
42 *Ibid.*, p. 764.
43 *Ibid.*, p. 765.
44 *Ibid.*, p. 766.
45 *Ibid.*
46 *Ibid.*, p. 768.
47 Hunter to Portland, *HRNSW*, Vol. 3, p. 237.
48 Flinders, *Terra Australis*, Vol. 1, p. 104.
49 Hunter to Portland, 6 July 1797, *HRNSW*, Vol. 3, p. 278.
50 Appendix A, 'Narrative of the Shipwreck of Captain Hamilton', *HRNSW*, Vol. 3, p. 760.
51 *Ibid.*

Chapter Ten
1 Governor Hunter to the Duke of Portland, 1 March 1798, *HRNSW*, Vol. 3, p. 364.
2 *Gratefull to Providence*, Vol. 2, p. 195, 1 September 1798.
3 Matthew Flinders, 'Land Grants and Leases (Registers)', 1792-1865 NRS 13836 [7/445], Register 1, Page 95; Reel 2560, search.records.nsw.gov.au.
4 *Flinders' Biographical Tribute to Trim*, pp. 1–2.
5 *Gratefull to Providence*, Vol. 2, p. 195, 1 September 1798.
6 'Survey of Reliance', *HRNSW*, p. 308.
7 Flinders, *Terra Australis*, Vol. 1, p. cv.
8 J. Shortland Jr to J. Shortland Sr, *HRNSW*, Vol. 3, pp. 481–82.
9 Governor Hunter to Joseph Banks, 1 August 1797, SLNSW, Banks Papers, Series 38.06.
10 Ibid.

11 Governor Hunter to the Duke of Portland, 1 March 1798, *HRNSW*, Vol. 3, p. 363.
12 'The Discovery of Bass Strait', *HRNSW*, Vol. 3, p. 314.
13 *Ibid.*, p. 326.
14 Flinders, *Terra Australis*, Vol. 1, p. cxiii.
15 Governor Hunter to the Duke of Portland, 10 January 1798, *HRNSW*, Vol. 3, p. 345.
16 Ibid., 1 March 1798, *HRNSW*, Vol. 3, p. 364.
17 Flinders, *Terra Australis*, Vol. 1, p. cxix.
18 'Mr. Bass's Journal', *HRNSW*, Vol. 3, p. 323.
19 Flinders, *Terra Australis*, Vol. 1, p. cxv.
20 Governor Hunter to the Duke of Portland, 1 March 1798, *HRNSW*, Vol. 3, p. 364.
21 Flinders, *Terra Australis*, Vol. 1, p. cxx.
22 Hunter to Portland, 1 March 1798, *HRNSW*, Vol. 3, p. 365.
23 Flinders, *Terra Australis*, Vol. 1, p. cxxi.
24 *Ibid.*, p. cxii.
25 *Ibid.*
26 *Ibid.*, p. cxxviii.
27 *Ibid.*, p. cxxxiv.
28 *Ibid.*, p. clxiv. The name was Anglicised later to Chappell.
29 Scott, *The Life of Captain Matthew Flinders, R.N.*, p. 166.
30 Francis Galton, *Family Faculties Manuscript*, 1884.
31 Flinders, *Terra Australis*, Vol. 1, p. cxxxv.
32 *Ibid.*, p. cxxx.
33 *Ibid.*, p. cxxxv.
34 *Ibid.*, p. cxxxvi
35 *Ibid.*, p. cxxxvii
36 *Ibid.*, p. cxviii
37 Hunter to Portland, 1 March 1798, *HRNSW*, Vol. 3, p. 363.
38 Hunter to Banks, 12 March, 25 July 1798, SLNSW, Banks Papers, Series 38.09.
39 Published in March 1799.
40 Banks to Under Secretary Nepean, 1 5 May 1798, *HRNSW*, Vol. 3, p. 382.
41 *Ibid.*

Chapter Eleven

1 Flinders, *Terra Australis*, Vol. 1, p. cxl.
2 D'Arcy Wentworth, b. 1762 (Portadown, Ireland); d. 7 July 1827 (Homebush, New South Wales).
3 Thomas Jamison, b. 1753 (Ireland), d. 27 January 1811 (London).
4 Old Bailey Session Papers, 1787 to 1788, pp. 15–20, cited in Kathleen Mary Dermody, 'D'Arcy Wentworth 1762–1827: A Second Chance', PhD thesis, Australian National University, April 1990, pp. 23–24.
5 Hunter to Portland, 1 July 1798, *HRA*, p. 159.
6 Bungaree, b. circa 1775 (Broken Bay, New South Wales), d. 24 November 1830 (Garden Island, Sydney). Flinders wrote his name as 'Bongaree'.
7 Flinders, *Terra Australis*, Vol. 1, p. cxciv.
8 Log of Reliance, 15 May 1798, AJCP Reel No: 5736, NLA.
9 Charles Bishop, 'Memoranda on the ship Nautilus, Amboyna to Port Jackson and thence to Macao, 23 Oct. 1796 to 9 Sept. 1799', SLNSW, C 192, p. 58.
10 Collins, *An Account of the English Colony*, Vol. 1, p. 448.
11 Charles Bishop, b. circa 1765, d. 1810.
12 'Mutiny on the Barwell', *HRNSW*, Vol. 3, 20 August 1798, p. 453.
13 Hunter to Nepean, *HRNSW*, Vol. 3, 3 September 1798, pp. 474–75.
14 *Ibid.*
15 Flinders, *Terra Australis*, Vol. 1, p. cxxviii.
16 *Ibid.*
17 Collins, *An Account of the English Colony*, Vol. 2, p. 129.
18 'Government and General Order', 3 October 1798, *HRNSW*, Vol. 3, p. 495.
19 Collins, *An Account of the English Colony*, Vol. 2, p. 130.
20 *Ibid.*
21 Flinders, *Terra Australis*, Vol. 1, p. clxxxiii.
22 *Ibid.*, p. cxxxix.
23 *Ibid.*, p. cxl.
24 *Ibid.*, p. cxli.
25 *Ibid.*, p. cxlvi.
26 *Ibid.*, p. cxlix.

27 *Ibid.*, p. cli
28 Collins, *An Account of the English Colony*, Vol. 2, p. 167.
29 Flinders, *Terra Australis*, Vol. 1, p. cliii.
30 *Ibid.*, p. cliv.
31 Collins, *An Account of the English Colony*, Vol. 2, p. clxix.
32 *Ibid.*, p. clxvii.
33 *Ibid.*, p. clvi
34 *Ibid.*, p. clxxvii.
35 *Ibid.*, p. clx.
36 'Narrative of an Expedition in the Colonial Sloop *Norfolk*, by Matthew Flinders, 2nd l't, H.M.S. Reliance', *HRNSW*, Vol. 3, p. 786.
37 Charles Bishop, 'Memoranda on the Ship Nautilus', SLNSW, C 192, p. 99.
38 Flinders, *Terra Australis*, Vol. 1, p. clxi.
39 *Ibid.*, p. clxvii.
40 'Narrative of *Norfolk*', *HRNSW*, Vol. 3, p. 795.
41 Flinders, *Terra Australis*, Vol. 1, p. clxix.
42 *Ibid.*
43 *Ibid.*, p. clxx.
44 *Ibid.*
45 *Ibid.*
46 *Ibid.*, p. clxxii.
47 'Narrative of *Norfolk*', *HRNSW*, Vol. 3, p. 801.
48 Their names were actually Heemskerck and Zeehaen.
49 'Narrative of *Norfolk*', *HRNSW*, Vol. 3, p. 804.
50 Antoine Raymond Joseph de Bruni, chevalier d'Entrecasteaux, b. 8 November 1737, d. 21 July 1793.
51 Commodore Sir John Hayes, baptised 11 February 1768, d. 3 July 1831.
52 Flinders, *A Voyage to Terra Australis*, Vol. 1, p. clxxxv.
53 *Ibid.*, p. clxxxvii.
54 'Narrative of *Norfolk*', *HRNSW*, Vol. 3, p. 813.
55 Flinders, *Terra Australis*, Vol. 1, p. cxciii.
56 Hunter to Nepean, 15 August 1799, *HRNSW*, Vol. 3, p. 704.
57 Galton, *Record of Family Faculties*, 1884.
58 *Gratefull to Providence*, Vol. 2, p. 195, 1 September 1798.
59 *Gratefull to Providence*, Vol. 2, p. 196, 11 October 1798.

Chapter Twelve

1 Flinders, *Terra Australis*, Vol. 1, p. cciv.
2 Banks to Under Secretary Nepean, 1
5 May 1798, *HRNSW*, Vol. 3, p. 382.
3 *Ibid.*
4 Hunter to Portland, 21 February 1799,
HRNSW, Vol. 3, p. 551.
5 Isaac Nichols, b. 29 July 1770 (Calne,
Wiltshire), d. 8 November 1819
(Sydney).
6 'The King agst the sd Isaac Nicholls',
HRA, Series 1, Vol. 2, p. 284.
7 Flinders to Governor Hunter, 30 April
1799, *HRA*, Series 1, Vol. 2, p. 333.
8 Matthew Flinders to Ann Chappelle,
16–19 March 1799, Flinders Papers,
60/017, FLI/25, NMM.
9 *Ibid.*
10 Christopher Smith to Flinders, 30 June
1799, Flinders Papers, 60/017, FLI/1,
NMM; Flinders to Christopher Smith,
14 February 1800, Flinders Papers,
60/017, FLI/4, NMM.
11 Flinders to Bass, 15 February 1800,
SLNSW, MLMSS 7046.
12 *Ibid.*
13 *Ibid.*
14 *Ibid.*
15 Michael Burge, 'Matthew Flinders –
Hello, Sailor!', australiancountrylife.
wordpress.com, 28 June 2015.
16 Flinders to Bass, 15 February 1800,
SLNSW, MLMSS 7046.
17 Hunter to Portland, 1 May 1799, *HRA*,
Series 1, Vol. 2, p. 354.
18 Hunter to Portland, 15 November 1799,
HRA, Series 1, Vol. 2, p. 395.
19 *Ibid.*, p. 394.
20 *Ibid.*
21 Flinders, *Terra Australis*, Vol. 1,
p. cxciii.
22 Collins, *An Account of the English Colony*,
Vol. 2, p. 225.
23 *Ibid.*, p. 226.
24 Flinders, *Terra Australis*, Vol. 1, p. cxcv.
25 *Matthew Flinders' Journal in the Norfolk
Sloop*, 1799, SLNSW, C 211/2, p. 6.
26 *Ibid.*
27 *Ibid.*, p. 7.
28 *Ibid.*

29 *Flinders' Biographical Tribute to Trim.*
30 By John Hawkesworth, editing Cook's
notes in 1775.
31 Collins, *An Account of the English Colony*,
Vol. 2, p. 231.
32 Flinders, *Terra Australis*, p. 196. The place
where the skirmish took place is now
known as South Point.

Chapter Thirteen

1 Collins, *An Account of the English Colony*,
Vol. 2, p. 245.
2 *Ibid.*, p. 232.
3 *Ibid.*, p. 234.
4 *Ibid.*
5 *Ibid.*
6 *Ibid*, p. 242.
7 Today it is known as the Pumicestone
Passage. It is actually a narrow bay
separating Bribie Island from the
mainland, and at its northern end is the
resort town of Caloundra.
8 It is now known as Woody Point; the
suburb of Redcliffe is further north.
9 *Flinders' Journal in the Norfolk*, 1799,
SLNSW, C 211/2, p. 13.
10 John St Pierre, *Matthew Flinders: The First
English Explorer in Moreton Bay 1799*,
Redcliffe Historical Society, 1997, p. 6.
11 Collins, *An Account of the English Colony*,
Vol. 2, p. 239.
12 *Ibid.*, p. 240.
13 *Ibid.*, p. 241.
14 Flinders, *Terra Australis*, p. cxcvii.
15 Collins, *An Account of the English Colony*,
Vol. 2, p. 243.
16 *Ibid.*, p. 245.
17 *Ibid.*
18 *Ibid.*, p. 248.
19 *Ibid.*
20 *Ibid.*, p. 251.
21 *Ibid.*, p. 250.
22 *Ibid.*, p. 251.
23 *Ibid.*, p. 252.
24 Flinders, *Terra Australis*, p. cxcviii.
25 *Ibid.*, p. cc.
26 Augustus John Hervey, b. 1724, d. 1779.
27 *Gratefull to Providence*, Vol. 2, p. 217.
The letters were dated 3 and
10 September 1799.

28 Flinders, *Terra Australis*, p. ccii.

29 Acting Governor King to Portland, 10 March 1801, *HRA*, Series 1, Vol. 3, p. 14.

Chapter Fourteen

1 Flinders to Matthew Flinders Sr, 13 April 1801, Lincolnshire Archives Office, Flinders Papers 3/6.

2 Governor Hunter to Secretary Nepean, 15 August 1799, *HRNSW*, Vol. 3, p. 702.

3 Estensen, *Life of Matthew Flinders*, p. 193.

4 Hunter to Portland, 2 January 1800, *HRA*, Series 1, Vol. 2, pp. 416, 420.

5 King v. Powell, Freebody, Metcalf, Timms and Butler for wantonly killing two native men of this territory, *R. v. Powell* [1799] NSWKR 7; [1799] NSWSupC 7.

6 Hunter to Portland, 20 April 1800, *HRA*, Series 1, Vol. 2, p. 412.

7 *Ibid.*, p. 404.

8 *Ibid.*, pp. 401–02.

9 Lyn Stewart, *Blood Revenge: Murder on the Hawkesbury 1799*, Rosenberg Publishing, 2015.

10 Hunter to Portland, 20 April 1800, *HRA*, Series 1, Vol. 2, pp. 401–02.

11 Flinders to Pultney Malcolm, 17 January 1800, Flinders Papers, La Trobe Collections, State Library of Victoria.

12 'Land Grants', 6 February 1800, *HRNSW*, Vol. 4., p. 48.

13 *Leichhardt Historical Journal*, no. 24, 2014.

14 Flinders to Christopher Smith, 14 February 1800, Flinders Papers, 60/017, FLI/4, NMM

15 *Ibid.*

16 *Ibid.*

17 Flinders to Bass, 15 February 1800, SLNSW, MLMSS 7046.

18 *Ibid.*

19 Flinders to Joseph Banks, 12 July 1804, *HRNSW*, Vol. 5, pp. 397–98.

20 Hunter to Captain Waterhouse, 12 February 1800, *HRNSW*, Vol. 4, p. 53.

21 Waterhouse to Evan Nepean, 27 August 1800, *HRNSW*, Vol. 4, p. 119.

22 *Matthew Flinders' Biographical Tribute to Trim*, p. 2.

23 Isle Penantipode, discovered by HMS *Reliance*, H. Waterhouse, commander, 25 March 1800, Papers of Matthew Flinders (as filmed by the AJCP), 1698–1852: [M444, M3033-M3037]/Series FLI 15 - 20/Subseries FLI 15, NLA.

24 Waterhouse to Evan Nepean, 27 August 1800, *HRNSW*, Vol. 4, p. 119.

25 *Gratefull to Providence*, Vol. 2, p. 210, 1 November 1799.

26 *Ibid.*, p. 209, 3 September 1799.

27 *Ibid.*, 1 May 1801. John Flinders died at the asylum in March 1834, aged fifty-two.

28 Flinders, *Terra Australis*, p. i.

29 Draft Instructions 'To Governor King ...', circa February 1800, SLNSW, Banks Papers, Series 39.057.

30 James Grant, baptised 6 September 1772 (Forres, Morayshire, Scotland), d. 11 November 1833 (St Servan, France).

31 *Observations on the Coasts of Van Diemen's Land, on Bass's Strait, etc,* John Nichols, 1801.

32 Flinders to Banks, 6 September 1800, SLNSW, Banks Papers, Series 65.01.

33 *Ibid.*

34 Flinders to Ann Chappelle, 25 September 1800, Flinders Papers, 60/017, FLI/25, NMM.

35 *Ibid.*

36 'State and Defects of His Majesty's Ship Reliance', *HRNSW*, Vol. 4, p. 227. The *Reliance* would be employed in harbour service for the next decade and a half.

37 *Gratefull to Providence*, Vol. 2, p. 221, 1 November 1800.

38 *Matthew Flinders' Biographical Tribute to Trim*, p. 1.

39 *Gratefull to Providence*, Vol. 2, p. 231, 30 June 1801.

40 Flinders Sr to Flinders, 23 November 1800, Lincolnshire Archives Office, Flinders Papers 3/3.

41 Flinders to Flinders Sr, 29 November 1800, Flinders Papers, 60/017, FLI/25, NMM.

42 Banks to Flinders, 16 November 1800, Flinders Papers, 60/017, NMM.

43 George John Spencer, 2nd Earl Spencer, b. 1 September 1758, d. 10 November 1834.

44 Nicolas Thomas Baudin, b. 17 February 1754, d. 16 September 1803.

45 Flinders, *Terra Australis*, p. 4.
46 *Ibid.*
47 Later Vice-Admiral Robert Merrick Fowler, b. 1778 (Horncastle, Lincolnshire), d. 25 May 1860 (Whitchurch-on-Thames, Oxfordshire).
48 Isabella Tyler, 'Biographical Outline of Capt. & Mrs Flinders 1852', NLA, NMM.
49 Flinders to Ann Chappelle, 18 December 1800, Flinders Papers, 60/017, FLI/25, NMM.
50 *Ibid.*
51 *Ibid.*, 16 January 1801.
52 Flinders to Banks, 24 January 1801, SLNSW, Banks Papers, Series 65.02.
53 *Gratefull to Providence*, Vol. 2, p. 225, 2 February 1801.
54 *Ibid.*, p. 228.
55 Flinders, *Terra Australis*, p. 5.
56 Lincolnshire Archives Office, Flinders Papers 3/5.
57 Flinders to Ann Chappelle, 6 April 1801, Flinders Papers 60/017 FLI/25, NMM.
58 *Ibid.*
59 *Ibid.*
60 Statement by W.A. Standert, Matthew Flinders' agent, Flinders Papers 60/017, FLI/6, NMM.
61 Flinders to Flinders Sr, 13 April 1801, Lincolnshire Archives Office, Flinders Papers 3/6.
62 *Ibid.*, loose pages from Flinders' private letterbook, flinders.rmg.co.uk, FLI04.
63 Isabella Tyler, 'Biographical Outline of Capt. & Mrs Flinders 1852', Flinders Papers 60/017, FLI 107, NMM.
64 *Ibid.*
65 *Gratefull to Providence*, Vol. 2, p. 228, 1 May 1801.

Chapter Fifteen

1 Flinders to Ann Flinders, 5 July 1801, copy by W.M.F. Petrie in Flinders Papers, La Trobe Collection.
2 Ann Flinders to Elizabeth Franklin, 17 April 1801, CY Reel 2457, Mitchell Library Af 2/1-14, 16, State Library of NSW.
3 Flinders to Flinders Sr, 9 May 1801, Lincolnshire Archives, Flinders 3/7.
4 William Henry Smyth, 'A Brief Memoir of Captain Matthew Flinders, RN', Box 81/3(d), La Trobe Collection.
5 Tyler, 'Biographical Outline of Capt. & Mrs Flinders 1852'.
6 Flinders to Henrietta Flinders, 10 May 1801, Flinders Papers 60/017, FLI/4, NMM.
7 Scott, *The Life of Captain Matthew Flinders, R.N.*, p. 170.
8 Tyler, 'Biographical Outline'.
9 Flinders to Banks, 29 April 1801, SLNSW, Banks Papers, Series 65.15.
10 Banks to Flinders, 1 May 1801, ibid., 65.16.
11 *Ibid.*
12 Scott, *The Life of Captain Matthew Flinders, R.N.*, p. 176.
13 'House of Commons, Monday', *Bath Chronicle and Weekly Gazette*, 11 June 1801, p. 3.
14 Banks to Flinders, 21 May 1801, SLNSW, Banks Papers, Series 65.20.
15 Royal Collection Trust, George III Calendar, Cook to unnamed (probably Philip Stephens, the Secretary to the Admiralty), 1 August 1772, GEO/MAIN/1359.
16 Flinders to Banks, 24 May 1801, SLNSW, Banks Papers, Series 65.21.
17 Flinders, *Terra Australis*, p. 7.
18 *Ibid.*, p. 136.
19 Flinders to Banks, 3 June 1801, SLNSW, Banks Papers, Series 65.22.
20 Matthew Flinders, *Journal on HMS 'Investigator'*, Vol. 1, 1801–1802, July 1801, SLNSW, Safe 1/24, p. 46.
21 Banks to Flinders, 5 June 1801, SLNSW, Banks Papers, Series 65.23.
22 Flinders to Banks, 6 June 1801, ibid., 65.24.
23 Robert Brown, b. 21 December 1773, d. 10 June 1858.
24 Ferdinand Lukas Bauer, b. 20 January 1760, d. 17 March 1826.
25 William Westall, b. 12 October 1781, d. 22 January 1850.
26 Richard Westall to Flinders, 13 March 1802, flinders.rmg.co.uk, FLI01.
27 Robert Brown, *Prodromus florae Novae-Hollandiae et insulae Van-Diemen*, 1812.

28 John Crosley, b. 1762, d. 1817.

29 Peter Good to William Townsend Aiton, 14 January 1801, SLNSW, Banks Papers, Series 63.16.

30 Scott, *The Life of Captain Matthew Flinders R.N.*, pp. 181–82.

31 Banks to Flinders, June 1801, SLNSW, Banks Papers, Series 65.26.

32 Flinders, *Journal on HMS 'Investigator'*, p. 52

33 Tyler, 'Biographical Outline', Flinders Papers 60/017, FLI 107, NMM.

34 Flinders to Ann Flinders, 30 June 1801, copy by W.M.F. Petrie in Flinders Papers, La Trobe Collection.

35 *Ibid.*, 5 July 1801.

36 Flinders, *Journal on HMS 'Investigator'*, p. 55.

37 *Ibid.*

38 *Gratefull to Providence*, Vol. 2, p. 231, 30 June 1801.

39 Flinders to Flinders Sr, 10 July 1801, flinders.rmg.co.uk, FLI04.

40 *Ibid.*

41 Flinders to Ann Flinders, 12 July 1801, *ibid.*

42 Banks to Flinders, June 1801, SLNSW, Banks Papers, Series 65.26.

43 Carolyn Webb, 'Australian Collectors Snap Up Rare Relics of Explorer Matthew Flinders', *The Sydney Morning Herald*, 16 December 2021.

44 Flinders, *Terra Australis*, p. lxxvi.

45 *Ibid.*

46 *Ibid.*

47 *Ibid.*

48 *Ibid.*

49 Ibid., p. 16.

50 French Passport for Matthew Flinders in HMS *Investigator*, flinders.rmg.co.uk, FLI03.

Chapter Sixteen

1 Flinders, *Terra Australis*, pp. 60–61.

2 *Ibid.*, p. 17.

3 *Gratefull to Providence*, Vol. 2, p. 234, 1 December 1801.

4 *Ibid.*, p. 235.

5 Flinders, *Terra Australis*, p. 36

6 Flinders, *Journal on HMS 'Investigator'*, Vol. 1, SLNSW, Safe 1/24, p. 110.

7 Flinders, *Terra Australis*, p. 28.

8 *Ibid.*, p. 29.

9 *The Journal of Peter Good, Gardener on Matthew Flinders Voyage to Terra Australis 1801–03*, edited by Phyllis I. Edwards, *Bulletin of the British Museum (Natural History)*, Historical series, vol 9, 1981, 7 September 1801, p. 40.

10 *Ibid.*

11 *Matthew Flinders' Biographical Tribute to Trim*, p. 1.

12 Flinders, *Terra Australis*, p. 32.

13 *Ibid.*, p. 38.

14 Flinders to Flinders Sr, 22 October 1801, Flinders Correspondence 3/10, Lincolnshire Archives.

15 *Gratefull to Providence*, Vol. 2, p. 238, 31 January 1802.

16 Flinders to Ann Flinders, 3 November 1801, flinders.rmg.co.uk, FLI25.

17 *Ibid.*

18 Flinders, *Terra Australis*, p. 45.

19 *Ibid.*, p. 46.

20 Flinders to Willingham Franklin, 27 November 1801, Matthew Flinders – Private letters, Vol. 1, 1801–1806, State Library of NSW, Safe 1/55.

21 *Ibid.*

22 Flinders, *Terra Australis*, pp. 47–48.

23 Pieter Nuyts or Nuijts, b. 1598, d. 11 December 1655.

24 Flinders, *Terra Australis*, pp. 48–49.

25 *Ibid.*, p. 49.

26 *Ibid.*, p. 54.

27 *Ibid.*, p. 57.

28 *Ibid.*, pp. 57–58.

29 *The Journal of Peter Good*, 25 December 1801, p. 51.

30 Richard Westall to Flinders, 13 March 1802, flinders.rmg.co.uk, FLI01.

31 Flinders, *'Investigator' Journal*, Vol. 1, SLNSW, Safe 1/24, p. 240.

32 Flinders to Ann Flinders, 31 May 1802, flinders.rmg.co.uk, FLI25.

33 Retter & Sinclair, *Letters to Ann*.

34 Flinders to Ann Flinders, 31 May 1802, flinders.rmg.co.uk, FLI25.

35 Flinders, *Terra Australis*, pp. 60–61.

36 *Ibid.*

37 Samuel Smith, *Journal on Board Investigator and Porpoise under Captain*

Matthew Flinders, SLNSW SAFE/C 222 (Safe 1/254).

38 *Ibid.*

39 *Ibid.*, p. 67.

40 *Ibid.*, p. 68.

41 Robert Brown to Joseph Banks, 30 May 1802, HRNSW, Vol. IV, p. 777.

42 *Ibid* to C.F. Greville, 30 May 1802, *ibid.*, p. 776.

43 Flinders, *Terra Australis*, p. 74.

44 *Ibid.*, p. 79.

45 *Ibid.*, p. 80.

46 *Ibid.*, p. 81.

47 *Ibid.*, p. 83.

48 *The Journal of Peter Good*, 13 January 1802, p. 55.

49 Flinders, *Terra Australis*, p. 86.

50 Samuel Smith, Journal .

51 Flinders, *Terra Australis*, p. 87.

52 *Ibid.*, p. 96.

53 *Ibid.*, p. 97.

54 *Ibid.*, p. 98.

55 *Ibid.*, p. 104.

56 *Ibid.*, p. 132.

Chapter Seventeen

1 Flinders, *Terra Australis*, p. 133.

2 *Ibid.*

3 *Ibid.*

4 *The Journal of Peter Good*, 21 February 1802, p. 62.

5 *Ibid.*

6 Flinders, *Terra Australis*, p. 135.

7 *Ibid.*

8 Samuel Smith journal.

9 *The Journal of Peter Good*, 22 February 1802, p. 62.

10 Flinders, *Terra Australis*, p. 137.

11 *Ibid.* p. 139.

12 Flinders to Ann Flinders, 31 May 1802, flinders.rmg.co.uk, FLI25.

13 *Gratefull to Providence*, Vol. 2, p. 238.

14 Flinders, *Terra Australis*, p. 142.

15 *Ibid.*, p. 148.

16 *Ibid.*, p. 146.

17 *Ibid.*

18 *Ibid.*, p. 145.

19 Miriam Estensen, *The Life of Matthew Flinders*, p. 193.

20 Flinders, *Terra Australis*, p. 155.

21 *Ibid.*, p. 156.

22 *The Journal of Peter Good*, 9 March 1802, p. 65.

23 *Ibid.*, 10 March 1802, p. 66.

24 Flinders, *Terra Australis*, p. 158.

25 *Ibid*, p. 159.

26 The Flinders Ranges were named by South Australia's Governor Gawler in 1839.

27 Flinders, *Terra Australis*, p. 166.

28 *Ibid.*, p. 160.

29 *Matthew Flinders' Biographical Tribute to Trim*, p. 3.

30 *Ibid.*

31 Charles Philip Yorke (12 March 1764 – 13 March 1834).

32 Flinders, *Terra Australis*, p. 169.

33 *Ibid.*

34 *Ibid.*, p. 170.

35 *The Journal of Peter Good*, 23 March 1802, p. 69.

36 Flinders, *Terra Australis*, p. 171.

37 *Ibid.*, p. 181.

38 *The Journal of Peter Good*, 2 April 1802, p. 71.

39 *Ibid.*, 3 April 1802, ibid.

40 Flinders, *Terra Australis*, p. 183.

41 Also recorded as Richard Staley.

42 Flinders, *Terra Australis*, p. 184.

43 *Ibid.*, p. 186.

44 Flinders, *'Investigator' journal*, Vol. 1, p. 432.

45 Flinders, *Terra Australis*, p. 188.

46 François Péron, b. 1775, d. 1810.

47 Louis-Claude Desaulses De Freycinet, b. 1779, d. 1842.

48 Pierre Leveque, b. 1746, d. 1814.

49 Leslie R. Marchant & J.H. Reynolds, 'Baudin, Nicolas Thomas (1754–1803)', *Australian Dictionary of Biography*, Australian National University, 1966.

50 *The Journal of Peter Good*, 8 April 1802, p. 73.

51 Samuel Smith *Journal*, 9 April 1802.

52 Nicolas Baudin, Journal de mer (entry dated 19 Germinal–9 Apr. 1802), Archives nationales, Paris, Marine 5JJ 3640, 5 vols. From Jean Fornasiero & John West-Sooby, 'A Cordial Encounter? The Meeting of Matthew Flinders and Nicolas Baudin (8–9 April, 1802)', News and Papers from the George Rudé Seminar, h-france.net.

53   *Ibid.*

54   Alphonse de Fleurieu, the descendant of Count Charles-Pierre Claret de Fleurieu (the French navigator and former Minister of Marine who drew up the itinerary for the Baudin expedition), compiled papers about the meeting of Baudin and Flinders. These papers are held in the library of the Royal Geographical Society of South Australia, within the State Library of South Australia (RGMS 107c). The comment relating to the *Lady Nelson* is at RGMS 107c, note 6.

55   Flinders, *Terra Australis*, p. 190.

56   *Ibid.*, p. 200.

57   Later Baron Jacques Félix Emmanuel Hamelin, b. 13 October 1768, d. 23 April 1839.

58   Flinders, *Terra Australis*, p. 190.

59   Baudin, *Journal de mer*, 18 Germinal–8 Apr. 1802.

60   Rebecca Brice, 'Sacre Bleu! French Invasion Plan for Sydney', abc.net.au, 10 December 2012.

61   Diana S. Jones, 'The Baudin Expedition in Australian waters (1801–1803): The Faunal Legacy', museum.wa.gov.au.

62   Flinders, *Terra Australis*, p. 202.

63   John Black, b. 31 October 1778, d. circa May 1802.

64   Flinders, *Terra Australis*, p. 205.

65   *Ibid.*, p. 211.

66   *Ibid.*

67   Smith, *Journal*, 28 April 1802.

68   Flinders, *Terra Australis*, p. 213.

69   *The Journal of Peter Good*, 27 April 1802, p. 75.

70   Flinders, *Terra Australis*, p. 215.

71   *Ibid.*, p. 216.

72   *Ibid.*, p. 221.

73   *Ibid.*, p. 218.

74   Flinders, *Terra Australis*, p. 227.

75   *Gratefull to Providence*, Vol. 2, p. 240.

Chapter Eighteen

1   Flinders to Ann Flinders, 20 July 1802, flinders.rmg.co.uk, FLI25.

2   He left Sydney on the ship *Hunter* in November 1801.

3   *The Journal of Peter Good*, 9 May 1802, p. 79.

4   Flinders, *A Voyage to Terra Australis*, Vol. 1, p. 227.

5   *Ibid.*, p. 228.

6   'Contract with Messrs. Bass and Bishop for the Importation of Pork, 9th October, 1801, *HRA*, Series 1, Vol. 3, pp. 337–38.

7   Governor King to Lord Hobart, 15 November 1802, *ibid.*, p. 724.

8   Flinders, *A Voyage to Terra Australis*, Vol. 1, pp. 231–32.

9   Commander Flinders to the Admiralty, 11 May 1802, *HRNSW*, Vol. 4. pp. 747–51.

10   Flinders to Ann Flinders, 31 May 1802, flinders.rmg.co.uk, FLI25.

11   *Ibid.*

12   *Ibid.*

13   *Ibid.*

14   *Ibid.*

15   Flinders, *Terra Australis*, Vol. 1, p. 236.

16   Flinders to Banks, 20 May 1802, *HRNSW*, Vol. 4. pp. 755–57.

17   Flinders to Ann Flinders, 31 May 1802, flinders.rmg.co.uk, FLI25.

18   Also recorded as McIntyre.

19   On 22 November 1801.

20   Smith, *Journal*, 20 July 1802.

21   King to Banks, 5 June 1802, *HRNSW*, Vol. 4, p. 784.

22   Commander Flinders to the Admiralty, 11 May 1802, *HRNSW*, Vol. 4, p. 749.

23   Sir Joseph Banks to Governor King, 8 April 1803, *HRNSW*, Vol. 5, p. 835.

24   Flinders to Ann Flinders, 20 July 1802, flinders.rmg.co.uk, FLI25.

25   *Ibid.*

26   Flinders, *Terra Australis*, p. 230.

27   Governor King to Baudin, 23 June 1802, *HRNSW*, Vol. 4, p. 949.

28   Flinders, *Terra Australis*, p. 230.

29   Baudin to Mr Harris, 23 September 1802, *HRNSW*, Vol. 4, p. 959.

30   'Péron's Report on Port Jackson', from Scott, *The Life of Captain Matthew Flinders, R.N.*, p. 439.

31   *Matthew Flinders' Biographical Tribute to Trim*, p. 4.

32 *Ibid.*

33 Flinders, *A Voyage to Terra Australis*, Vol. 1, p. 231.

34 John Murray, b. circa 1775, d. 1807.

35 Nanbarry, b. circa 1780, d. 1821. Flinders record his name as Nanbaree.

36 Flinders, *A Voyage to Terra Australis*, p. 235.

37 Daniel Southwell, Southwell Papers, British Library, London, MS 16381.

38 Keith Vincent Smith, 'Nanbarry', dictionaryofsydney.org, 2010.

39 Watkin Tench, *A Complete Account of the Settlement at Port Jackson*, p. 19.

40 *Ibid.*, p. 34.

41 Tooth knocked out – 'Irra badiang', Robert Brown, Georges River Vocabulary, 1803, Botanic Library, Natural History Museum, London, MS B3V, ff 258–259.

42 Illustration in Collins, *An Account of the English Colony*, Vol. I, 1798.

43 Flinders, *Terra Australis*, Vol. 1, p. 237.

44 Flinders, *'Investigator' Journal*, Vol. 1, p. 508.

45 Flinders, *Journal on HMS 'Investigator'*, Vol. 2, pp. 10–11.

46 *Ibid.*, p. 19

47 *The Journal of Peter Good*, 31 July 1802, p. 82.

48 Flinders, *Journal on HMS 'Investigator'*, Vol. 2, pp. 26–27.

49 *The Journal of Peter Good*, 31 July 1802, p. 82.

50 Ibid.

51 Admiral Sir Roger Curtis, b. 4 June 1746, d. 14 November 1816.

52 Flinders, *'Investigator' Journal*, Vol. 2, p. 37.

53 *The Journal of Peter Good*, 5 August 1802, p. 84.

54 Flinders, *'Investigator' Journal*, Vol. 2, p. 39.

55 *Ibid.*, p. 78.

56 Flinders, *Terra Australis*, Vol. 2, pp. 21–22.

57 *Ibid.*, p. 26.

58 *The Journal of Peter Good*, 15–16 August 1802, p. 86.

59 *Ibid.*

60 *Ibid.*

61 Ida Lee, *The Logbooks of the Lady Nelson*, Grafton & Co, 1915, p. 174.

62 Flinders, *Terra Australis*, Vol. 2, p. 313.

63 The place was renamed Port Clinton in 1892 to avoid confusion with the town of Bowen, 400 kilometres to the north-west.

64 Flinders, *Terra Australis*, Vol. 2, p. 99.

65 Flinders, *'Investigator' Journal*, Vol. 2, p. 93.

66 *Ibid.*, p. 94.

67 *Matthew Flinders' Biographical Tribute to His Cat Trim*, p. 7.

68 Flinders, *Journal on HMS 'Investigator'*, Vol. 2, p. 96.

69 *Ibid.*, p. 143.

70 *Ibid.*, p. 132.

71 *The Journal of Peter Good*, 15 September 1802, p. 92.

72 Flinders, *Terra Australis*, Vol. 2, p. 73.

73 *Ibid.*, p. 77.

74 *The Journal of Peter Good*, 15 September 1802, pp. 94–95.

75 Flinders, *Terra Australis*, Vol. 2, p. 88.

76 *Ibid.*, p. 87.

77 *Ibid.*, p. 91.

78 *Ibid.*, p. 92

79 *Ibid.*, p. 96

80 Flinders to Ann Flinders, 18 October 1802, 60/017, FLI/25, NMM.

Chapter Nineteen

1 Flinders, *A Voyage to Terra Australis*, Vol. 2, p. 138.

2 *Ibid.*, p. 99.

3 *The Journal of Peter Good*, 20–21 October 1802, p. 96.

4 Flinders, *A Voyage to Terra Australis*, Vol. 2, p. 100.

5 'General Chart of Terra Australis or Australia: showing the parts explored between 1798 and 1803 by M. Flinders Commr. of H.M.S. Investigator', G. & W. Nicol, 1814, SLNSW, 153830.

6 Flinders, *Terra Australis*, Vol. 2, p. 104.

7 Flinders, *'Investigator' Journal*, Vol. 2, p. 212.

8 Murray Island, also known as Mer Island or Maer Island.

9   Flinders, *Terra Australis*, Vol. 2, p. 109.

10 *Ibid.*, p. 110.

11 *The Journal of Peter Good*, 30 October 1802, pp. 97–98.

12 Flinders, *A Voyage to Terra Australis*, Vol. 2, p. 110.

13 *The Journal of Peter Good*, 30 October 1802, p. 98.

14 Flinders, *Terra Australis*, Vol. 2, p. 135.

15 *Ibid.*, p. 118.

16 *Ibid.*

17 *Ibid.*, p. 119.

18 *Ibid.*, p. 120.

19 *Ibid.*, p. 123.

20 *Ibid.*

21 Jan Etienne Gonsal, also known as Jean Etienne Gonzal, d. 4 July 1770.

22 Pieter Arend Leupe, *The Voyages of the Dutch: To the Zuidland or New Holland in the 17th and 18th centuries*, G. Hulst van Keulen, 1868.

23 Willem Janszoon (sometimes abbreviated to Jansz), b. circa 1570, d. circa 1630.

24 Usually spelt as the *Duyfken*.

25 Flinders, *A Voyage to Terra Australis*, Vol. 2, p. 129.

26 Brown to Banks, March 1802, *HRNSW*, Vol. 5. p.82.

27 Flinders, *A Voyage to Terra Australis*, Vol. 2, p. 132.

28 Edward Duyker 'Mary Beckwith', dictionaryofsydney.org.

29 Flinders, *Terra Australis*, Vol. 2, p. 138.

30 *Ibid.*, p. 141.

31 In later years, explorers with HMS *Beagle* in 1841, A.C. Gregory's North Australian Expedition in 1856, and members of William Landsborough's expedition searching for Burke and Wills in 1861 also carved their names on the tree. The tree was damaged by a storm in 1888; the autographed timber is now at Brisbane's Museum of Lands, Mapping and Surveying.

32 Flinders, *Terra Australis*, Vol. 2, p. 143.

33 *Ibid.*, p. 142.

34 Smith, *Journal* 8 September 1802.

35 *Matthew Flinders' Biographical Tribute to Trim*, p. 7.

36 Flinders, *'Investigator' Journal*, Vol. 2, p. 272.

37 Flinders, *Terra Australis*, Vol. 2, pp. 153-4.

38 *The Journal of Peter Good*, 4 December 1802, p. 104.

39 Flinders, *Terra Australis*, Vol. 2, p. 161.

40 *Ibid.*, p. 189. Now known as the Maluku Islands of Indonesia.

41 *Ibid.*, p. 173.

42 *The Journal of Peter Good*, 24 December 1802, p. 107.

Chapter Twenty

1   Flinders, *Terra Australis*, Vol. 2, p. 196.

2   *Ibid.* p. 180.

3   *Ibid.*, p. 182.

4   *The Journal of Peter Good*, 4 January 1803, p. 109.

5   Flinders, Terra Australis, Vol. 2, p. 189.

6   *Ibid.*, p. 196.

7   *The Journal of Peter Good*, 21 January 1803, p. 111.

8   Ibid.

9   Flinders, *Terra Australis*, Vol. 2, p. 197.

10 *Ibid.*, p. 198.

11 *Ibid.*, p. 197.

12 *Ibid.*, p. 189.

13 *Ibid.*, p. 229.

14 *Ibid.*, p. 206.

15 *Ibid.*, p. 208.

16 *Ibid.*

17 *Ibid.*, p. 209.

18 *Ibid.*, p. 227.

19 *Ibid.*, pp. 228–29.

20 *Ibid.*, p. 229.

21 *The Journal of Peter Good*, 17 February 1803, p. 118.

22 Flinders, *Terra Australis*, Vol. 2, p. 230.

23 *Ibid.*, p. 238.

24 *Ibid.*, p. 247.

25 Flinders to Ann Flinders, 28 March 1803, SLNSW, Matthew Flinders – Private letters, Vol. 1, 1801–1806, Safe 1/55.

26 Flinders, Journals of the Investigator, Porpoise and Cumberland, 1801-3, SLNSW, ML Safe 1/26, 6 March 1803.

27 Flinders to Ann Flinders, 28 March 1803, SLNSW, Safe 1/55.

28 Flinders to Osborne Standert, 6 April 1803, *ibid.*

29 Flinders to Sir Joseph Banks, 28 March 1803, *ibid.*

30 Flinders to Thomas Franklin, 4 April 1803, *ibid.*

31 Flinders to Henrietta Flinders, 2 April 1803, *ibid.*

32 Flinders to Christopher Smith, 1 April 1803, *ibid.*

33 The location of Tryal Rocks was not confirmed until 1969. The reef is located in the Indian Ocean off the north-west coast of Australia, fourteen kilometres north-west of the outer edge of the Montebello Islands.

34 *The Journal of Peter Good*, 17 May 1803, p. 122.

35 Flinders, *Terra Australis*, Vol. 2, p. 266.

36 Flinders, *'Investigator' Journal*, Vol. 2, p. 512.

37 Flinders, *Terra Australis*, Vol. 2, p. 269.

38 Flinders to Hugh Bell, 27 May 1803, SLNSW, Safe 1/55.

39 *Ibid.*, 29 May 1803.

40 Flinders, *Terra Australis*, Vol. 2, p. 270–71.

Chapter Twenty-one

1 Flinders to Elizabeth Flinders, 10 June 1803, SLNSW. He crossed these words out.

2 'Ship News', *The Sydney Gazette and New South Wales Advertiser*, 12 June 1803, p. 4.

3 'Copy of a letter from Matthew Flinders, esq commander of His Majesty's Sloop Investigator, to his excellency the Governor', *ibid.*

4 George Howe, b. 1769 (St Kitts), d. 11 May 1821 (Sydney).

5 Flinders to Elizabeth Flinders, 10 June 1803, SLNSW, Safe 1/55.

6 *Ibid.*

7 *Ibid.*

8 'Weekly Occurrences', *Sydney Gazette*, 19 June 1803, p. 1.

9 Flinders to Ann Flinders, 25 June 1803, SLNSW, Matthew Flinders – Private letters.

10 Flinders, *Terra Australis*, Vol. 2, pp. 273–74.

11 James Inman, b. 1776 (Garsdale, now in Cumbria), d. 7 February 1859 (Portsmouth).

12 Flinders, *Terra Australis*, Vol. 2, p. 275.

13 Flinders to Ann Flinders, 25 June 1803, SLNSW, Private letters.

14 Sometimes reported as Robert Cummings.

15 Flinders, *Terra Australis*, Vol. 2, p. 277.

16 Flinders to Ann Flinders, 25 June 1803.

17 *Ibid.*

18 *Ibid.*

19 *Ibid.*

20 William Westall to Joseph Banks, 31 January 1804, SLNSW, Banks Papers, Series 23.44.

21 *Matthew Flinders' Biographical Tribute to Trim*, p. 7.

22 Flinders, *Terra Australis*, Vol. 2, p. 277.

23 Flinders to George Bass, undated, Private letters.

24 *Ibid.*

25 Flinders to Mrs Kent, undated, Private letters.

26 Flinders, *Terra Australis*, Vol. 2, p. 296.

27 Flinders to Mrs McArthur, 6 July 1803, Private letters.

28 *Ibid.*, 28 July 1803.

29 Flinders, *Terra Australis*, Vol. 2, p. 279.

30 Robert Allen to Joseph Banks, 6 August 1803, *HRNSW*, Vol. 5, p. 181.

31 'Ship News', *The Sydney Gazette*, 14 August 1803, p. 4.

32 Flinders, *Terra Australis*, Vol. 2, p. 298.

33 *Ibid.*

34 *Ibid.*, p. 299.

35 William Westall's account, from James Stanier Clarke, *Naufragia: Or Historical Memoirs of Shipwrecks and of the Providential Deliverance of Vessels*, Vol. 1, 1805, p. 385.

36 *Ibid.*, p. 386.

37 *Ibid.*

38 Flinders, *Terra Australis*, Vol. 2, p. 300.

39 'Commander Flinders's Account of the Wreck of the *Porpoise* and *Cato*,' *HRA*, Series 1, Vol. 4, p. 402.

40 Flinders, *Terra Australis*, Vol. 2, p. 302.

41 *Ibid.*, p. 305.

42 Smith, *Journal*, August 1803.

43 Flinders, *A Voyage to Terra Australis*, Vol. 2, p. 304.

44 *Ibid.*, p. 312.

45 *Ibid.*, p. 305.

46 *Matthew Flinders' Biographical Tribute to His Cat Trim*, p. 7.

47 Flinders, *Terra Australis*, Vol. 2, p. 306.

48 *Ibid.*, p. 307, quoting Palmer's account published in the Calcutta newspaper *The Orphan*, 3 February 1804.

49 Governor King to Sir Joseph Banks, 14 August 1804, *HRNSW*, Vol. 5, p. 450.

50 Flinders, *Terra Australis*, Vol. 2, p. 308.

51 *Matthew Flinders' Biographical Tribute to Trim*, p. 7.

52 Flinders, *Terra Australis*, Vol. 2, p. 309.

53 *Ibid.*, p. 311.

54 Scott, *The Life of Captain Matthew Flinders, R.N.*, p. 293.

55 Flinders, *Terra Australis*, Vol. 2, p. 309.

56 *Ibid.*

57 *Ibid.*, p. 310.

58 *Ibid.*, p. 315.

59 *Ibid.*, p. 316.

60 *Ibid.*, p. 318.

61 *Ibid.*, p. 321.

62 Papers of Matthew Flinders (as filmed by the AJCP), 1698–1852: [M444, M3033-M3037]/Series ADM 7/File Adm 7/708/Narrative of Matthew Flinders of his imprisonment by the French, p. 27.

63 Flinders, *Terra Australis*, Vol. 2, p. 322.

Chapter Twenty-two

1 Governor King to Captain Flinders, 17 September 1803, *HRNSW*, Vol. 5, p. 222.

2 Ibid.

3 'Ship News', *The Sydney Gazette*, 18 September 1803, p. 4.

4 *Ibid.*

5 Flinders to Governor King, 24 September to 10 October, *HRNSW*, Vol. 5, p. 240.

6 *Ibid.*

7 *Ibid.*

8 Flinders, *Terra Australis*, p. 328.

9 *Ibid.*

10 *Ibid*, p. 327.

11 *Ibid.*

12 *Ibid.*, pp. 328–29.

13 *Ibid.*, p. 347.

14 *Ibid.*, p. 351.

15 *Ibid.*, p. 325.

16 *Ibid.*, p. 353.

17 Papers of Matthew Flinders (as filmed by the AJCP), 1698-1852 : [M444, M3033-M3037]/Series ADM 7/File Adm 7/708/Narrative of Matthew Flinders of his imprisonment by the French, p. 54.

18 Estensen, *The Life of Matthew Flinders*, p. 315.

19 Flinders, *Terra Australis*, Vol. 2, p. 354.

20 *Ibid.*, p. 355.

21 General Charles Mathieu Isidore Decaen, b. 13 April 1769 (Caen, France), d. 9 September 1832 (Barre, France).

22 On 17 September 1803.

23 Flinders, *Terra Australis*, Vol. 2, p. 360.

24 *Ibid.*

25 Scott, *The Life of Captain Matthew Flinders, R.N.*, p. 323.

26 Flinders, *Terra Australis*, Vol. 2, p. 361.

27 *Ibid.*

28 *Ibid.*, p. 362.

Chapter Twenty-three

1 Decaen Papers, Municipal Library of Caen, Normandy, Vol. 10, as quoted in Scott, *The Life of Captain Matthew Flinders, R.N.*, p. 345.

2 Flinders, *Terra Australis*, Vol. 2, p. 363.

3 Decaen Papers, as quoted in Scott, *The Life of Captain Matthew Flinders, R.N.*, pp. 324–28.

4 *Ibid.*

5 Flinders, *Terra Australis*, Vol. 2, p. 363.

6 Decaen Papers, as quoted in Scott, *The Life of Captain Matthew Flinders, R.N.*, p. 345.

7 Flinders, *Terra Australis*, Vol. 2, p. 489.

8 Flinders to General Decaen, in Flinders, Private Journal, 17 December 1803 to 8 July 1814, SLNSW, CY 227, Safe 1/58, 20 December 1803.

9 Flinders, *A Voyage to Terra Australis*, Vol. 2, p. 366.

10 Governor King to Lord Hobart, 9 November 1802, *HRA*, Series 1, Vol. 3, p. 628.

11 Governor King to Sir Joseph Banks, 9 May 1803, *HRNSW*, Vol. 5, p. 134.

12 François Péron's Report on Port Jackson, translated into English in Scott, *The Life of Captain Matthew Flinders, R.N.*, pp. 436, 464.

13 Governor King to Lord Hobart, *HRA*, Series 1, Vol. 4, p. 358.

14 Flinders, *Terra Australis*, Vol. 2, p. 416.

15 *Ibid.*, p. 367.

16 *Ibid.*, p. 369.

17 *Ibid.*, p. 370.

18 Decaen Papers, as quoted in Scott, *The Life of Captain Matthew Flinders, R.N.*, p. 345.

19 Flinders, *Terra Australis*, Vol. 2, p. 372.

20 *Ibid.*, p. 375.

21 *Ibid.*, p. 377.

22 Mrs Ann Tyler to Matthew Flinders, 19 September 1803, flinders.rmg.co.uk, FLI01.

23 Flinders, Private Journal, 31 December 1803.

24 Denis Decrès, b. 18 June 1761, d. 7 December 1820.

25 Decaen to Denis Decrès, January 1804, Decaen Papers, Vol. 10, as quoted in Scott, *The Life of Captain Matthew Flinders, R.N.*, p. 345.

26 Flinders, *Terra Australis*, Vol. 2, p. 383.

27 *Ibid.*, p. 384.

28 *Matthew Flinders' Biographical Tribute to Trim*, p. 8.

29 Jacques Bergeret, b. 15 May 1771, d. 26 August 1857.

30 Flinders, *Terra Australis*, Vol. 2, p. 384.

31 *Ibid.*, p. 385.

32 *Ibid.*

33 *Ibid.*, p. 388.

34 Matthew Flinders to Sir Joseph Banks, 12 July 1804, SLNSW, Banks Papers, Series 65.31.

35 Flinders to Rev. W. Tyler, 26 April 1804, flinders.rmg.co.uk, FLI25.

36 Sir Joseph Banks to Mrs Anne [sic] Flinders, 4 June 1804, *ibid.*, FLI26.

37 Flinders, Private Journal, 31 March 1803.

38 Flinders, *Terra Australis*, Vol. 2, p. 389.

39 *Ibid.*, p. 391.

40 *Matthew Flinders' Biographical Tribute to Trim*, p. 8.

Chapter Twenty-four

1 Flinders to Rev. W. Tyler, 20 May 1804, flinders.rmg.co.uk, FLI25.

2 *Ibid.*, 26 April 1804.

3 *Ibid.*, 20 May 1804.

4 *Ibid.*

5 Emmanuel Halgan, b. 31 December 1771, d. 20 April 1852.

6 Charles-Alexandre Léon Durand, Comte de Linois, b. 27 January 1761, d. 2 December 1848.

7 Flinders, *Terra Australis*, Vol. 2, p. 391.

8 *Ibid.*

9 Flinders, Private Journal, 14–15 May 1804.

10 *Ibid.*, 18 May 1804.

11 *Ibid.*

12 Flinders, *A Voyage to Terra Australis*, Vol. 2. p. 460.

13 Matthew Flinders to Sir Joseph Banks, 12 July 1804, SLNSW, Banks Papers, Series 65.31.

14 *Ibid.*

15 *Ibid.*

16 Banks to Flinders, 18 June 1805, *ibid.*, Series 65.35.

17 Banks to Institut de France, 22 August 1804, *HRNSW*, Vol. 5, p. 455.

18 Banks to Governor King, 22 August 1804, *HRNSW*, Vol. 5, p. 455.

19 Flinders to Governor King, 8 August 1804, *HRA*, Vol. 5, p. 438.

20 'Ship News', *The Sydney Gazette and New South Wales Advertiser*, 7 April 1805, p. 3.

21 *HRA*, Vol. 5, p. 439.

22 *Ibid.*, p. 442.

23 Rif Winfield, *British Warships in the Age of Sail 1793–1817*, Seaforth, 2008, p. 399.

24 Flinders to Ann Flinders, 13 July, 23 August, 4 November 1804, SLNSW, Safe 1/55.

25 *Ibid.*

26 *Ibid.*

27 Flinders to Samuel Flinders, 25 August 1804, *ibid.*

28 *Ibid.*

29 M. Flinders, *Australia or Terra Australis*, Charts of the Hydrographic Department (as filmed by the AJCP) [microform]: pre-1825 :[M406], 1770-1824./File 1.Y46/2.

30 John Alexander Ferguson, *Bibliography of Australia: 1784–1830*. Vol. 1, National Library of Australia, 1975, p. 77.

31 James Wilson, *A Missionary Voyage to the Southern Pacific Ocean*, London Missionary Society, 1799, p. lxxxvii.

32 Letter from Matthew Flinders originally enclosing a chart of 'New Holland' (Australia), 23 August 1804, re-dated 4 November 1804, Board of Longitude Papers, RGO 14/51, p. 172, University of Cambridge, cudl.lib.cam.ac.uk.

33 Admiral of the Fleet Sir George Cockburn, b. 22 April 1772, d. 19 August 1853.

34 Francis Galton, *Inquiries into Human Faculty and Its Development*, Macmillan, 1883.

35 Flinders, Private Journal, 13 October 1804.

36 *Ibid.*

37 Flinders, *Terra Australis*, Vol. 2, p. 401.

38 Flinders, Private Journal, 17 January 1805.

39 Flinders to Ann Flinders, 31 December 1804, SLNSW, Safe 1/55.

40 Charles Thomi Pitot De La Beaujardière, b. 1779, d. 1821.

41 Flinders to Banks, 16 May 1805, *HRNSW*, Vol. 5, p. 623.

42 Flinders, Private Journal, 25 July 1805.

43 *Ibid.*

Chapter Twenty-five

1   Flinders, Private Journal, 22 June 1806.

2   Flinders, *Terra Australis*, Vol. 2, p. 456.

3   Flinders, Private Journal, 18 August 1805.

4   *Ibid.*

5   *Ibid.*

6   Flinders, *A Voyage to Terra Australis*, Vol. 2, p. 418.

7   Flinders, Private Journal, 21 August 1805.

8   *Ibid.*, 25 August 1805

9   *Ibid.*

10 *Ibid.*, 9 October 1805.

11 Flinders to John Aken, 23 November 1805, SLNSW, Flinders – Private letters, Vol. 1, Safe 1/55.

12 Flinders, Private Journal, 24 October 1805.

13 *Ibid.*, 13 October 1805.

14 *Ibid.*, 22 October 1805.

15 Matthew to Ann Flinders, 20 November 1805, SLNSW, Safe 1/55.

16 *Ibid.*

17 Banks to Flinders, 18 June 1805, SLNSW, Banks Papers, Series 65.35.

18 Charles Baudin, b. 21 July 1784, d. 7 June 1854.

19 Flinders, Private Journal, 30 August 1805.

20 Flinders to Samuel Flinders, 29 March 1806, SLNSW, Safe 1/55.

21 Flinders to Joseph Banks, 20 March 1806, SLNSW, Banks Papers, Series 65.38.

22 Scott, *The Life of Captain Matthew Flinders, R.N.*, p. 366.

23 Flinders to Captain Gaml. M Ward, 19 May 1806, SLNSW, Safe 1/55.

24 Flinders to John Aken, 7 December 1806, SLNSW, Safe 1/56.

25 Flinders, *Terra Australis*, Vol. 2, p. 456.

26 Flinders to Joseph Banks, 8 December 1806, *HRNSW*, Vol. 6, p. 207.

27 Flinders to William Bligh, 23 March 1807, SLNSW, Safe 1/54.

28 *Ibid.*

29 On 26 January 1808.

30 Toussaint Antoine de Chazal (1770-1854)

31 Portrait of Captain Matthew Flinders, RN, 1774–1814, by Toussaint Antoine De Chazal De Chamerel, Art Gallery of South Australia, Adelaide.

32 Flinders, Private Journal, 11 January 1807.

33 Estensen, *The Life of Matthew Flinders*, p. 393.

34 Flinders to Ann Flinders, 31 May 1807, SLNSW, Safe 1/56.

35 *Ibid.*, 30 June 1807, flinders.rmg.co.uk, FLI 25.

36 Flinders, *Terra Australis*, Vol. 2, p. 460.

37 Decaen to Denis Decrès, 20 August 1807, from M. Albert Pitot, *Esquisses Historiques de l'Ile-de-France*, E. Pezzani, 1899, as quoted in Scott, *The Life of Captain Matthew Flinders, R.N.*, p. 372.

38 Flinders, *Terra Australis*, Vol. 2, p. 419.

39 Paul David Labauve D'Arifat.

40 Flinders, Private Journal, 16 March 1808.

41 *Ibid.*, 20 April 1808.

42 *Ibid.*, 24 July 1808.
43 Flinders to Charles Baudin, 25 July
    1808, quoted in Sidney John Baker, *My
    Own Destroyer*, Currawong Publishing,
    1962, p. 135.
44 Flinders, Private Journal, 7 January 1809.
45 Flinders to Joseph Banks, *HRNSW*,
    Vol. 7, p. 52.
46 Flinders, *Terra Australis*,Vol. 2, p. 470.
47 Hugh Hope, b. 1782, d. 1822.
48 Gilbert Elliot-Murray-Kynynmound,
    1st Earl of Minto, b. 23 April 1751, d. 21
    June 1814.
49 Hugh Hope to Flinders, 23 December
    1809, flinders.rmg.co.uk, FLI01.
50 Flinders, *A Voyage to Terra Australis*,Vol. 2,
    p. 475.
51 Hugh Hope to Flinders, 10 March
    1810, flinders.rmg.co.uk, FLI01.
52 Flinders, *Terra Australis*,Vol. 2, p. 479.
53 *Ibid.*
54 *Ibid.*, p. 480.
55 Flinders, Private Journal, 5 June 1810.
56 Joseph Banks to Ann Flinders, 12 June
    1810, flinders.rmg.co.uk, FLI26.
57 Flinders, *Terra Australis*,Vol. 2, p. 485.

Chapter Twenty-six
1   Ann Flinders to Thomas Pitot, 19
    September 1814, SLNSW, Safe 1/57.
2   Flinders, *Terra Australis*,Vol. 2, p. 494.
3   Joseph Banks to Ann Flinders, flinders.
    rmg.co.uk, FLI26.
4   Tyler, 'Biographical Outline'.
5   Vice-Admiral Sir Thomas Bertie, b. 3
    July 1758, d. 13 June 1825.
6   Flinders, *A Voyage to Terra Australis*,Vol. 2,
    p. 494.
7   John Franklin to Flinders, 1 November
    1810, flinders.rmg.co.uk, FLI01.
8   Scott, *The Life of Captain Matthew
    Flinders, R.N.*, p. 382.
9   Flinders, Private Journal, 25 October
    1810.
10 Flinders, *Terra Australis*,Vol. 2, p. 495.
11 *Ibid.*
12 Captain Kent to Flinders, rmg.co.uk,
    FLI01.
13 *Ibid.*
14 Flinders, Private Journal, 7 November
    1810.
15 Joseph Banks to Matthew Flinders,
    28 October 1810, rmg.co.uk, FLI01.
16 Flinders, Private Journal, 16 November
    1810.
17 *Ibid.*, 20 November 1810.
18 *Ibid.*, 28 December 1810.
19 *Ibid.*, 3 January 1811.
20 Flinders to Charles Desbassayns,
    2 February 1811, SLNSW, Flinders –
    Private letters,Vol. 3, Safe 1/57.
21 Flinders, Private Journal, 13 January
    1811.
22 *Ibid.*, 14 January 1811.
23 William Wilberforce, b. 24 August 1759,
    d. 29 July 1833.
24 Flinders, Private Journal, 11 May 1811.
25 *Ibid.*, 9 June 1811.
26 *Ibid.*, 11 June 1811.
27 Flinders, *A Voyage to Terra Australis*,Vol. 2,
    p. 495.
28 Flinders to James Wiles, 2 July 1811,
    SLNSW, Flinders – Private letters,Vol. 3,
    Safe 1/57.
29 Flinders, Private Journal, 8 August 1811.
30 *Ibid.*, 27 January 1811.
31 George Nicol, b. circa 1740, d. 25 June
    1828.
32 Flinders to Madame D'Arifat, 7 March
    1812, SLNSW, Safe 1/57.
33 Flinders, Private Journal, 10 August 1811.
34 *Ibid*, 11 August 1811.
35 *Ibid.*, 30 September 1811.
36 *Ibid*, 11 October 1811.
37 Flinders, *A Voyage to Terra Australis*,Vol. 1,
    p. iii.
38 Flinders, Private Journal, 14 December
    1811.
39 *Ibid.*, 28 December 1811.
40 Flinders to Samuel Flinders, 27
    December 1811, SLNSW, Safe 1/57.
41 *Ibid.*
42 Flinders, Private Journal, 30 December
    1811.
43 *Ibid.*, 30 March 1812.
44 Scott, *The Life of Captain Matthew
    Flinders R.N.*, p. 413.
45 Flinders, Private Journal, 25 March 1812.
46 Tyler, 'Biographical Outline'.
47 Flinders, Private Journal, 1 April 1812.
48 Flinders to Mrs Elizabeth Flinders,
    2 April 1812, SLNSW, Safe 1/57.

49 Tyler, 'Biographical Outline'.
50 Flinders to Charles Desbassayns, 14 July 1812, SLNSW, Safe 1/57.
51 Flinders to James Wiles, 2 July 1811, *ibid.*
52 Flinders, Private Journal, 1 January 1813.
53 *Ibid.*, 2 March 1813.
54 *Ibid.*, 30 March 1813.
55 *Ibid.*, 27 February 1814.
56 *Ibid.*, 9 March 1814.
57 *Ibid.*, 17 March 1814.
58 *Ibid.*, 21 May 1814.
59 Tyler, 'Biographical Outline'.
60 Ann Flinders to Thomas Pitot, 19 September 1814, SLNSW, Safe 1/57.
61 Tyler, 'Biographical Outline'.
62 Ann Flinders to Thomas Pitot, 19 September 1814, SLNSW, Safe 1/57.
63 Estensen, *The Life of Matthew Flinders*, p. 470.
64 Scott, *The Life of Captain Matthew Flinders, R.N.*, p. 396.
65 Rob Mundle, *Flinders: The Man Who Mapped Australia*, Hachette, 2012, p. 364.
66 Tyler, 'Biographical Outline'.
67 *Ibid.*
68 'A Memoir of Captain Flinders, Apparently in His Wife's Handwriting'.
69 Tyler, 'Biographical Outline'.
70 'A Memoir of Captain Flinders, Apparently in His Wife's Handwriting'.

Epilogue
1 A speech by the Duke of Cambridge at Australia House, published 8 July 2014, royal.uk.
2 Flinders to Sir Joseph Banks, 20 March 1806, SLNSW, Banks Papers, Series 65.38.
3 Gillian Dooley, 'Matthew Flinders: The Man Behind the Map of Australia', transactionsvic.blogspot.com, 14 October 2015.
4 Geoffrey C. Ingleton, *Matthew Flinders: Navigator and Chartmaker*, Genesis Publications, 1986, p. 425.
5 'A Memoir of Captain Flinders, Apparently in His Wife's Handwriting'.
6 Ann Flinders to Charles Hursthouse, 28 December 1814, SLNSW, Flinders – Private letters, Vol. 3, Safe 1/57.
7 Retter & Sinclair, *Letters to Ann*, quoting Matthew Flinders Petrie.

8 Account from William Whitelaw to Ann Flinders for marble tablets, 21 April 1815, flinders.rmg.co.uk, FLI06.
9 Ann Flinders to Thomas Pitot, 19 September 1814, SLNSW, Flinders – Private letters, Vol. 3, Safe 1/57.
10 Macquarie to Henry Goulburn, 21 December 1817, *HRA*, Series I, Vol. IX, p. 747.
11 *The Sydney Gazette*, 4 February 1815, p. 1.
12 Matora 'Cora' Gooseberry, b. circa 1777, d. 30 July 1852.
13 *The Sydney Gazette*, 7 July 1829, p. 2.
14 *Ibid.*
15 'Death of King Boongarie', *The Sydney Gazette*, 27 November 1830, p. 2.
16 John Franklin to Ann Flinders, 27 December 1842. flinders.rmg.co.uk, FL127.
17 Anne Flinders (Philomathes), *The Connexion Between Revelation and Mythology Illustrated and Vindicated*, W.H. Dalton, 1845.
18 William Petrie, b. 21 January 1821, d. 16 March 1908.
19 'Ann Chappelle Flinders', findagrave.com.
20 Sir William Matthew Flinders Petrie, b. 3 June 1853 (Charlton, London), d. 28 July 1942 (Jerusalem).
21 'Matthew Flinders', *The Age*, 23 October 1925, p. 3.
22 Carolyn Webb, 'Australian Collectors Snap Up Rare Relics of Explorer Matthew Flinders', *The Age*, 16 December 2021.
23 James Vyver, 'Collection of Books Used by Matthew Flinders during 1801 Circumnavigation Donated to the National Archives of Australia', abc.com.au, 1 March 2023.
24 A speech by the Duke of Cambridge at Australia House, published 8 July 2014, royal.uk.
25 Nick Miller, 'Matthew Flinders Found: London Dig Unearths Grave of Great Explorer', *The Sydney Morning Herald*, 25 January 2019.
26 Lynne Harrison, 'Donington Hero Matthew Flinders Is Coming Home', spaldingtoday.co.uk, 17 October 2019.
27 *Ibid.*

Index